T0235015

Lecture Notes in Mathematics

Volume 2298

This series reports on new developments in all areas of mathematics and their applications - quickly, informally and at a high level. Mathematical texts analysing new developments in modelling and numerical simulation are welcome. The type of material considered for publication includes:

1. Research monographs
2. Lectures on a new field or presentations of a new angle in a classical field
3. Summer schools and intensive courses on topics of current research.

Texts which are out of print but still in demand may also be considered if they fall within these categories. The timeliness of a manuscript is sometimes more important than its form, which may be preliminary or tentative.

Titles from this series are indexed by Scopus, Web of Science, Mathematical Reviews, and zbMATH.

Maria Fragoulopoulou • Atsushi Inoue •
Martin Weigt • Ioannis Zarakas

Generalized B*-Algebras and Applications

 Springer

Maria Fragoulopoulou
Department of Mathematics
National and Kapodistrian
University of Athens
Athens, Greece

Atsushi Inoue
Department of Applied Mathematics
Fukuoka University
Fukuoka, Japan

Martin Weigt
Department of Mathematics and Applied
Mathematics
Nelson Mandela University
Port Elizabeth, South Africa

Ioannis Zarakas
Department of Mathematics
Hellenic Army Academy
Vari, Greece

ISSN 0075-8434   ISSN 1617-9692   (electronic)
Lecture Notes in Mathematics
ISBN 978-3-030-96432-0   ISBN 978-3-030-96433-7   (eBook)
https://doi.org/10.1007/978-3-030-96433-7

Mathematics Subject Classification: 46H20, 46H35, 46K10, 47L60, 46H30, 46K05, 46L60

This Springer imprint is published by the registered company Springer Nature Switzerland AG
The registered company address is: Gewerbestrasse 11, 6330 Cham, Switzerland

Ἐν δὲ τούτοις καὶ πρὸ τούτων οἱ κα-
λούμενοι Πυθαγόρειοι τῶν μαθημάτων
ἁψάμενοι πρῶτοι ταῦτά τε προήγαγον,
καὶ ἐντραφέντες ἐν αὐτοῖς τὰς τούτων
ἀρχὰς τῶν ὄντων ἀρχὰς ᾠήθησαν εἶναι
πάντων.

Ἀριστοτέλους
Τῶν μετὰ τὰ Φυσικά
Α, 985b 23-26
see [G.S. Kirk, J.E. Raven, *The Presocratic
Philosophers* (Cambridge Univ. Press,1957/
with corrections, Cambridge, 1973), pp.
236–237]

Contemporaneously with these philosophers [sc. Leucippus and Democri-
tus], and before them, the Pythagore-
ans, as they are called, devoted them-
selves to mathematics; they were the
first to advance this study, and having been brought up in it they thought
its principles were the principles of all
things.

W. Jaeger (ed.)
Aristotelis Metaphysica
book 1, 985b 23-26
see [W. Jaeger (ed.), *Aristotelis Metaphysica*
(Oxford Univ. Press, Oxford, 1957), *pp. 13,14*]

To the fond memory of the late Professor
G.R. ALLAN
and to Professor
P.G. DIXON
with much respect

Preface

In 1967, G.R. Allan initiated and studied a class of locally convex algebras with continuous involution, called GB^*-algebras (an abbreviation for "Generalized B^*-algebras"). The structure of a GB^*-algebra $A[\tau]$ is defined by a certain collection $\mathscr{B}^*_A$ of subsets of the underlying locally convex $*$-algebra $A[\tau]$, which, being partially ordered by inclusion, attains a maximal member, denoted by B_0. Among the first results, Allan showed an algebraic Gelfand–Naimark type theorem for commutative GB^*-algebras. Namely, he proved that *a commutative GB^*-algebra $A[\tau]$ is algebraically $*$-isomorphic to a $*$-algebra of extended-complex valued continuous functions on a compact Hausdorff space* $\mathfrak{M}_0$. The latter is, in fact, the Gelfand space of the C^*-subalgebra $A[B_0] := \{\lambda x : \lambda \in \mathbb{C}, x \in B_0\}$ of $A[\tau]$, endowed with the gauge function of B_0. This C^*-algebra is the key tool for investigating the structure of a GB^*-algebra.

In 1970, P.G. Dixon extended Allan's definition of a GB^*-algebra to include topological $*$-algebras that are not locally convex, thus enriching the set of examples of GB^*-algebras. Dixon then showed that the locally convex noncommutative GB^*-algebras are realized by closed operators on a Hilbert space. Namely, he gave a noncommutative algebraic Gelfand–Naimark type theorem for GB^*-algebras, which determines them among unbounded operator algebras.

Typical examples of GB^*-algebras are C^*-algebras, pro-C^*-algebras (i.e., inverse limits of C^*-algebras), C^*-like locally convex $*$-algebras initiated by A. Inoue and K.-D. Kürsten, the Arens algebra

$$L^{\omega}[0, 1] = \bigcap_{1 \leq p < \infty} L^p[0, 1]$$

(Allan) equipped with the topology of the L^p-norms, $1 \leq p < \infty$ and the algebra $\mathfrak{M}[0, 1]$ of all measurable functions on $[0, 1]$ (modulo equality a.e.), endowed with the topology of convergence in measure, which is not necessarily locally convex (Dixon).

Thus, GB^*-algebras generalize the celebrated C^*-algebras, which consist entirely of bounded operators. As noted, a GB^*-algebra $A[\tau]$ consists mainly of unbounded operators. The bounded operators that it may contain are all concentrated in the C^*-subalgebra $\mathcal{A}[B_0]$ of $\mathcal{A}[\tau]$, mentioned above.

The main body of the present monograph is based on the theory of GB^*-algebras as developed by Allan and Dixon. In addition, it is augmented by related results of, among others, S.J. Bhatt, W. Kunze, G. Lassner, M. Oudadess, K. Schmüdgen, A.W. Wood, and the authors.

It is very well known that our physical world mostly consists of unbounded operators. Among them, the most well-known are the Hamiltonian operator H representing the observable energy, and the operators P, Q representing the observable momentum and observable position, respectively. The following algebraic equations involving the previous operators

$$[P, Q] = PQ - QP = -i\hbar I, \quad H = \frac{P^2}{2m} + \frac{m\omega^2 Q^2}{2},$$

correspond to the one-dimensional harmonic oscillator, where i is the imaginary unit, $\hbar$ is the Planck's constant, I is the identity operator, and m, ω the mass and frequency of the oscillator, respectively. According to quantum mechanics, a physical observable is represented by a self-adjoint linear operator, while certain algebraic relations, as before, correspond to a physical system, whose mathematical image is an operator *-algebra in an inner product space. In the case of the one-dimensional harmonic oscillator, the respective *-algebra is the one generated by the essentially self-adjoint operators, H, P, Q.

The journey to the quantum mechanical relation $PQ - QP = -i\hbar I$, known as the *Heisenberg* (or canonical) *commutation relation*, was not straightforward. Moreover, the conclusion that the quantum mechanical observables are not represented by numbers but by operators is considered to be *one of the greatest achievements of science*.

From the above it is clear and quite natural why scientists were led to the study of unbounded operator algebras, among them being the GB^*-algebras. This is the reason why the latter have been investigated, in different directions, by various authors.

Athens, Greece   Maria Fragoulopoulou
Fukuoka, Japan   Atsushi Inoue
Port Elizabeth, South Africa   Martin Weigt
Vari, Greece   Ioannis Zarakas

Contents

Chapter 1
Introduction

A Generalized B^*-algebra (for short GB^*-algebra) is a symmetric pseudo-complete locally convex $*$-algebra $\mathcal{A}[\tau]$, with an identity element, whose strucure is determined by a collection $\mathcal{B}^*_\mathcal{A}$ of certain subsets of $\mathcal{A}$ that enjoy algebraic and topological properties (Definition 3.3.2).

The aim of the book in hands, is to exhibit the results of the fathers of the subject G.R. Allan and P.G. Dixon, as carried out in [3–5, 48–50]. Related results by other authors are also included, providing useful information and indicating how far the investigation of GB^*-algebras can reach. There is a small number of new results presented in Sects. 6.3 and 6.4. There are results that are not included in the monograph, although they are connected with GB^*-algebras; some of them can be found in [19, 20, 38, 44, 56, 121, 153–155].

As was remarked in the Preface, GB^*-algebras are realized by unbounded operators acting on a dense subspace of a Hilbert space (see e.g., Theorems 6.3.5 and 6.3.11). For the most known unbounded linear operators P, Q, H mentioned in Preface, the reader is referred to [29, 57, 108, 138, 142].

In this regard, notice that until the theory of quantum was recognized, it passed from various stages, that could not be interpreted by the views of classical physics of that time. This was first stated, in 1900, by the German physicist Max Planck, aiming to clarify the laws of the 'thermic radiation' that are taken experimentally, but cannot be interpreted theoretically through classical physics. For blending the theoretical computations, with the experimental data, Planck stated the theory of quantum according to which the light, and more general, the electromagnetic radiation is emitted and absorbed not continuously, but in elementary amounts, which he named quantum of light. Later Einstein used the name photon instead of quantum.

The quantum theory interpreted various phenomena like photoelectric, fluorescence, etc. From the end of 1924 to the beginning of 1925 Heisenberg and Schrödinger introduced independently equivalent, although seemingly not similar, interpretations for the empirical quantization laws of Bohr and Sommerfeld. These laws were based on the matrix, respectively wave mechanics. Later on, it turned out

© The Author(s), under exclusive license to Springer Nature Switzerland AG 2022
M. Fragoulopoulou et al., *Generalized B*-Algebras and Applications*, Lecture Notes in Mathematics 2298, https://doi.org/10.1007/978-3-030-96433-7_1

that these two approaches are equivalent and correspond to two different realizations of the momentum and position operators P, Q of quantum mechanics. For a very recent algebraic approach to quantum theories, see [138, Chapter 1].

But, GB^*-algebras occur also, among the so-called *unbounded Hilbert algebras* [78–83, 157], that are very important for the Tomita Takesaki theory for unbounded operator algebras developed in [85], by the second named author; on the other hand, they contribute to the rising of the so-called EW^*-algebras; see Chap. 5, as well [49, 83, 85], that also play a decisive role in the aforementioned Tomita Takesaki theory.

All the preceding combined with the desire of each one of us to make the structure of GB^*-algebras more familiar to a wider spectrum of researchers, gave us the motive and impetus for attempting the writing of this monograph.

The whole essay is divided in seven chapters. Chapter 2 offers a preparatory stage for the introduction and basic properties of GB^*-algebras. Namely, Chap. 2 gives a general theory of locally convex algebras with separately continuous multiplication, aiming to a proper definition of spectrum for elements of this sort of algebras and to the investigation of its properties, in comparison to the usual theory of spectrum in the classical case. A motivation for this comes from the spectral theory of a closed operator T on a Banach space E, where the spectrum of T is given by those complex numbers λ, such that the operator $\lambda I - T$ has no bounded inverse (see, for instance, [145]), where I is the restriction of the identity operator of E on the domain of T. Thus, a proper definition of a bounded element (Allan) in a locally convex algebra should be given first. This was done in Definition 2.2.1 and its choice is confirmed from the theory that rises from it, based on the definition of the spectrum of an element (see Sect. 2.3), as it can be seen from Theorem 2.3.7 and its corollaries. In Theorem 2.3.13, the new concept of spectrum is related with the usual concept of the spectrum of an element, when the considered locally convex algebra $A[\tau]$ has continuous inversion. When $A[\tau]$ is also pseudo-complete (a weaker notion than completeness, that plays an essential role in the whole theory of GB^*-algebras), then Allan bounded elements are just characterized by the boundedness of the usual spectrum (Corollary 2.3.14). Similar ideas to the previous ones were considered earlier by L. Waelbroeck [148], in 1957, but in a more specific framework and under the assumption of quasi-completeness and commutativity of the topological algebras involved. Moreover, a comparable definition to that of G.R. Allan for a bounded element, but in an m-convex algebra, was given by S. Warner [149, p. 197, Definition 3], in 1956. Finally, in Sects. 2.4 and 2.5 a functional calculus for a pseudo-complete locally convex algebra, respectively the carrier space of a commutative pseudo-complete locally convex algebra $A[\tau]$ are discussed, where in both cases the set A_0 of (Allan-)bounded elements of $A[\tau]$ plays an important role; in the second case A_0 coincides with $A[B_0]$, therefore it is a C^*-algebra (cf. Lemma 3.3.7(ii) and Theorem 3.3.9(i)), although, in the general case, A_0 is not even a subspace. In this regard, an interesting result is given by Corollary 6.4.5.

Chapter 3 deals with the basic theory of GB^*-algebras. More precisely, Sect. 3.1 concerns hermitian and symmetric locally convex $*$-algebras (see Definition 3.1.6), where the respective concepts are given by using, in fact, (Allan) bounded elements (cf. Definitions 2.2.1 and 2.3.1). Note that the classical definitions of hermiticity and symmetry are completely algebraic, given through the usual spectrum $sp(x)$ of an element x in a $*$-algebra $\mathcal{A}$ (see discussion before Definition 3.1.3). Every symmetric locally convex $*$-algebra is "classically" symmetric and as in the algebraic case, every symmetric pseudo-complete locally convex $*$-algebra is hermitian (Corollary 3.1.7); symmetry and pseudo-completeness are among the main ingredients of a GB^*-algebra (Definition 3.3.2). In Sect. 3.2, another of the main ingredients of a GB^*-algebra $\mathcal{A}[\tau]$, the collection $\mathfrak{B}^*_\mathcal{A}$ arises, in an attempt of characterizing the C^*-condition in a normed $*$-algebra $\mathcal{A}[\|\cdot\|]$ by using not the properties of the given norm $\|\cdot\|$, but the properties of $\mathcal{A}[\|\cdot\|]$, as a locally convex $*$-algebra (see, e.g., Theorem 3.2.9). Sect. 3.3 discusses and comments on the definitions of a GB^*-algebra given by Allan and Dixon (cf., e.g., Remark 3.3.4 and Definitions 3.3.5, 3.3.6). At the same time, it unfolds the fundamental structure of such an algebra and presents several examples. Section 3.4 gives a functional representation of a commutative GB^*-algebra $\mathcal{A}[\tau]$, which is an algebraic analogue of the commutative Gelfand–Naimark theorem for C^*-algebras, due to G.R. Allan. The decisive role of the C^*-subalgebra $\mathcal{A}[B_0]$ of $\mathcal{A}[\tau]$ plays an essential role, in this regard (Theorem 3.4.9). In Sect. 3.5, the subclass of GB^*-algebras, called C^*-like locally convex algebras, is discussed; this was introduced by A. Inoue and K.-D. Kürsten, in 2002 (cf. [88]). The main result in this section states that every C^*-like locally convex algebra $\mathcal{A}[\tau]$ is a GB^*-algebra, with B_0 the unit ball of the "bounded part" $\mathcal{A}_b$ of $\mathcal{A}[\tau]$ (see Theorem 3.5.3 and the indicated by ▶ discussion in Remark 3.5.7(3)). An important fact in this direction is the comparison of the $*$-subalgebras $\mathcal{A}[B_0]$, $\mathcal{D}(p_\Lambda)$ of $\mathcal{A}[\tau]$ in the case of a GB^*-algebra $\mathcal{A}[\tau]$ and their coincidence with $\mathcal{A}_b$, when $\mathcal{A}[\tau]$ is a C^*-like locally convex $*$-algebra. For this (cf. Corollary 3.5.4 and Proposition 3.5.8, together with all results between them).

Chapter 4 considers various aspects of the theory of commutative GB^*-algebras. Namely, Sect. 4.1 treats a functional calculus analogous to that of commutative C^*-algebras (cf., for instance, Theorem 4.1.2(Allan)). Such a deal is natural, after an algebraic commutative Gelfand–Naimark type theorem has been proved for GB^*-algebras (Theorem 3.4.9). The preceding functional calculus is extended to noncommutative GB^*-algebras, in Chap. 6 (Theorem 6.1.3 (Dixon)). In Sect. 4.2, it is proved that every commutative GB^*-algebra admits an abundance of positive linear functionals that separate its points (Corollary 4.2.6, Allan). Unlike to the C^*-algebras theory, there is only a 'partial analogue' of the last result in the noncommutative case, due to Dixon (Theorem 6.3.4). It is well known that every C^*-algebra has a unique C^*-norm. Section 4.3 investigates whether a similar situation happens for GB^*-algebras. In this regard, one defines an equivalence between two GB^*-topologies, in virtue of the $\mathfrak{B}^*$ collections corresponding to the given topologies; this means that the C^*-algebras generated by the maximal elements of the $\mathfrak{B}^*$-collections, under consideration, coincide. It is proved that any two locally convex $*$-algebra topologies, that make a commutative $*$-algebra with identity a GB^*-algebra, are equivalent in the preceding sense (Corollary 4.3.10). This result of G.R. Allan was extended later to the noncommutative case by P.G. Dixon (see Corollary 6.3.7). Furthermore, it is shown that on every commutative GB^*-

algebra $\mathcal{A}[\tau]$ there is a finest locally convex $*$-algebra topology "equivalent" to τ (Theorem 4.3.13). The last Sect. 4.4 presents an example of P.G. Dixon, according to which there is a $*$-algebra of functions with no GB^*-topology.

In Chap. 5, GB^*-algebras of closable operators in a Hilbert space are discussed. Section 5.1 introduces the basic definitions and properties of unbounded operator algebras called $\mathcal{O}^*$-algebras and unbounded $*$-representations of a $*$-algebra. The notion of closedness (resp. self-adjointness and integrability) of an $\mathcal{O}^*$-algebra is defined and studied in analogy with the notion of a closed (resp. self-adjoint) operator. Section 5.2 introduces several locally convex topologies on $\mathcal{O}^*$-algebras. The uniform topologies τ_u, τ^u and τ_*^u and the quasi-uniform topologies τ_{qu} and τ_{qu}^*, which are generalizations of the operator-norm topology, in the bounded case, are defined and they are different to one another (see (5.2.3)). Moreover, other topologies called weak, strong, strong*, quasi-strong and quasi-strong* are defined and the relations among all the preceding topologies are investigated (see Proposition 5.2.1(1)). Section 5.3, deals with a symmetric $\mathcal{O}^*$-algebra called extended C^*-algebra (resp. extended W^*-algebra), whose bounded part is a C^*-algebra (resp. a von Neumann algebra). In this regard, it is proved that a closed $\mathcal{O}^*$-algebra $\mathcal{M}$ is an extended C^*-algebra, if and only if, $\mathcal{M}[\tau_u]$ is a locally convex GB^*-algebra of Dixon (Theorem 5.3.4). In Sect. 5.4, the concept of an unbounded Hilbert algebra, which is a generalization of a standard Hilbert algebra, is introduced. Given a Hilbert algebra $\mathfrak{A}_0$, we use the noncommutative integration theory of I.E. Segal to define the noncommutative Arens algebra $L_2^\omega(\mathfrak{A}_0)$, which is an unbounded Hilbert algebra, maximal among all unbounded Hilbert algebras containing $\mathfrak{A}_0$ (Theorem 5.4.4). In case $\mathfrak{A}_0$ has an identity element, then $L_2^\omega(\mathfrak{A}_0)$ is a Fréchet GB^*-algebra equipped with the locally convex topology defined by the family of L^p norms $\{\| \cdot \|_p\}_{2 \leq p < \infty}$ (Theorem 5.4.10). Furthermore, an extended W^*-algebra is constructed by the left regular representation of the unbounded Hilbert algebra $L_2^\omega(\mathfrak{A}_0)$ and the left regular representation of the Hilbert algebra $\mathfrak{A}_0$ (Theorem 5.4.5). Note that the unbounded Hilbert algebras play an important role in the Tomita–Takesaki theory in algebras of unbounded operators (see [85]).

In Chap. 6, our aim is to present a noncommutative Gelfand–Naimark type theorem for GB^*-algebras. For this purpose, Sect. 6.1 discusses a noncommutative functional calculus (Theorem 6.1.3 (Dixon)) similar to the commutative one discussed in Sect. 4.1. Section 6.2 introduces positive elements in a GB^*-algebra $\mathcal{A}[\tau]$. Denoting by $\mathcal{A}^+$ the set of all positive elements in $\mathcal{A}[\tau]$, it is proved that $\mathcal{A}^+$ is a convex cone (Theorem 6.2.5), τ-closed under a certain condition; the latter is fulfilled by a barrelled and bornological topology T defined on $\mathcal{A}$, finer than τ (see Theorems 6.2.6, 6.2.11 and 6.2.13). In Sect. 6.3, after discussing the algebraic version of the Gelfand–Naimark–Segal construction and defining the direct sum $*$-representation of a family of closed $*$-representations, we prove that every GB^*-algebra has enough positive linear functionals in order to separate its points (Theorem 6.3.4). All these lead to an 'algebraic' extension of the celebrated noncommutative Gelfand–Naimark theorem for C^*-algebras in the context of GB^*-algebras (Dixon); see Theorem 6.3.5. In this regard, a natural question is whether one can obtain an extension of the latter result, where the constructed faithful $*$-

representation is bicontinuous on its image. This is achieved by employing the so called $A\mathcal{O}^*$-algebras, introduced and studied by G. Lassner in [103], K. Schmüdgen [133] and others. In this regard, see Theorem 6.3.11. Among the biproducts of the latter theorem, Corollary 6.3.16 characterizes an $A\mathcal{O}^*$-algebra $\mathcal{A}[\tau]$ in the context of Fréchet GB^*-algebras by the fact that each self-adjoint continuous linear functional on $\mathcal{A}[\tau]$ is expressed as a difference of two (continuous) positive linear functionals on $\mathcal{A}[\tau]$. The last Sect. 6.4 deals with an analogue of Ogasawara's commutativity condition theorem in C^*-algebras, within the context of GB^*-algebras. The corresponding result is due to M. Oudadess [118] (see Theorem 6.4.1). Two byproducts of Theorem 6.4.3, that are also due to M. Oudadess and give a more general version of Ogasawara's commutativity condition theorem, are Corollaries 6.4.4 and 6.4.5 that provide interesting new results for a GB^*-algebra.

The last two chapters are devoted to applications of the developed theory in the previous five chapters. Thus, in Chap. 7 and, in particular, in Sect. 7.1, we investigate the structure and the representation theory of the completion of a given C^*-algebra $\mathcal{A}[\|\cdot\|]$ with identity, with respect to a locally convex $*$-algebra topology τ, coarser than the topology of the given C^*-norm and such that the multiplication in $\mathcal{A}[\tau]$ is jointly continuous. Such a study started in 2006, with [12] and continued by [62], in 2007. If we denote by $\widetilde{\mathcal{A}}[\tau]$ the aforementioned completion, it was proved that this is a GB^*-algebra over the closed unit ball $\mathcal{U}(\mathcal{A})$ of $\mathcal{A}[\|\cdot\|]$, if and only if, $\mathcal{U}(\mathcal{A})$ is τ-closed (see Corollary 7.1.3). Even more, $\widetilde{\mathcal{A}}[\tau]$ is a GB^*-algebra, if and only if, there exists a faithful $(\tau - \tau_{s^*})$-continuous $*$-representation π of $\widetilde{\mathcal{A}}[\tau]$, such that $\tau \prec r_\pi$, where τ_{s^*} is the strong* topology on $\mathcal{L}^\dagger(\mathcal{D}(\pi))$ and r_π an unbounded C^*-seminorm in $\widetilde{\mathcal{A}}[\tau]$ (see Theorem 7.1.7). If the multiplication on $\mathcal{A}$ is not jointly continuous with respect to τ, then $\widetilde{\mathcal{A}}[\tau]$ is not necessarily a locally convex $*$-algebra, but it has the structure of a partial $*$-algebra. The investigation of structure and representation theory of $\widetilde{\mathcal{A}}[\tau]$, in this case, was studied in [12, Section 3] and [62, Section 3]. Partial $*$-algebras consist of unbounded operators and admit a product operation that is not defined for all pairs of elements. For a complete account in their theory and a rich literature, see [7]. Section 7.2 studies continuity of positive linear functionals on a GB^*-algebra. Unlike to the C^*-case, positive linear functionals on a GB^*-algebra are not always continuous (see Theorem 7.2.1, its corollaries and Example 7.2.4). In Sect. 7.3, it is proved that as in the C^*-case, every GB^*-algebra admits a bounded approximate identity (Theorem 7.3.1, Bhatt). In Sect. 7.4, it is shown that the inverse limit of an inverse system of GB^*-algebras is again a GB^*-algebra (cf. Corollary 7.4.7, Kunze). Moreover, every closed ideal of a GB^*-algebra $\mathcal{A}[\tau]$ is a $*$-ideal and the quotient algebra $\mathcal{A}/I$, under the quotient topology, is a GB^*-algebra with underlying C^*-algebra the quotient $\mathcal{A}[B_0]/(I \cap \mathcal{A}[B_0])$ (Theorem 7.4.8, Kunze); recall that $\mathcal{A}[B_0]$ is the underlying C^*-algebra of $\mathcal{A}[\tau]$. Finally, Sect. 7.5 deals with various types of the well known Vidav–Palmer theorem for GB^*-algebras, due to A.W. Wood [156] (see, for example, Theorems 7.5.43, 7.5.49, 7.5.51). For stating such results, a proper definition of the numerical range of an element in a locally convex $*$-algebra is needed and its basic properties are investigated. At the same time an equivalent variant of the usual definition of a GB^*-algebra has been used, given by A.W. Wood, for reaching the obtained generalizations of a non-normed Vidav–Palmer theorem (see, for instance, Theorem 7.5.49 and Theorem 7.5.51).

Applications carry on in Chap. 8, which exclusively deals with tensor products of GB^*-algebras. Such a matter was first considered in [64]. For other sources of topological tensor products of unbounded operator algebras, see [1, 65, 71]. There is a physical justification for using tensor products. They are viewed to describe two quantum systems as one joint system [2], while the physical significance of tensor products always depends on the applications, which may involve wave functions, spin states, oscillators etc.; see e.g., [29, 70].

In Sect. 8.1, a background material is presented concerning standard notions and known results on topological tensor products. Moreover, some examples of GB^*-algebras are exhibited that are realized as tensor products of GB^*-algebras. The Example 8.1.4 leads us to Definition 8.2.5 of the tensor product of two GB^*-algebras. Section 8.2 deals with $\mathfrak{B}^*$ collections (see Definitions 3.3.1, 3.3.2) in the case of topological tensor products of (pseudo-complete) locally convex *-algebras. In Sect. 8.3, we testify Definition 8.2.5 of the GB^*-tensor product in several cases. For instance, if ε is the injective locally convex tensor topology, Corollary 8.3.6 shows that if $\mathcal{A}_1[\tau_1]$, $\mathcal{A}_2[\tau_2]$ are complete locally convex *-algebras with identities, such that either $\mathcal{A}_1$ or $\mathcal{A}_2$ is commutative, then $\mathcal{A}_1 \widehat{\otimes}_\varepsilon \mathcal{A}_2$ is a tensor product GB^*-algebra, if and only if, $\mathcal{A}_1[\tau_1]$ and $\mathcal{A}_2[\tau_2]$ are GB^*-algebras. In fact, commutativity is needed only in the 'if' part. Justifications of the usage of commutativity are given at the introduction of Sect. 8.3. Section 8.4, touches nuclear GB^*-algebras, where some examples are presented and a few characterizations are mentioned together with some questions on this interesting matter. For some further information on this topic, see [152]. The final Sect. 8.5 provides some applications, concerning mainly existence and properties of (unbounded) *-representations on tensor product GB^*-algebras.

We have tried to make this book as self-contained as possible and have tried to give rather detailed proofs of the results, for the convenience of the readers. In cases that some proofs are omitted, full references are given.

For the reading of this book, some familiarity is assumed with general topology [97, 100], functional analysis [73, 74, 91, 113, 123, 128, 129, 131], Banach and Banach *-algebras [32, 43, 72, 115, 120, 127], C^*-algebras and von Neumann algebras [45, 46, 52, 114, 122, 130, 144], basic theory of topological (*-)algebras [15, 60, 72, 111, 112, 136, 158] and basic theory of bounded and unbounded operators [7, Chapter 1], [66, pp. 237–248], [95, 114, 136, 144]. Our suggestions are only indicative, since almost all mentioned fields have to present a very rich literature.

The last 31 years, seven books have been published on unbounded operator algebras and their representation theory. The readers, who are interested in being more acquainted with unbounded operator theory and its applications, may chronologically be referred to [7, 53, 66, 85, 136, 138] and [86].

Finally, we would like to express our cordial thanks to all of our colleagues and collabotators, with whom either we had useful discussions on the topics of the present book, or we were receivers of their important comments, having them as listeners in our seminars, on relevant topics.

Last, but not least, we want to address our heartfelt thanks to our colleague Professor Emerita Maria Papathanassiou (who is also an archeologist), for providing us with the ancient Greek text of Aristotle, as well with its translation in English (see front matter).

Chapter 2
A Spectral Theory for Locally Convex Algebras

Chapter 2 acts as a preparative stage to the basic definitions and fundamentals of our main theme, which is the development of the theory of GB^*-algebras, as algebras of unbounded operators. More precisely, the notion of a 'bounded element' (Definition 2.2.1) in a locally convex algebra is introduced (Allan), as well the concept of spectrum of an element (Definition 2.3.1). In addition, the carrier space of a commutative complete locally convex algebra is defined (cf. Sect. 2.5) and a Gelfand-type theory is developed.

2.1 Basic Definitions and Notation

All the algebras considered throughout this book are complex, linear, associative. If an algebra $\mathcal{A}$ has an identity element, this will be denoted by e. *We emphasize that all topological spaces are considered Hausdorff*, unless indicated otherwise.

In this section we exhibit the basic definitions and basic notations that will be used in this book.

The symbols $\mathbb{N}, \mathbb{R}, \mathbb{C}$ stand for the natural, real and complex numbers, respectively. The *imaginary unit* is denoted by i and the *conjugate* of a complex number α by $\overline{\alpha}$.

Let X be a topological space with two topologies τ, τ'. In order to indicate that τ *is coarser than* τ', or equivalently τ' *is finer than* τ, we shall write $\tau \prec \tau'$. In order to indicate that τ, τ' *are equivalent*, in the usual sense, we shall write $\tau \approx \tau'$; see also Definition 4.3.1 and Remark 4.3.2.

If $X[\tau]$ is a topological space endowed with a topology τ and S a subset of X, *the closure of S* will be denoted by $\overline{S}$ or for distinction, by $\overline{S}^{\tau}$.

If $\mathcal{A}$ is an algebra with identity, denote by $G_{\mathcal{A}}$, the *group of the invertible elements of $\mathcal{A}$*.

© The Author(s), under exclusive license to Springer Nature Switzerland AG 2022
M. Fragoulopoulou et al., *Generalized B*-Algebras and Applications*, Lecture Notes in Mathematics 2298, https://doi.org/10.1007/978-3-030-96433-7_2

If $\mathcal{A}$ is an algebra without identity, the symbol $\mathcal{A}_1$ will stand for its *unitization*. That is, $\mathcal{A}_1 = \mathcal{A} \oplus \mathbb{C}$, with linear operations defined coordinatewise and multiplication by

$$(x, \alpha)(y, \beta) := (xy + \alpha y + \beta x, \alpha\beta), \ \forall \, x, y \in \mathcal{A} \text{ and } \alpha, \beta \in \mathbb{C}.$$

For the identity of the unitization $\mathcal{A}_1$ of $\mathcal{A}$, we shall use the symbol $e_1 \equiv (0, 1)$.

Take again an algebra $\mathcal{A}$ without identity. If x is in $\mathcal{A}$, employing the *circle operation* $\circ$, we shall say that x is *quasi-invertible*, if there are elements $y, z \in \mathcal{A}$ with

$$x \circ y = 0 = z \circ x, \quad \text{where } x \circ y := x + y + xy$$

(alternatively, you can use in the previous equality $-xy$ instead of $+xy$). If x is *quasi-invertible* in $\mathcal{A}$, then $y = z$ is unique, it is called the *quasi-inverse* of x and it is denoted by x° [60, 111]. It is easily seen, that $x \in \mathcal{A}$ is quasi-invertible with quasi-inverse x°, if and only if, $e + x$ is invertible in the unitization $\mathcal{A}_1$ of $\mathcal{A}$ with inverse $e + x^\circ$. Readily, the same is true if $\mathcal{A}$ has an identity. The symbol $G_{\mathcal{A}}^q$, will stand for the *group of quasi-invertible elements* of $\mathcal{A}$.

We first give some definitions and terminology needed in what follows. For the basic theory of topological vector spaces and general topological algebras with or without involution, the reader is referred, respectively to [74, 113, 128, 131] and [15, 60, 72, 111, 112].

Definition 2.1.1 By a *topological algebra* $\mathcal{A}[\tau]$ we shall mean a topological vector space, which is also an algebra, such that the ring multiplication is separately continuous. $\mathcal{A}[\tau]$ is said to be a *locally convex algebra*, if it is a topological algebra, whose underlying topological vector space is a locally convex space. In this case, the topology τ of $\mathcal{A}$ may be described by an *upwards directed family of* (nonzero) *seminorms*, or otherwise by a *saturated family of seminorms* [74, p. 96]. Such a family will be usually denoted by $\Gamma = \{p\}$ or $\{p_\nu\}_{\nu \in \Lambda}$, with Λ an *upwards directed index set*; sometimes, for distinction, we shall use the symbol $\Gamma_{\mathcal{A}}$ instead of just Γ. When each seminorm p_ν, $\nu \in \Lambda$ is also *submultiplicative* (briefly, *m-seminorm*) i.e.,

$$p_\nu(xy) \le p_\nu(x)p_\nu(y), \ \forall \, x, y \in \mathcal{A} \text{ and } \nu \in \Lambda,$$

then $\mathcal{A}[\tau]$ is called an *m-convex algebra*, where *m*-convex is an abbreviation of "multiplicatively-convex" (the term '*m*-convex' was used in [158], but other authors have used instead, the term 'locally *m*-convex'; see, e.g., [15, 37, 111, 112]). A seminorm p with the property $p(x) = 0$, if and only if, $x = 0$, is called a *norm* and it is usually denoted by $\| \cdot \|$. An *m*-seminorm, which is also a norm, is called *algebra norm*. An algebra $\mathcal{A}$ endowed with an algebra norm $\| \cdot \|$, is called *normed algebra* and it is denoted by $\mathcal{A}[\| \cdot \|]$. A complete normed algebra is said to be a *Banach algebra*.

▶ Concerning Definition 2.1.1, we shall (almost) always refer to Γ (and/or to Γ_A), as a *defining family of seminorms for the topology* τ. The same will apply, when we speak for *a family of $*$-seminorms*, too.

If $A[\tau]$ is a topological algebra without identity, its *unitization* A_1 endowed with the product topology, denoted by τ_1, is also a topological algebra, for which we shall use the symbol $A_1[\tau_1]$. In the case, when $A[\|\cdot\|]$ is a normed (resp. Banach) algebra, its unitization is denoted by $A_1[\|\cdot\|_1]$, where $\|(x,\lambda)\|_1 := \|x\| + |\lambda|$, for all $(x,\lambda) \in A_1 \equiv A \times \mathbb{C}$.

▶ *Note that since we are working on algebras*, we shall use, *without any particular mention*, the term *norm*, instead of the term *algebra norm*. Moreover, we shall always assume that *the seminorms in* Γ, as above, *will be mutually non-equivalent*, in the usual sense of equivalence between two seminorms.

2.2 The Set of Bounded Elements. Radius of Boundedness

Definition 2.2.1 (Allan) Let $A[\tau]$ be a locally convex algebra. An element x of A is called (Allan-) *bounded* and for simplicity just *bounded*, if there exists a nonzero complex number λ, such that the set $\{(\lambda x)^n : n \in \mathbb{N}\}$ is a bounded subset of $A[\tau]$.

For some historical comments on the concept introduced in Definition 2.2.1, see Introduction, just before the description of the results of Chap. 3. The set of all bounded elements of $A[\tau]$ will be denoted by A_0. It is straightforward to see that every element of a normed algebra is bounded in the previous sense. Also if $A[\tau]$ has an identity e, then e is bounded.

Definition 2.2.2 For a locally convex algebra $A[\tau]$, let $(\mathfrak{B}_0)_A$ denote the collection of all subsets B of A, which fulfill the following properties:

(1) B is absolutely convex and $B^2 \subset B$;
(2) B is bounded and closed.

In case no confusion arises as to which algebra the previous collection of sets is regarded, *we use the symbol* $\mathfrak{B}_0$. For every $B \in \mathfrak{B}_0$, we use $A[B]$ to denote the subalgebra of A generated by B. Based on the properties that each set $B \in \mathfrak{B}_0$ possesses, we have that

$$A[B] = \{\lambda x : \lambda \in \mathbb{C}, \, x \in B\}. \tag{2.2.1}$$

It is easily proven that for every $B \in \mathfrak{B}_0$ the Minkowski functional (or *gauge function*) $\| \cdot \|_B$ on $\mathcal{A}[B]$, that is, the function

$$\|x\|_B = \inf \left\{ t > 0 : x \in tB \right\}, \ x \in \mathcal{A}[B], \tag{2.2.2}$$

defines a norm with which $\mathcal{A}[B]$ becomes a normed algebra. If not explicitly stated otherwise, $\mathcal{A}[B]$ will always be assumed to carry this norm topology.

We note that the topology τ, which $\mathcal{A}[B]$ carries as a subalgebra of $\mathcal{A}$, is weaker than $\| \cdot \|_B$. Indeed let $(x_n)_{n \in \mathbb{N}}$ be a sequence in $\mathcal{A}[B]$, such that $x_n \to x$, with respect to $\| \cdot \|_B$. Then, since B is τ-bounded, for each 0-neighbourhood in $\mathcal{A}[\tau]$, say U, there is an $\varepsilon > 0$, such that $B \subset \varepsilon U$. Due to convergence of the sequence (x_n) to x, with respect to $\| \cdot \|_B$, we have that there is $n_0 \in \mathbb{N}$, such that $\|x_n - x\|_B < \frac{1}{\varepsilon}$, for every $n \in \mathbb{N}$, $n \geq n_0$. Hence, $x_n - x \in \frac{1}{\varepsilon} B$, $n \geq n_0$; thus, $x_n - x \in U$, for all $n \geq n_0$. Therefore, $x_n \to x$ with respect to τ.

The intimate relation between the set $\mathcal{A}_0$, of all bounded elements in $\mathcal{A}$, and the normed algebras $\mathcal{A}[B]$, $B \in \mathfrak{B}_0$ is given by the following proposition, for which the next definition is helpful.

Definition 2.2.3 A subcollection $\mathfrak{B}_1$ of $\mathfrak{B}_0$ is called *basic* if for every $B \in \mathfrak{B}_0$, there is some $B_1 \in \mathfrak{B}_1$, such that $B \subset B_1$.

Proposition 2.2.4 *Let* $\mathcal{A}[\tau]$ *be a locally convex algebra and* $\mathfrak{B}_1$ *a basic subcollection of* $\mathfrak{B}_0$. *Then,* $\mathcal{A}_0 = \bigcup \left\{ \mathcal{A}[B] : B \in \mathfrak{B}_1 \right\}$.

Proof Let $x \in \mathcal{A}[B]$, for some $B \in \mathfrak{B}_1$. Then, for $\lambda > \|x\|_B$ we have that $\frac{1}{\lambda} x \in B$. Since $B^2 \subset B$, by induction we have $B^n \subset B$, for $n \in \mathbb{N}$, so that $\left\{ (\frac{1}{\lambda} x)^n : n \in \mathbb{N} \right\} \subset B$. Therefore, the subset $\left\{ (\frac{1}{\lambda} x)^n : n \in \mathbb{N} \right\}$ of $\mathcal{A}$ is bounded, since B is bounded and so $x \in \mathcal{A}_0$.

In the inverse direction, if $x \in \mathcal{A}_0$ and $\lambda \in \mathbb{C}$, $\lambda \neq 0$, then the set $S \equiv \left\{ (\lambda x)^n : n \in \mathbb{N} \right\}$ is bounded in $\mathcal{A}[\tau]$. It is easy to show that the closure of the absolutely convex hull of S belongs to $\mathfrak{B}_0$. Therefore, there is some $B \in \mathfrak{B}_1$, such that $S \subset B$ and so $x \in \mathcal{A}[B]$. $\qquad \square$

If for a locally convex algebra $\mathcal{A}[\tau]$, all the normed subalgebras $\mathcal{A}[B]$, $B \in \mathfrak{B}_0$ are Banach algebras, then $\mathcal{A}[\tau]$ is called *pseudo-complete* .

Proposition 2.2.5 *If* $\mathcal{A}[\tau]$ *is sequentially complete, then* $\mathcal{A}[\tau]$ *is pseudo-complete.*

Proof Let $B \in \mathfrak{B}_0$ and $(x_n)_{n \in \mathbb{N}} \subset \mathcal{A}[B]$, such that the sequence $(x_n)_{n \in \mathbb{N}}$ is $\| \cdot \|_B$-Cauchy. Since τ is weaker than $\| \cdot \|_B$ on $\mathcal{A}[B]$, $(x_n)_{n \in \mathbb{N}}$ is τ-Cauchy and therefore, from the assumption of $\mathcal{A}[\tau]$ being sequentially complete, we have that there is an $x \in \mathcal{A}$, such that $x_n \to x$, with respect to τ. Let $\varepsilon > 0$. Since $(x_n)_{n \in \mathbb{N}}$ is $\| \cdot \|_B$-Cauchy, there is a $n_0 \in \mathbb{N}$, with $\|x_m - x_n\|_B < \frac{\varepsilon}{4}$, for all $m, n \geq n_0$. Then, since $x_m - x_{n_0} \to x - x_{n_0}$, with respect to τ and B is τ-closed, we have that $x - x_{n_0} \in \frac{\varepsilon}{4} B$ and thus $x \in \mathcal{A}[B]$. So, for

$$m \geq n_0, \ x_m - x = x_m - x_{n_0} + x_{n_0} - x \in \frac{\varepsilon}{4} B + \frac{\varepsilon}{4} B \subset \frac{\varepsilon}{2} B,$$

since B is convex. Therefore, $\|x_m - x\|_B < \varepsilon$, for all $m \geq n_0$. Hence, $\mathcal{A}[B]$ is sequentially complete, therefore a Banach algebra. $\square$

Proposition 2.2.6 *If $\mathfrak{B}_0$ contains a basic subcollection $\mathfrak{B}_1$, such that $\mathcal{A}[B]$ is a Banach algebra for every $B \in \mathfrak{B}_1$, then $\mathcal{A}[\tau]$ is pseudo-complete.*

Proof Let $B \in \mathfrak{B}_0$ and $(x_n)_{n \in \mathbb{N}}$ be a $\|\cdot\|_B$-Cauchy sequence in $\mathcal{A}[B]$. There is some $B_1 \in \mathfrak{B}_1$, such that $B \subset B_1$ and hence $\|x\|_{B_1} \leq \|x\|_B$, for $x \in \mathcal{A}[B]$. Therefore, $(x_n)_{n \in \mathbb{N}}$ is a $\|\cdot\|_{B_1}$-Cauchy sequence in $A[B_1]$, hence there is an element $x \in A[B_1]$, such that $x_n \to x$ with respect to $\|\cdot\|_{B_1}$. Following similar arguments to those of the proof of Proposition 2.2.5 we have that $x \in \mathcal{A}[B]$ and $x_n \to x$ with respect to $\|\cdot\|_B$. Therefore, $\mathcal{A}[B]$ is a Banach algebra and since B is an arbitrary element of $\mathfrak{B}_0$, we have that $\mathcal{A}[\tau]$ is pseudo-complete. $\square$

In the example that follows we are going to show that the converse of Proposition 2.2.5 is not valid. Towards this direction, the notion of convergence in the sense of Mackey will be proven useful (cf. [74, Chapter 3, §5] and/or [131, Chapter IV, 3.]). We recall that in a topological vector space $E[\tau]$, a sequence $(x_n)_{n \in \mathbb{N}}$ is *a Mackey Cauchy sequence* if there is some sequence $(\varepsilon_n)_{n \in \mathbb{N}}$ of positive numbers, which tends to 0 and a bounded subset B of E, such that

$$\forall\, n \in \mathbb{N},\ x_n - x_m \in \varepsilon_n B,\ \text{for } m > n.$$

Furthermore, the sequence $(x_n)_{n \in \mathbb{N}}$ is *Mackey convergent* to an element $x \in E$, if there is some sequence $(\varepsilon_n)_{n \in \mathbb{N}}$ of positive numbers, which tends to 0 and a bounded subset B of E, such that

$$x_n - x \in \varepsilon_n B,\ \forall\, n \in \mathbb{N}.$$

The topological vector space $E[\tau]$ is *Mackey complete*, if every Mackey Cauchy sequence in E is Mackey convergent. We note that *if a locally convex space $E[\tau]$ is metrizable and sequentially complete, then it is also Mackey complete.*

Indeed, a metrizable locally convex space $E[\tau]$ is bornological (see Definition 4.3.14(2) in Sect. 4.2 and [131, p. 61, 8.1]). But then, $E[\tau]$ becomes a Mackey space (ibid., p. 132, 3.4). This means that the topology τ is the Mackey topology $\tau(E, E')$, E' the dual of $E[\tau]$. Namely, $\tau(E, E')$ is the topology of uniform convergence on the absolutely convex $\sigma(E', E)$-compact subsets of E' (cf. Definitition 4.3.11 and (4.3.12) in Sect. 4.3). But, a metrizable and sequentially complete locally convex space is complete, hence $E[\tau]$ is Mackey complete.

The example that follows illustrates a locally convex algebra, which is metrizable and pseudo-complete, but not Mackey complete and so a fortiori not sequentially complete.

Example 2.2.7 Let $\mathcal{A}$ be the algebra of all polynomials in one variable with complex coefficients. We endow $\mathcal{A}$ with the topology τ of uniform convergence on compact subsets of the positive real line $\mathbb{R}^+$. Then, clearly $\mathcal{A}[\tau]$ is metrizable. Moreover, $\mathcal{A}_0$ is the set of all constant functions. Yet, the family $(\mathfrak{B})_{\mathcal{A}}$ of subsets of $\mathcal{A}$ has a greatest member, say B_0, which consists of all constant functions of absolute value not exceeding 1. Therefore, $\mathcal{A}[B_0]$ is a Banach algebra (see (2.2.1) and (2.2.2)) and so from Proposition 2.2.6 we have that $\mathcal{A}[\tau]$ is pseudo-complete.

Consider a sequence of polynomials $(P_n)_{n \in \mathbb{N}}$ in $\mathcal{A}$, such that

$$P_n(x) = \sum_{r=0}^{n} (-1)^{r+1} \frac{x^{2r+1}}{(2r+1)!}. \tag{2.2.3}$$

Then, $P_n \to \sin(x)$, with respect to τ and the set

$$B = \left\{ P \in \mathcal{A} : |P(x)| \leq e^x, \ \forall\, x \in \mathbb{R}^+ \right\}$$

is clearly τ-bounded. Since,

$$|P_n(x) - \sin(x)|e^{-x} \to 0, \quad \text{uniformly on } \mathbb{R}^+,$$

we deduce that $(P_n)_{n \in \mathbb{N}}$ is a Mackey Cauchy sequence. Nevertheless, $(P_n)_{n \in \mathbb{N}}$ is not convergent to any element of $\mathcal{A}$. Hence, $\mathcal{A}[\tau]$ *is not Mackey complete.*

Note that this last implication is based on the fact that *if a sequence* $(x_n)_{n \in \mathbb{N}}$ *in a topological vector space* $E[\tau]$ *is Mackey convergent to an element, say* $x \in E$, *then* $x_n \to x$, *with respect to* τ. Indeed, since $x_n \to x$ in the sense of Mackey, there is a sequence of positive real numbers $(\varepsilon_n)_{n \in \mathbb{N}}$, which tends to 0 and a bounded subset V of E, such that $x_n - x \in \varepsilon_n V$, for all $n \in \mathbb{N}$. Let U be a 0-neighbourhood in E. There is $\varepsilon > 0$, such that $V \subset \varepsilon U$, hence $\varepsilon_n V \subset \varepsilon_n \varepsilon U$, $n \in \mathbb{N}$. Since $\varepsilon_n \to 0$, there is $n_0 \in \mathbb{N}$, such that $\varepsilon_n < \frac{1}{\varepsilon+1}$, for every $n \geq n_0$. Suppose, without loss of generality, that U is also balanced. Then, we have that $\varepsilon_n V \subset \varepsilon_n \varepsilon U \subset U$, for every $n \geq n_0$. Therefore, $x_n - x \in U$, for all $n \geq n_0$, thus $x_n \to x$ with respect to τ.

Proposition 2.2.8

(1) If $\mathcal{A}[\tau]$ is a pseudo-complete algebra and $\mathcal{B}$ is a closed subalgebra of $\mathcal{A}$, then $\mathcal{B}[\tau\!\restriction_{\mathcal{B}}]$ is also pseudo-complete.
(2) If $\mathcal{A}[\tau]$ has no identity, then its unitization $\mathcal{A}_1[\tau_1]$ is pseudo-complete, if and only if, $\mathcal{A}[\tau]$ is pseudo-complete.
(3) If $\mathcal{A}[\tau]$ has an identity, say e, then the collection $\left\{ B : B \in \mathfrak{B}_0, e \in B \right\}$ is a basic subcollection of $\mathfrak{B}_0$.

Proof

(1) Let $(\mathfrak{B}_0)_{\mathcal{A}}$, $(\mathfrak{B}_0)_{\mathcal{B}}$ be the corresponding collections of subsets for $\mathcal{A}[\tau]$ and $\mathcal{B}[\tau\!\restriction_{\mathcal{B}}]$ respectively. Given the fact that $\mathcal{B}$ is a closed subalgebra of $\mathcal{A}$ it is straightforward that $(\mathfrak{B}_0)_{\mathcal{B}} \subset (\mathfrak{B}_0)_{\mathcal{A}}$. Hence, the result follows.
(2) If $\mathcal{A}_1[\tau_1]$ is pseudo-complete, then from (1) we have that $\mathcal{A}[\tau]$ is pseudo-complete. In the inverse direction let $\mathcal{A}[\tau]$ be pseudo-complete. For $B \in (\mathfrak{B}_0)_{\mathcal{A}_1}$ we will show that $\mathcal{A}_1[B]$ is a Banach algebra with respect to the norm $\|\cdot\|_B$. We first note that if $(\lambda, x) \in B$, then $\lambda \in D = \{z \in \mathbb{C} : |z| \leq 1\}$: indeed

since $B^2 \subset B$, by induction we have that $(\lambda, x)^n = (\lambda^n, x_n) \in B$, for all $n \in \mathbb{N}$, where x_n is an element of $\mathcal{A}$. Since B is bounded the fact that $(\lambda^n, x_n) \in B$ is possible only for those complex numbers λ that belong to D.

Consider now $B_1 = \frac{1}{3} D + \frac{1}{3} B$. It is easy to show that $B_1 \in (\mathfrak{B}_0)_{\mathcal{A}_1}$. Let $B_2 = \{x \in \mathcal{A} : (\lambda, x) \in B, \text{ for some } \lambda \in \mathbb{C}\}$. Then, $\frac{1}{3} B_2 \subset B_1 \cap \mathcal{A}$. If C denotes the closed absolutely convex hull of $\frac{1}{3} B_2$, then $C \subset B_1 \cap \mathcal{A}$ and $C \in (\mathfrak{B}_0)_{\mathcal{A}}$. Hence, by the assumption of pseudo-completeness for $\mathcal{A}[\tau]$ we have that $\mathcal{A}[C]$ is a Banach algebra. Let $\mathcal{A}[C]_1[\| \cdot \|_C^1]$ be the unitization of $\mathcal{A}[C]$, with norm $\|(\lambda, x)\|_C^1 = |\lambda| + \|x\|_C$, for $\lambda \in \mathbb{C}$, $x \in \mathcal{A}[C]$. If $B_e \equiv U(\mathcal{A}[C]_1)$ is the closed unit ball of $\mathcal{A}[C]_1$, with respect to the norm $\| \cdot \|_C^1$, it can easily be shown that $\frac{1}{6} B \subset B_e$ and therefore $\mathcal{A}_1[B] \subset \mathcal{A}[C]_1$. From the latter inclusion we have that, on $\mathcal{A}_1[B]$, the topology of the norm $\| \cdot \|_C^1$ is weaker than that of $\| \cdot \|_B$. Indeed, let us consider a sequence $(x_n)_{n \in \mathbb{N}} \subset \mathcal{A}_1[B]$, such that $x_n \to x$, with respect to $\| \cdot \|_B$ and $\varepsilon > 0$. Then,

$$\exists \, m_0 \in \mathbb{N} : \frac{1}{m_0} < \varepsilon \text{ and } n_0 \in \mathbb{N} \text{ with } \|x_n - x\|_B < \frac{1}{6m_0}, \, \forall \, n \geq n_0.$$

Therefore,

$$x_n - x \in \frac{1}{m_0} B_e, \, \forall \, n \geq n_0, \text{ so that } \|x_n - x\|_C^1 \leq \frac{1}{m_0} < \varepsilon, \text{ for } n \geq n_0.$$

Hence, $x_n \to x$, with respect to the norm $\| \cdot \|_C^1$ on $\mathcal{A}[C]_1$.

Since $\tau \prec \| \cdot \|_C$ on $\mathcal{A}[C]$ we have that the norm topology $\| \cdot \|_C^1$ on $\mathcal{A}[C]_1$ is stronger than the product topology induced on $\mathcal{A}[C]_1$ by the product topology from $\mathcal{A}[C] \times \mathbb{C}$. Based on this relation between the two topologies and since $B \in (\mathfrak{B}_0)_{\mathcal{A}_1}$ is closed, with respect to the product topology on $\mathcal{A}_1$, we have that B is $\| \cdot \|_C^1$-closed in $\mathcal{A}[C]_1$. Aiming to show that $\mathcal{A}_1[B]$ is a Banach algebra, let $(x_n)_{n \in \mathbb{N}}$ be a $\| \cdot \|_B$-Cauchy sequence in $\mathcal{A}_1[B]$. Since $\| \cdot \|_C^1 \prec \| \cdot \|_B$ on $\mathcal{A}_1[B]$, we have that $(x_n)_{n \in \mathbb{N}}$ is a $\| \cdot \|_C^1$-Cauchy sequence in $\mathcal{A}[C]_1$. Since $\mathcal{A}[C]_1[\| \cdot \|_C^1]$ is a Banach algebra, there exists an $x \in \mathcal{A}[C]_1$, such that $x_n \to x$, with respect to $\| \cdot \|_C^1$. Let $\varepsilon > 0$. Then,

$$\exists \, n_0 \in \mathbb{N} : \|x_n - x_m\|_B < \frac{\varepsilon}{4}, \, \forall \, m, n \geq n_0.$$

Hence, $x_n - x_{n_0} \in \frac{\varepsilon}{4} B$, for all $n \geq n_0$. Given that $x_n - x_{n_0} \to x - x_{n_0}$ with respect to $\| \cdot \|_C^1$ and that B is $\| \cdot \|_C^1$-closed, we have that $x - x_{n_0} \in \frac{\varepsilon}{4} B$. Therefore,

$$x_n - x = (x_n - x_{n_0}) + (x_{n_0} - x) \in \frac{\varepsilon}{4} B + \frac{\varepsilon}{4} B \subset \frac{\varepsilon}{2} B, \, \forall \, n \geq n_0.$$

So, $\|x_n - x\|_B < \varepsilon$, for all $n \geq n_0$, i.e. $x_n \to x$ with respect $\| \cdot \|_B$. Therefore, $\mathcal{A}_1[B]$ is a Banach algebra and thus $\mathcal{A}_1[\tau_1]$ is pseudo-complete.

(3) For a subset $B \in \mathfrak{B}_0$ let S be the closed absolutely convex hull of $B \cup \{e\}$. Then it is clear that $S \in \mathfrak{B}_0$, $B \subset S$ and $e \in S$. The result then immediately follows.

$\square$

▶ If $\mathcal{A}$ is a locally convex algebra with identity e, we shall *denote the basic subcollection* of Proposition 2.2.8(3) *by* $(\mathfrak{B})_{\mathcal{A}}$ *(or simply by* $\mathfrak{B}$ *if no confusion arises).* In what follows, *for simplicity of notation, we retain the same symbol,* i.e., $\mathfrak{B}$, *to refer also to the collection of subsets of Definition 2.2.2, for the case where $\mathcal{A}$ does not have an identity.*

Theorem 2.2.10 below, will be proven very helpful in various arguments in the sections of this chapter. In the course of the proof of Theorem 2.2.10, the following general result, i.e. Proposition 2.2.9, is used, which we record for sake of completeness. For the proof of Proposition 2.2.9, the reader is referred to [131, §4.2, p. 83, Theorem].

Proposition 2.2.9 *Let E, F be locally convex spaces, such that E is barrelled, or let E, F be topological vector spaces, such that E is a Baire space. Then, every bounded subset H of $L(E, F)$ with respect to the topology of pointwise convergence on E, where $L(E, F)$ denotes all continuous linear maps on E into F, is equicontinuous.*

Theorem 2.2.10 *If $\mathcal{A}[\tau]$ is a commutative and pseudo-complete locally convex algebra, then $\mathfrak{B}$ is outer-directed by inclusion, that is for $B, C \in \mathfrak{B}$, there is $F \in \mathfrak{B}$, such that $B \cup C \subset F$.*

Proof Let us first consider the case where $\mathcal{A}$ has an identity e. For $B, C \in \mathfrak{B}$ and for every $b \in B$ consider the linear map $L_b : \mathcal{A}[C] \rightarrow \mathcal{A}$, such that $L_b(a) = ba$, $a \in \mathcal{A}[C]$. From the separate continuity of multiplication in $\mathcal{A}[\tau]$ we have that L_b is continuous. Moreover, since B is bounded we have that the set $\Omega = \{L_b : b \in B\}$ is bounded with respect to the topology of pointwise convergence on $\mathcal{A}[C]$. Hence, since $\mathcal{A}[C]$ is a Banach algebra and $\mathcal{A}[\tau]$ a locally convex algebra, from Proposition 2.2.9 we have that Ω is equicontinuous. Therefore, if V is a 0-neighbourhood in $\mathcal{A}[\tau]$ there is a 0-neighbourhood U in $\mathcal{A}[C]$, such that $L_b(U) \subset V$, for every $b \in B$. Since C is a bounded subset of $\mathcal{A}[C]$ there is $\varepsilon > 0$, such that $C \subset \varepsilon U$. Thus, for every $b \in B$, we have that $bC \subset L_b(\varepsilon U) \subset \varepsilon V$, that is BC is a bounded subset of $\mathcal{A}[\tau]$. Furthermore, since $\mathcal{A}$ is commutative, $(BC)^2 = B^2 C^2$, hence $(BC)^2 \subset BC$. Therefore, the closed absolutely convex hull of BC belongs to $\mathfrak{B}_0$. Therefore, by Proposition 2.2.8, there is an $F \in \mathfrak{B}$, such that $BC \subset F$. Since $e \in B \cap C$ we conclude that $B \cup C \subset BC \subset F$.

In case $\mathcal{A}$ has no identity, for $B, C \in \mathfrak{B}$ we consider B_1, C_1 to be the closed absolutely convex hulls of the sets $B \cup \{(\lambda, 0) : \lambda \in D\}$ and $C \cup \{(\lambda, 0) : \lambda \in D\}$ respectively, where D denotes the closed unit disk in $\mathbb{C}$. Then $B_1, C_1 \in (\mathfrak{B}_0)_{\mathcal{A}_1}$

and so, as seen above, there is an $F \in (\mathfrak{B})_{A_1}$, such that $B_1 \cup C_1 \subset F$. Therefore, $B \cup C \subset F \cap A$, where $F \cap A \in (\mathfrak{B})_A$.   □

Corollary 2.2.11 *If $A[\tau]$ is a commutative and pseudo-complete locally convex algebra, then A_0 is a subalgebra of A.*

Proof From Proposition 2.2.4 we have that $A_0 = \cup\{A[B] : B \in \mathfrak{B}\}$. Hence, for $x, y \in A_0$ there are $B, C \in \mathfrak{B}$ such that $x \in A[B]$ and $y \in A[C]$. By Theorem 2.2.10, there is an $F \in \mathfrak{B}$ such that $B \cup C \subset F$. Therefore, $x + y, xy \in A[F]$ and thus $x + y, xy \in A_0$.   □

Remark 2.2.12 Note that if $A[\tau]$ is not commutative, then A_0 need not even be a linear subspace. But if $A[\tau]$ is pseudo-complete and $x, y \in A$, such that $xy = yx$, then taking the closed subalgebra B of $A[\tau]$ generated by x, y, B is a commutative pseudo-complete locally convex algebra (see Proposition 2.2.8(i)), therefore Corollary 2.2.11 implies that $xy, x+y \in B_0$, consequently $xy, x+y \in A_0$ too.

Corollary 2.2.13 *Let $A[\tau]$ be a commutative and pseudo-complete locally convex algebra and $B_0 = \cup\{B : B \in \mathfrak{B}\}$. Then, B_0 is absolutely convex, $B_0^2 \subset B_0$ and B_0 is absorbent in A_0. Hence, the Minkowski functional $\| \cdot \|_{B_0}$ of B_0 is a submultiplicative seminorm on A_0.*

Proof Let $0 \neq x \in A_0$. Then there is $B \in \mathfrak{B}$ such that $x \in A[B]$. Hence, $x = \lambda y$, for some $\lambda \in \mathbb{C}$, $y \in B$. Since B is balanced we have that $\frac{1}{|\lambda|} x = \frac{\lambda}{|\lambda|} y \in B \subset B_0$. Therefore, B_0 is absorbent in A_0. The properties that $B_0 \subset B_0^2$ and that of B_0 being absolutely convex result from the fact that every $B \in \mathfrak{B}$ enjoys the same properties and from Theorem 2.2.10.   □

For a locally convex algebra $A[\tau]$ a useful tool is the notion of the radius of boundedness for an element $x \in A$. The *radius of boundedness* $\beta(\cdot)$ of x is defined by the relation

$$\beta(x) := \inf\left\{\lambda > 0 : \left\{(\frac{1}{\lambda} x)^n : n \in \mathbb{N}\right\} \text{ is bounded}\right\},$$

where $\inf \emptyset = +\infty$. In Sect. 2.3 we are going to see the relation between the spectral radius and the radius of boundedness of an element in a locally convex algebra. In the proposition, which follows some basic properties of $\beta(x)$ are listed. The proofs of these properties are obvious, so that are omitted.

Proposition 2.2.14 *Let $A[\tau]$ be a locally convex algebra and $x \in A$. Then, the following hold:*

(1) $\beta(x) \geq 0$ *and* $\beta(\lambda x) = |\lambda|\beta(x)$, *for* $\lambda \in \mathbb{C}$, *with the convention that* $0 \cdot \infty = 0$.
(2) $\beta(x) < +\infty$, *if and only if,* $x \in A_0$.
(3) $\beta(x) = \inf\left\{\lambda > 0 : (\frac{1}{\lambda} x)^n \to 0, \text{ for } n \to \infty\right\}$.

(4) *For $x \in \mathcal{A}_0$, if $\lambda \in \mathbb{C}$, such that $|\lambda| > \beta(x)$, then $(\frac{1}{\lambda} x)^n \to 0$, for $n \to \infty$. If $0 < |\lambda| < \beta(x)$, then the set $\{(\frac{1}{\lambda} x)^n : n \in \mathbb{N}\}$ is unbounded.*

Proposition 2.2.15 *Let $\mathcal{A}[\tau]$ be a locally convex algebra. Then, the restriction of β to $\mathcal{A}_0$ is the Minkowski functional $\| \cdot \|_{B_0}$ of $B_0 = \bigcup \{B : B \in \mathfrak{B}\}$.*

Proof Let $x \in \mathcal{A}_0$ and $\lambda > 0$, such that the set $S = \{(\frac{1}{\lambda} x)^n : n \in \mathbb{N}\}$ is bounded. Then the closed absolutely convex hull of S, say B, belongs to $\mathfrak{B}$. Thus, $B \subset B_0$ and therefore $\frac{1}{\lambda} x \in B_0$. Hence, $\|x\|_{B_0} \leq \beta(x)$.

For the reverse inequality, let $\mu > 0$ such that $x \in \mu B_0$. Then, there is some $B \in \mathfrak{B}$, such that $\frac{1}{\mu} x \in B$. Since $B^2 \subset B$, by induction we have that $\{(\frac{1}{\mu} x)^n : n \in \mathbb{N}\} \subset B$, hence $\{(\frac{1}{\mu} x)^n : n \in \mathbb{N}\}$ is bounded. Therefore,

$$\left\{ \mu > 0 : x \in \mu B_0 \right\} \subset \left\{ \mu > 0 : \left\{ (\frac{1}{\mu} x)^n : n \in \mathbb{N} \right\} \text{ is bounded} \right\},$$

from which follows that $\beta(x) \leq \|x\|_{B_0}$.   □

Based on Corollary 2.2.13 and on Proposition 2.2.15 we have the following result.

Corollary 2.2.16 *Let $\mathcal{A}[\tau]$ be a commutative and pseudo-complete locally convex algebra. Then, the restriction of β to $\mathcal{A}_0$ is a submultiplicative seminorm.*

Corollary 2.2.17 *Let $\mathcal{A}[\tau]$ be a locally convex algebra and $x \in \mathcal{A}_0$. Then, $\beta(x) = \inf \{\|x\|_B : B \in \mathfrak{B}, x \in \mathcal{A}[B]\}$.*

Proof Let $\alpha > 0$, such that $x \in \alpha B_0$. Then, there is $B \in \mathfrak{B}$, such that $x \in \alpha B$. Hence,

$$\inf \{\alpha > 0 : x \in \alpha B_0\} \geq \inf \{\|x\|_B : B \in \mathfrak{B}, x \in \mathcal{A}[B]\}$$

and thus from Proposition 2.2.15 we have that

$$\beta(x) \geq \inf \{\|x\|_B : B \in \mathfrak{B}, x \in \mathcal{A}[B]\}.$$

On the other hand, if $B \in \mathfrak{B}$ with $x \in \mathcal{A}[B]$, it is clear that for every $\varepsilon > 0$: $\|x\|_B + \varepsilon \geq \inf \{\alpha > 0 : x \in \alpha B_0\}$. Therefore, $\inf \{\|x\|_B : B \in \mathfrak{B}, x \in \mathcal{A}[B]\} \geq \beta(x)$.   □

The next proposition provides us with two other relations with which the radius of boundedness of an element can be expressed. For a locally convex algebra $\mathcal{A}[\tau]$, let $\mathcal{A}'$ denote the topological dual of $\mathcal{A}$ and let Γ_τ be a family of seminorms defining the locally convex topology τ of $\mathcal{A}$. Consider the following formulae

$$\beta'(x) = \sup\left\{\limsup_{n\to\infty}|f(x^n)|^{\frac{1}{n}} : f \in \mathcal{A}'\right\} \text{ and}$$

$$\beta''(x) = \sup\left\{\limsup_{n\to\infty}|p(x^n)|^{\frac{1}{n}} : p \in \Gamma_\tau\right\}, \ x \in \mathcal{A}.$$

Proposition 2.2.18 *Let $\mathcal{A}[\tau]$ be a locally convex algebra, and $x \in \mathcal{A}$. Then,*

$$\beta(x) = \beta'(x) = \beta''(x).$$

Proof Let $x \notin \mathcal{A}_0$. Then, $\beta(x) = +\infty$ and hence, trivially, $|f(x)| \leq \beta(x)$, for every $f \in \mathcal{A}'$.

Consider now $f \in \mathcal{A}'$ and suppose that $x \in \mathcal{A}_0$ and $\lambda > 0$ with $\lambda > \beta(x)$. Then, from Proposition 2.2.14(4), we have $(\frac{1}{\lambda} x)^n \to 0$, for $n \to \infty$. Hence, there is $n_0 \in \mathbb{N}$, such that $|f(\frac{1}{\lambda^n} x^n)| \leq 1, \forall n \geq n_0$. Therefore, $\limsup\limits_{n\to\infty}|f(x^n)|^{\frac{1}{n}} \leq \lambda$, for each $f \in \mathcal{A}'$ and $\lambda > \beta(x)$. Thus, $\beta'(x) \leq \beta(x)$, for all $x \in \mathcal{A}$. The same arguments as above hold if in place of the functional f we have a seminorm $p \in \Gamma_\tau$. Hence, $\beta''(x) \leq \beta(x), x \in \mathcal{A}$.

Next we show that $\beta(x) \leq \beta'(x)$. For $\beta'(x) = +\infty$ this is trivial, so we suppose that $\beta'(x) < +\infty$. In this case, for $\lambda > 0$ with $\lambda > \beta'(x)$, there is some $n_0 \in \mathbb{N}$, such that for any

$$f \in \mathcal{A}', \ |f((\frac{1}{\lambda} x)^n)| < 1, \ \forall n \geq n_0.$$

Therefore, the set $\{(\frac{1}{\lambda} x)^n : n \in \mathbb{N}\}$ is weakly bounded and thus τ-bounded by [128, p. 67, Theorem 1]. Hence, $\lambda \geq \beta(x)$ and so $\beta(x) \leq \beta'(x)$.

The result will be proven once we show that $\beta'(x) \leq \beta''(x)$. Indeed, if $f \in \mathcal{A}'$, then from the continuity of f there is some $p \in \Gamma_\tau$ and $M > 0$, such that

$$|f(x^n)|^{\frac{1}{n}} \leq M^{\frac{1}{n}} p(x^n)^{\frac{1}{n}}, \forall n \in \mathbb{N}.$$

Therefore,

$$\limsup_{n\to\infty}|f(x^n)|^{\frac{1}{n}} \leq \limsup_{n\to\infty}|p(x^n)|^{\frac{1}{n}} \leq \beta''(x),$$

so $\beta'(x) \leq \beta''(x)$. $\qquad\square$

2.3   Spectrum and Spectral Radius

The spectrum of an element x *of an arbitrary algebra* $\mathcal{A}$, will be denoted by $sp(x)$ or by $sp_{\mathcal{A}}(x)$, when more than one algebra is involved; more precisely, if $\mathcal{A}$ has an identity e

$$sp(x) := \{\lambda \in \mathbb{C} : \lambda e - x \notin G_{\mathcal{A}}\}.$$

If $\mathcal{A}$ has no identity, then

$$sp_{\mathcal{A}}(x) := sp_{\mathcal{A}_1}(x).$$

Throughout this book $\mathbb{C}^*$ will denote *the one point compactification of* $\mathbb{C}$. A partial algebraic structure is defined on $\mathbb{C}^*$ as follows:

$$\infty + \lambda = \infty, \ \lambda \in \mathbb{C}; \ \infty \cdot \lambda = \infty, \ \lambda \in \mathbb{C}^* \backslash \{0\}; \ \text{and} \ \overline{\infty} = \infty.$$

Now, in case of a locally convex algebra $\mathcal{A}[\tau]$, G.R. Allan [4, Definition (3.1)] defined a new spectrum of an element $x \in \mathcal{A}$ as follows:

Definition 2.3.1 Let $\mathcal{A}[\tau]$ be a locally convex algebra with an identity e. The *spectrum of an element* $x \in \mathcal{A}$, denoted by $\sigma(x)$ (or by $\sigma_{\mathcal{A}}(x)$, when more than one algebra is involved), is the subset of $\mathbb{C}^*$ defined by

$$\sigma_{\mathcal{A}}(x) = \{\lambda \in \mathbb{C} : \lambda e - x \text{ has no inverse in } \mathcal{A}_0\} \cup \{\infty \Leftrightarrow x \notin \mathcal{A}_0\}.$$

In case $\mathcal{A}$ has no identity then $\sigma_{\mathcal{A}}(x) := \sigma_{\mathcal{A}_1}(x, 0)$.

Concerning the latter, since from Proposition 2.2.8(2), $\mathcal{A}[\tau]$ is pseudo-complete, if and only if, $\mathcal{A}_1[\tau_1]$ is pseudo-complete, we can assume without loss of generality that the locally convex algebra $\mathcal{A}[\tau]$ has an identity e.

The complement of $\sigma_{\mathcal{A}}(x)$ in $\mathbb{C}^*$ is denoted by $\rho_{\mathcal{A}}(x)$ (or simply by $\rho(x)$, if no confusion arises) and is called the *resolvent set of* x.

If $\mathcal{A}[\tau]$ is a locally convex $*$-algebra, it is clear that

$$\sigma_{\mathcal{A}}(x^*) = \{\overline{\lambda} : \lambda \in \sigma_{\mathcal{A}}(x)\}, \ \forall \, x \in \mathcal{A}.$$

Proposition 2.3.2 *Let* $\mathcal{A}[\tau]$ *be a locally convex algebra and* $x \in \mathcal{A}$. *If* $\mathcal{C}$ *denotes a maximal commutative subalgebra that contains* x, *then* $e \in \mathcal{C}$ *and* $\sigma_{\mathcal{A}}(x) = \sigma_{\mathcal{C}}(x)$. *Also if* $\mathcal{A}[\tau]$ *is pseudo-complete, then* $\mathcal{C}[\tau\upharpoonright_{\mathcal{C}}]$ *is pseudo-complete.*

Proof Clearly the facts that $e \in \mathcal{C}$ and that $\mathcal{C}$ is closed follow from the assumption of $\mathcal{C}$ being a maximal commutative subset of $\mathcal{A}$. Therefore, from Proposition 2.2.8(1) we have that $\mathcal{C}[\tau\upharpoonright_{\mathcal{C}}]$ is pseudo-complete if $\mathcal{A}[\tau]$ is supposed to be pseudo-complete. If $\lambda \notin \sigma_{\mathcal{A}}(x)$ and $\lambda \neq \infty$, then $(\lambda e - x)^{-1} \in \mathcal{A}_0$. Since $x \in \mathcal{C}$ and $\mathcal{C}$ is maximal commutative it is straightforward that $(\lambda e - x)^{-1} \in \mathcal{C}$. So, $(\lambda e - x)^{-1}$ is a bounded element in $C[\tau\restriction_C]$. Hence, $\lambda \notin \sigma_C(x)$. In case $\infty \notin \sigma_A(x)$ then $x \in A_0$. Thus, x is a bounded element in C, hence $\infty \notin \sigma_C(x)$. Therefore, $\sigma_C(x) \subset \sigma_A(x)$. The other inclusion is trivial. □

Towards the development of a functional calculus for a pseudo-complete locally convex algebra that we are going to describe in Sect. 2.4, the definition of the following function is crucial.

Definition 2.3.3 Let $A[\tau]$ be a locally convex algebra and $x \in A$. The resolvent of x is the function $R(x)$ defined by $R(x)\lambda \equiv R_\lambda(x) = (\lambda e - x)^{-1}$, for all λ, such that the inverse of $\lambda e - x$ is defined. When there is no danger of confusion with respect to which element x the resolvent function is being considered, the symbol R_λ is used instead of $R_\lambda(x)$.

The resolvent of x is thus a function on a subset of $\mathbb{C}^*$, which takes values in A. We recall that a function f on a subset of $\mathbb{C}^*$ that takes values in a locally convex space $E[\tau]$ is holomorphic on some open set G of $\mathbb{C}$ if the limit $\lim\limits_{\lambda \to \lambda_0} \frac{f(\lambda) - f(\lambda_0)}{\lambda - \lambda_0}$ exists for every $\lambda_0 \in G$. Moreover, f is holomorphic at ∞ if the limit $\lim\limits_{\lambda \to 0} f(\frac{1}{\lambda})$ exists and the function

$$
g(\lambda) = \begin{cases} f(\frac{1}{\lambda}), & \text{if } \lambda \neq 0 \\ \lim\limits_{\lambda \to 0} f(\frac{1}{\lambda}), & \text{if } \lambda = 0 \end{cases}
$$

is holomorphic in some neighbourhood of 0. In order to obtain some useful results concerning the resolvent R_λ we are going to use the weak topology on A. The reason behind this choice can be seen by the following direct observation: if f is a holomorphic function that takes values in any locally convex algebra $A[\tau]$ and $x' \in A'$, then the function $\lambda \mapsto x'(f(\lambda))$ is a complex valued holomorphic function.

For any locally convex algebra $A[\tau]$ we have the following result concerning the weak topology $\sigma(A, A')$ on A.

Lemma 2.3.4 *Let $A[\tau]$ be a locally convex algebra. Then, A is also a locally convex algebra with respect to $\sigma(A, A')$.*

Proof It suffices to show that the multiplication in A is separately continuous with respect to $\sigma(A, A')$. Let $f \in A'$, $y \in A$. Consider the maps f_y, f^y from A into $\mathbb{C}$, such that $f_y(x) = f(yx)$, $f^y(x) = f(xy)$, $x \in A$. Then, from the separate continuity of multiplication in A, we have that f_y, $f^y \in A'$, from which the result is easily derived. □

Lemma 2.3.5 *Let $A[\tau]$ be a locally convex algebra and $x \in A$. Suppose that $R(x)$ is weakly holomorphic at μ ($\neq \infty$). Then, $R(x)$ has weak derivatives $R^{(n)}(x)$, $n \in \mathbb{N}$, at μ, given by $R_\mu^{(n)} = (-1)^n n! R_\mu^{n+1}$, $n \in \mathbb{N}$.*

Proof We note that for λ, μ in the domain of $R(x)$ we have that

$$
\begin{aligned}
R_\lambda - R_\mu &= (\lambda e - x)^{-1} - (\mu e - x)^{-1} \\
&= (\lambda e - x)^{-1}\big((\mu e - x) - (\lambda e - x)\big)(\mu e - x)^{-1} \\
&= -(\lambda - \mu)R_\lambda R_\mu.
\end{aligned}
$$

Therefore, based on Lemma 2.3.4 and the assumption for $R(x)$ being weakly holomorphic, hence weakly continuous at μ we have that

$$
\frac{R_\lambda - R_\mu}{\lambda - \mu} = -R_\lambda R_\mu \underset{\lambda \to \mu}{\to} -R_\mu^2, \quad \text{with respect to } \sigma(\mathcal{A}, \mathcal{A}').
$$

Hence, the result follows for $n = 1$. Let us suppose that the result holds for all $n = 1, 2, \ldots, m$. Then,

$$
\begin{aligned}
\frac{R_\lambda^{(m)} - R_\mu^{(m)}}{\lambda - \mu} &= (-1)^m m! (\lambda - \mu)^{-1}(R_\lambda^{m+1} - R_\mu^{m+1}) \\
&= (-1)^m m! (\lambda - \mu)^{-1}(R_\lambda - R_\mu)(R_\lambda^m + R_\lambda^{m-1} R_\mu + \cdots + R_\mu^m) \\
&= (-1)^{m+1} m! \sum_{r=0}^{m} R_\lambda^{r+1} R_\mu^{m+1-r}.
\end{aligned}
$$

We show that $R^{r+1}(x)$ is weakly continuous at μ, for $r = 0, 1, \ldots, m$. Towards this direction, given any $f \in \mathcal{A}'$ consider the complex valued function $\varphi(\lambda) = f(R_\lambda)$, which is holomorphic at μ by the assumption of the lemma. By the inductive hypothesis and taking into account the continuity of f we have that $\varphi^{(r)}(\lambda) = f(R_\lambda^{(r)})$ for λ in a neighbourhood of μ and for $r = 0, 1, \ldots, m$. The function $\varphi^{(r)}$ is necessarily continuous at μ for $r = 0, 1, \ldots, m$. Therefore, $R^{r+1}(x)$ is weakly continuous at μ for $r = 0, 1, \ldots, m$. Hence, we have that

$$
\frac{R_\lambda^{(m)} - R_\mu^{(m)}}{\lambda - \mu} \to (-1)^{m+1}(m+1)! R_\mu^{m+2}, \quad \text{for } \lambda \to \mu, \text{ with respect to } \sigma(\mathcal{A}, \mathcal{A}').
$$

Thus, the result of the lemma is established by induction.   □

Lemma 2.3.6 *Let $\mathcal{A}[\tau]$ be a locally convex algebra and $x \in \mathcal{A}$, such that $R(x)$ is holomorphic at ∞ with respect to the weak topology $\sigma(\mathcal{A}, \mathcal{A}')$. If*

$$
S_\lambda = \begin{cases} R_{\frac{1}{\lambda}}, & \lambda \neq 0, \\ \lim_{\lambda \to 0} R_{\frac{1}{\lambda}}, & \lambda = 0 \end{cases}
$$

then, in some neighbourhood of 0, S_λ has weak derivatives of all orders given by $S_\lambda^{(n)} = n! x^{n-1} (e - \lambda x)^{-(n+1)}$, for $n \in \mathbb{N}$ and $S_\lambda = \lambda (e - \lambda x)^{-1}$.

Proof For $\lambda \neq 0$ we have that $S_\lambda = R_{\frac{1}{\lambda}} = \left(\frac{1}{\lambda} e - x\right)^{-1} = \lambda (e - \lambda x)^{-1}$. Let l be the weak limit of $R_{\frac{1}{\lambda}}$, as $\lambda \to 0$ (note that the existence of l is ensured due to the assumption of $R(x)$ being weakly holomorphic at ∞). Then, $(e - \lambda x) S_\lambda = S_\lambda - x(\lambda S_\lambda)$ and the right hand side of the previous equation will tend weakly to l, as $\lambda \to 0$, based on Lemma 2.3.4. On the other hand, we have that

$$(e - \lambda x) S_\lambda = (e - \lambda x) \left(\frac{1}{\lambda} e - x\right)^{-1} = \lambda e \to 0 \text{ as } \lambda \to 0.$$

Therefore, $l = 0$; hence, $S_\lambda = \lambda (e - \lambda x)^{-1}$ holds true for $\lambda = 0$ too.

Now from the assumption that $R(x)$ is weakly holomorphic at ∞ we have that $S(x)$ is weakly holomorphic in some 0-neighbourhood, say N (similar to $R(x)$, the notation $S(x)$ denotes the function $S(x)(\lambda) \equiv S_\lambda$). Then, we have that

$$S_\lambda - S_\mu = R_{\frac{1}{\lambda}} - R_{\frac{1}{\mu}} = -\left(\frac{1}{\lambda} - \frac{1}{\mu}\right) R_{\frac{1}{\lambda}} R_{\frac{1}{\mu}}, \text{ for } \lambda \neq 0 \neq \mu \text{ in } N.$$

Hence, for the weak first derivative of $S(x)$, based on Lemma 2.3.4, we have that

$$S_\mu' = \lim_{\lambda \to \mu} \frac{S_\lambda - S_\mu}{\lambda - \mu} = \frac{1}{\mu^2} S_\mu^2 = \frac{1}{\mu^2} \left(\mu(e - \mu x)^{-1}\right)^2 = (e - \mu x)^{-2}.$$

Therefore, the formula for the weak derivatives of $S(x)$ in some 0-neighbourhood holds for $n = 1$. Suppose the result holds for $n = 1, \ldots, m$. Then we have that

$$
\begin{aligned}
\frac{S_\lambda^{(m)} - S_\mu^{(m)}}{\lambda - \mu} &= \frac{m! x^{m-1}}{\lambda - \mu} \left(\frac{1}{\lambda^{m+1}} R_{\frac{1}{\lambda}}^{m+1} - \frac{1}{\mu^{m+1}} R_{\frac{1}{\mu}}^{m+1}\right) \\
&= \frac{m! x^{m-1}}{\lambda - \mu} \left(\left(\frac{1}{\lambda^{m+1}} - \frac{1}{\mu^{m+1}}\right) R_{\frac{1}{\lambda}}^{m+1} + \frac{1}{\mu^{m+1}}\left(R_{\frac{1}{\lambda}}^{m+1} - R_{\frac{1}{\mu}}^{m+1}\right)\right) \\
&= \frac{m! x^{m-1}}{\lambda - \mu} (\mu - \lambda) \frac{(\mu^m + \mu^{m-1}\lambda + \cdots + \lambda^m)}{\mu^{m+1}\lambda^{m+1}} R_{\frac{1}{\lambda}}^{m+1} \\
&\quad + \frac{1}{\mu^{m+1}} m! x^{m-1}\left(-\frac{1}{\lambda\mu}\right)\left(\frac{1}{\lambda} - \frac{1}{\mu}\right)^{-1}\left(R_{\frac{1}{\lambda}}^{m+1} - R_{\frac{1}{\mu}}^{m+1}\right) \\
&\xrightarrow[\lambda \to \mu]{} \frac{-(m+1)!}{\mu^{m+2}} x^{m-1} R_{\frac{1}{\mu}}^{m+1} + \frac{1}{\mu^{m+1}} m! x^{m-1}\left(-\frac{1}{\mu^2}\right) \\
&\quad \cdot \left(-(m+1) R_{\frac{1}{\mu}}^{m+2}\right)
\end{aligned}
$$

$$= (m+1)! \frac{x^{m-1}}{\mu^{m+2}} R_{\frac{1}{\mu}}^{m+1} \left(\frac{1}{\mu} R_{\frac{1}{\mu}} - e \right) = (m+1)! \frac{x^{m-1}}{\mu^{m+2}} R_{\frac{1}{\mu}}^{m+1} x R_{\frac{1}{\mu}}$$

$$= (m+1)! x^m \frac{1}{\mu^{m+2}} R_{\frac{1}{\mu}}^{m+2} = (m+1)! x^m \left(e - \mu x \right)^{-(m+2)}.$$

Therefore, the result follows by induction.   □

Theorem 2.3.7 *Let $\mathcal{A}[\tau]$ be a locally convex algebra and $x \in \mathcal{A}$. Let $R(x)$ be the resolvent of x. Then, the following hold:*

(1) *If $R(x)$ is weakly holomorphic at μ, $\mu \in \mathbb{C}^*$, then $\mu \in \rho(x)$.*
(2) *If $\mu \in \rho(x)$, then there is a neighbourhood N of μ and $B \in \mathfrak{B}$, such that, for every $\lambda \in N \cap \rho(x)$, $R_\lambda \in \mathcal{A}[B]$. Also, for $\mu \neq \infty$, $R(x)$ is differentiable at μ relative to $\rho(x)$ in the sense of norm convergence in $\mathcal{A}[B]$.*
(3) *If $\mathcal{A}[\tau]$ is pseudo-complete and $\mu \in \rho(x)$, then the neighbourhood N and the set $B \in \mathfrak{B}$ as in (2) can be chosen, such that $N \subset \rho(x)$, $R_\lambda \in \mathcal{A}[B]$, for all $\lambda \in N$ and $R(x)$ is holomorphic at μ in the sense of norm convergence in $\mathcal{A}[B]$.*

Proof (1) Let $R(x)$ be weakly holomorphic at $\mu \neq \infty$. Then, there exists $\delta > 0$, such that $R(x)$ is weakly holomorphic at $\lambda \in \mathbb{C}$, for $|\lambda - \mu| < \delta$. If $f \in \mathcal{A}'$, consider the function $\phi(\lambda) = f(R_\lambda)$, $\lambda \in \mathbb{C}$. Note that, ϕ is holomorphic at λ, for every λ, such that $|\lambda - \mu| < \delta$. From Lemma 2.3.5 we have that $\phi^{(n)}(\mu) = f(R_\mu^{(n)}) = (-1)^n n! f(R_\mu^{n+1})$, $n \in \mathbb{N}$. Then, the Taylor expansion of ϕ about μ is given by

$$\phi(\lambda) = \sum_{k=0}^{+\infty} (-1)^k f(R_\mu^{k+1})(\lambda - \mu)^k, \text{ for } |\lambda - \mu| < \delta.$$

Therefore, from Cauchy's radius-of-convergence formula $\limsup_n |f(R_\mu^n)|^{\frac{1}{n}} \leq \frac{1}{\delta}$. Hence, from Proposition 2.2.18 we have that $\beta(R_\mu) \leq \frac{1}{\delta}$. Thus, $R_\mu \in \mathcal{A}_0$ (see Proposition 2.2.14(2)), so $\mu \in \rho(x)$.

In case now $R(x)$ is weakly holomorphic at ∞, we have

$$S_\lambda = \begin{cases} R_{\frac{1}{\lambda}}, & \lambda \neq 0 \\ \lim_{\lambda \to 0} R_{\frac{1}{\lambda}}, & \lambda = 0 \end{cases} \text{ is weakly holomorphic at } 0.$$

So, for $f \in \mathcal{A}'$, the function $\phi(\lambda) = f(S_\lambda)$ is holomorphic at λ, $|\lambda| < \delta$, for some $\delta > 0$. From Lemma 2.3.6 we have that $\phi^{(n)}(0) = f(n! x^{n-1})$. Therefore, the Taylor expansion of ϕ about 0 is given by

$$\phi(\lambda) = \sum_{k=0}^{+\infty} f(x^{k-1}) \lambda^k, \text{ for } |\lambda| < \delta.$$

Hence, we have that $\limsup_n |f(x^n)|^{\frac{1}{n}} \leq \frac{1}{\delta}$. Consequently, by Proposition 2.2.18, $\beta(x) \leq \frac{1}{\delta}$, thus $x \in \mathcal{A}_0$ and so $\infty \in \rho(x)$.

For (2) and (3), we consider two cases below.

- *Case $\mu \neq \infty$.* Let $\mu \in \rho(x)$, $\mu \neq \infty$. Then, R_μ exists and $R_\mu \in \mathcal{A}_0$. From Proposition 2.2.4 and the comment in ►, before Proposition 2.2.9, we have that there is $B \in \mathfrak{B}$, such that $R_\mu \in \mathcal{A}[B]$. Clearly $\|R_\mu\|_B > 0$. Let $\lambda \in \mathbb{C}$, with $|\lambda - \mu| < \frac{1}{\|R_\mu\|_B}$. Moreover, let s_n denote the n-th partial sum of the series

$$R_\mu - R_\mu^2(\lambda - \mu) + R_\mu^3(\lambda - \mu)^2 - \cdots \qquad (2.3.4)$$

For $m > n$ we have that

$$\|s_n - s_m\|_B \leq \|R_\mu\|_B \sum_{k=n+1}^{m} \|R_\mu\|_B^k |\lambda - \mu|^k \to 0, \quad \text{for} \ \ n, m \to +\infty.$$

Hence, $(s_n)_{n\in\mathbb{N}}$ forms a Cauchy sequence in $\mathcal{A}[B]$. Since $\tau \prec \|\cdot\|_B$ on $\mathcal{A}[B]$, for every seminorm $p \in \Gamma_\tau$, there is $C_p > 0$, such that $\|x\|_B \leq C_p p(x)$, $x \in \mathcal{A}[B]$. Therefore, for every $p \in \Gamma_\tau$ and λ, such that $\lambda \in N = \{z \in \mathbb{C} : |z - \mu| < \frac{1}{\|R_\mu\|_B}\}$ we have

$$p\big((\lambda e - x)s_n - e\big) = p\big((\lambda - \mu)s_n + (\mu e - x)s_n - e\big)$$
$$= p\big(e + (-1)^n (\lambda - \mu)^{n+1} R_\mu^{n+1} - e\big) \qquad (2.3.5)$$
$$\leq C_p |\lambda - \mu|^{n+1} \|R_\mu\|_B^{n+1} \to 0, \quad \text{for} \ \ n \to +\infty.$$

So, $\lim_n \big((\lambda e - x)s_n\big) = e$ and similarly $\lim_n \big(s_n(\lambda e - x)\big) = e$, with respect to τ. Therefore, for $\lambda \in N \cap \rho(x)$, we have that $s_n \to R_\lambda$, with respect to τ. Furthermore, since $(s_n)_{n\in\mathbb{N}}$ is a $\|\cdot\|_B$-Cauchy sequence in $\mathcal{A}[B]$ there is $M > 0$, such that $\|s_n\|_B \leq M$ for all $n \in \mathbb{N}$. Hence, $\frac{1}{M} s_n \in B$, for all $n \in \mathbb{N}$. Then, since $\frac{1}{M} s_n \to \frac{1}{M} R_\lambda$, with respect to τ, and given that B is τ-closed, we conclude that $R_\lambda \in \mathcal{A}[B]$. Then, we have that $s_n \to R_\lambda$, with respect to $\|\cdot\|_B$: indeed, by analogous computations as those of (2.3.5), we have that for $\lambda \in N$, $\|(\lambda e - x)s_n - e\|_B = |\lambda - \mu|^{n+1} \|R_\mu\|_B^{n+1} \to 0$, as $n \to \infty$. Therefore, $\|(\lambda e - x)(s_n - R_\lambda)\|_B \to 0$, for $n \to \infty$. Hence,

$$\|s_n - R_\lambda\|_B = \|R_\lambda(\lambda e - x)(s_n - R_\lambda)\|_B \leq \|R_\lambda\|_B \|(\lambda e - x)(s_n - R_\lambda)\|_B \to 0,$$

as $n \to \infty$. It is clear then from the series in (2.3.4) that R is differentiable at μ relative to $\rho(x)$ in the sense of norm convergence in $\mathcal{A}[B]$.

If, in addition, $\mathcal{A}[\tau]$ is pseudo-complete, then $\mathcal{A}[B]$ is a Banach algebra. Hence, for $\lambda \in N$, the series in (2.3.4) is necessarily convergent, so R_λ exists and belongs to $\mathcal{A}[B]$. Therefore, $\lambda \in \rho(x)$, so that $N \subset \rho(x)$.

- *Case $\mu = \infty$.* Suppose that $\infty \in \rho(x)$. Then, $x \in \mathcal{A}_0$, so there is some $B \in \mathfrak{B}$, such that $x \in \mathcal{A}[B]$. Let $N = \{\lambda \in \mathbb{C}^* : |\lambda| > \|x\|_B\}$. For $\lambda \in N \cap \rho(x)$, using similar arguments just as in the previous case, we have that $R_\lambda \in \mathcal{A}[B]$. Moreover, for $\lambda \neq \infty$, $\lambda \in N \cap \rho(x)$, we have that $R_\lambda = \sum_{n=1}^{+\infty} \lambda^{-n} x^{n-1}$, with respect to norm convergence in $\mathcal{A}[B]$. If $\mathcal{A}[\tau]$ is pseudo-complete, then R_λ exists and belongs to $\mathcal{A}[B]$ for $\lambda \in N$. Moreover, for $\lambda \in N$, $\lambda \neq \infty$, R is holomorphic at λ and

$$\|R_\lambda\|_B \leq \sum_{n=1}^{+\infty} \|\lambda^{-n} x^{n-1}\|_B = \frac{1}{|\lambda| - \|x\|_B}.$$

So, $\|R_\lambda\|_B \to 0$, for $\lambda \to \infty$. It is then easily deduced that R is holomorphic at ∞.

$\square$

Based on the previous proposition we can now establish the following corollaries.

Corollary 2.3.8 *Let $\mathcal{A}[\tau]$ be a locally convex algebra and $x \in \mathcal{A}$. Then, $\sigma(x) \neq \emptyset$. If $\mathcal{A}[\tau]$ is pseudo-complete, then $\sigma(x)$ is closed.*

Proof Let us suppose that $\sigma(x) = \emptyset$. Then from Theorem 2.3.7(2) we have that R is holomorphic on the whole complex plane $\mathbb{C}$. Also $x \in \mathcal{A}_0$ since $\infty \in \rho(x)$. Thus, by using the same argument just as at the end of the proof of Theorem 2.3.7, we have that $R_\lambda \to 0$, for $\lambda \to \infty$. Then from Liouville's theorem, as this is applied in locally convex spaces, we have that $R_\lambda = 0$, a contradiction. Moreover, in case $\mathcal{A}[\tau]$ is pseudo-complete, from Theorem 2.3.7(3) we obtain that $\sigma(x)$ is closed. $\square$

The following corollary *extends to the locally convex case the* Gelfand–Mazur *theorem.*

Corollary 2.3.9 *Let $\mathcal{A}[\tau]$ be a locally convex algebra, such that, for every $x \in \mathcal{A}$, there is some nonzero $\lambda \in \mathbb{C}$, such that $(\lambda x)^n \to 0$. If moreover $\mathcal{A}$ is a division algebra, then $\mathcal{A}$ is topologically isomorphic to $\mathbb{C}$.*

Proof We consider the map $\phi : \mathbb{C} \to \mathcal{A}[\tau] : \lambda \mapsto \lambda e$. It is clear that ϕ is a topological isomorphism onto its image. Based on the assumption that for every $x \in \mathcal{A}$ there is a nonzero $\lambda \in \mathbb{C}$, such that $(\lambda x)^n \to 0$, we derive that $\mathcal{A} = \mathcal{A}_0$. Then, by Corollary 2.3.8, we have that $\sigma(x) \cap \mathbb{C} \neq \emptyset$. For $\lambda \in \sigma(x) \cap \mathbb{C}$, the element $\lambda e - x$ has no inverse in $\mathcal{A}_0 = \mathcal{A}$. Hence, since $\mathcal{A}$ is a division algebra, $\lambda e = x$ and thus ϕ is onto. Therefore, the result follows. $\square$

Lemma 2.3.10 *Let $\mathcal{A}[\tau]$ be a pseudo-complete locally convex algebra, $x \in \mathcal{A}$ and K a compact subset of $\rho(x)$. Then, there is some $B \in \mathfrak{B}$, such that $R_\lambda \in \mathcal{A}[B]$, for every $\lambda \in K$.*

Proof Let C be a maximal commutative subalgebra of $\mathcal{A}$ that contains x. Being maximal commutative subalgebra, C is closed and $e \in C$. Since C is a closed subalgebra of $\mathcal{A}[\tau]$, $C[\tau \restriction_C]$ is pseudo-complete by Proposition 2.2.8(1). Furthermore, by Proposition 2.3.2, we have that $\rho(x) \equiv \rho_{\mathcal{A}}(x) = \rho_C(x)$, hence for each $\lambda \in \rho(x)$, $R_\lambda \in C$. Let now $\lambda \in K$. From Theorem 2.3.7(3) there is a neighbourhood N_λ of λ and a set $B \in \mathfrak{B}$, such that $R_\mu \in \mathcal{A}[B] \cap C$, with $\mu \in N_\lambda$. Since K is compact there are finitely many points $\lambda_1, \ldots, \lambda_n \in K$, such that the corresponding neighbourhoods $N_{\lambda_1}, \ldots, N_{\lambda_n}$ cover K. Let $B_1, \ldots, B_n$ be the respective subsets in $\mathfrak{B}$, such that $R_\mu \in A[B_i] \cap C$, for $i = 1, \ldots, n$, where $\mu \in N_{\lambda_i}$. Since C is closed and $e \in C$, it is straightforward to show that $B \cap C \in \mathfrak{B}_C$, for every $B \in \mathfrak{B}$. Hence, taking into account pseudo-completeness of C, we have that $A[B \cap C] = A[B] \cap C$ is a Banach algebra with respect to $\| \cdot \|_B$, for every $B \in \mathfrak{B}$. Therefore, by using analogous arguments as those in the proof of Theorem 2.2.10, it can be shown that the family $\{B \cap C : B \in \mathfrak{B}\}$ is outer-directed by inclusion. So, there is some $B \in \mathfrak{B}$, such that

$$(B_1 \cap C) \cup (B_2 \cap C) \cup \cdots \cup (B_n \cap C) \subset B \cap C.$$

Therefore, $B_i \cap C \subset B \cap C$, for $i = 1, \ldots, n$. Hence, $R_\lambda \in \mathcal{A}[B] \cap C$, for every $\lambda \in K$. $\qquad\square$

If $\mathcal{A}$ is an arbitrary algebra and $x \in \mathcal{A}$, the *spectral radius* of x will be denoted by $r(x)$, or for distinction by $r_{\mathcal{A}}(x)$, with

$$r(x) := \sup \{ |\lambda| : \lambda \in sp(x) \}.$$

▶ *In the case of a locally convex algebra $\mathcal{A}[\tau]$, considering $\sigma(x)$ in the place of $sp(x)$, we use the same symbols and the same formula for the spectral radius of an element x in $\mathcal{A}[\tau]$, with the convention $|\infty| = +\infty$.*

The relation between the spectral radius and the radius of boundedness, in a locally convex algebra, is given by the following

Theorem 2.3.11 *Let $\mathcal{A}[\tau]$ be a locally convex algebra and $x \in \mathcal{A}$. Then, $\beta(x) \leq r(x)$. In case $\mathcal{A}[\tau]$ is pseudo-complete, then $\beta(x) = r(x)$.*

Proof Let $r(x) < +\infty$, for otherwise the inequality is trivial. Then, $\infty \notin \sigma(x)$ and so $x \in \mathcal{A}_0$. Hence, from Proposition 2.2.4, there is some $B \in \mathfrak{B}$, such that $x \in \mathcal{A}[B]$. As in the proof of Theorem 2.3.7(2) we have that if

$$N = \{\mu \in \mathbb{C} : |\mu| > \|x\|_B\}, \quad \text{then } R_\lambda = \lambda^{-1}e + \lambda^{-2}x + \cdots, \quad \text{for } \lambda \in \rho(x) \cap N,$$

with respect to norm convergence in $\mathcal{A}[B]$. Then, for $f \in \mathcal{A}'$ the function $\phi(\lambda) = f(R_\lambda)$ is written as follows:

$$\phi(\lambda) = \lambda^{-1} f(e) + \lambda^{-2} f(x) + \cdots, \quad \text{for } \lambda \in N \cap \rho(x). \tag{2.3.6}$$

Moreover, by Theorem 2.3.7, ϕ is holomorphic at $\lambda \in \mathbb{C}$ with $|\lambda| > r(x)$. So, ϕ has a Laurent expansion in that region, which must coincide with the series in (2.3.6). Thus, we have that $\limsup_n |f(x^n)|^{\frac{1}{n}} \leq r(x)$ and so by Proposition 2.2.18, $\beta(x) \leq r(x)$, $x \in \mathcal{A}$.

Let us assume now that $\mathcal{A}[\tau]$ is pseudo-complete. We show that $r(x) \leq \beta(x)$, $x \in \mathcal{A}$. Suppose that $\beta(x) < +\infty$, for otherwise the inequality is trivial. Then, by Proposition 2.2.14(2), $x \in \mathcal{A}_0$. Let $\lambda \in \mathbb{C}$, such that $|\lambda| > \beta(x)$. By Corollary 2.2.17, there exists $B \in \mathfrak{B}$, such that $x \in \mathcal{A}[B]$ and $|\lambda| > \|x\|_B$. Then, as in the proof of Theorem 2.3.7(3), we conclude that R_λ exists in $\mathcal{A}[B]$ and so $\lambda \in \rho(x)$. Therefore, for every $\mu \in \sigma(x)$, we have that $|\mu| \leq \beta(x)$, from which the result follows. $\qquad\square$

In case of a pseudo-complete locally convex algebra $\mathcal{A}[\tau]$, the following result provides us with a relation between the spectrum $\sigma(x)$ of an element $x \in \mathcal{A}_0$ and the spectra $\sigma_{\mathcal{A}[B]}(x) = sp_{\mathcal{A}[B]}(x)$, for those $B \in \mathfrak{B}$, such that $x \in \mathcal{A}[B]$.

Proposition 2.3.12 *Let $\mathcal{A}[\tau]$ be a pseudo-complete locally convex algebra and $x \in \mathcal{A}_0$. Then, the following hold:*

(1) $\sigma_{\mathcal{A}}(x) = \bigcap \{\sigma_{\mathcal{A}[B]}(x) : B \in \mathfrak{B}, \ x \in \mathcal{A}[B]\}$;
(2) $r_{\mathcal{A}}(x) = \inf \{r_{\mathcal{A}[B]}(x) : B \in \mathfrak{B}, \ x \in \mathcal{A}[B]\}$.

Proof For the proof of both claims let $\mathcal{C}$ denote a maximal commutative subalgebra of $\mathcal{A}$ containing x.

(1) Since $\mathcal{A}[B] \subset \mathcal{A}$, for every $B \in \mathfrak{B}$, we have that

$$\sigma_{\mathcal{A}}(x) \subset \bigcap \{\sigma_{\mathcal{A}[B]}(x) : B \in \mathfrak{B}, \ x \in \mathcal{A}[B]\}.$$

Now let $\lambda(\neq \infty)$, such that $\lambda \notin \sigma_{\mathcal{A}}(x)$. Then, $R_\lambda \in \mathcal{A}_0$, so there is a $B_1 \in \mathfrak{B}$, such that $R_\lambda \in \mathcal{A}[B_1] \cap \mathcal{C}$ (Proposition 2.2.4). By the outer-directedness of $\{B \cap \mathcal{C} : B \in \mathfrak{B}\}$ (see Theorem 2.2.10) we have that there exists a $B \in \mathfrak{B}$, such that $R_\lambda \in \mathcal{A}[B] \cap \mathcal{C}$ and $x \in \mathcal{A}[B]$. Hence, $\lambda \notin \sigma_{\mathcal{A}[B]}(x)$ and so the inverse inclusion follows.

(2) From (1) we have that

$$r_{\mathcal{A}}(x) \leq \inf \{r_{\mathcal{A}[B]}(x) : B \in \mathfrak{B}, \ x \in \mathcal{A}[B]\}.$$

Consider $\mu \in \mathbb{C}$, such that $\mu > r_{\mathcal{A}}(x)$ and let $K = \{\lambda \in \mathbb{C}^* : |\lambda| \geq \mu\}$. Then, K is a compact subset of $\rho(x)$ and so from Lemma 2.3.10 we have that there is some $B \in \mathfrak{B}$, such that $R_\lambda \in \mathcal{A}[B]$, for all $\lambda \in K$. By using the same argument as in (1) we can assume that $x \in \mathcal{A}[B]$. Then, we have that $r_{\mathcal{A}[B]}(x) \leq \mu$. Hence,

$$\inf \left\{ r_{\mathcal{A}[B]}(x) : B \in \mathfrak{B}, \ x \in \mathcal{A}[B] \right\} \leq \mu, \ \forall \, \mu > r_{\mathcal{A}}(x),$$

from which the inverse inequality follows. $\qquad\square$

For a locally convex algebra $\mathcal{A}[\tau]$ in which inversion is continuous on the invertible elements of $\mathcal{A}$, the relation between $\sigma(x)$ and $sp(x)$, $x \in \mathcal{A}$ (for the respective definitions, see beginning of Sect. 2.3) is described by the following result. In the notation of the following theorem $\overline{sp(x)}$ stands for the closure of $sp(x)$ in $\mathbb{C}^*$.

Theorem 2.3.13 *Let $\mathcal{A}[\tau]$ be a locally convex algebra with continuous inversion and let $x \in \mathcal{A}$. Then, $sp(x) \subset \sigma(x) \subset \overline{sp(x)}$. Moreover, if $\mathcal{A}[\tau]$ is pseudo-complete, then $\sigma(x) = \overline{sp(x)}$.*

Proof It is clear from the respective definitions that $sp(x) \subset \sigma(x)$. For the second inclusion, assume that $\overline{sp(x)} \neq \mathbb{C}^*$, for otherwise the result is trivial and let $\mu \in \mathbb{C}$, such that $\mu \notin \overline{sp(x)}$. Then, there is some $\delta > 0$, such that if $\lambda \in \mathbb{C}$ with $|\lambda - \mu| < \delta$, then $\lambda \notin sp(x)$. Under the assumption of continuity of inversion on the invertible elements of $\mathcal{A}$, we have that $R(x)$ is continuous at μ. Then, since

$$R_\lambda - R_\mu = -(\lambda - \mu) R_\lambda R_\mu$$

(see beginning of the proof of Lemma 2.3.5), the function $R(x)$ is differentiable at μ, for every finite point μ in the complement of $\overline{sp(x)}$ in $\mathbb{C}^*$.

Now if $\infty \notin \overline{sp(x)}$, then there is $M > 0$, such that if $\lambda \neq \infty$ and $|\lambda| > M$, then $\lambda \notin \overline{sp(x)}$. For these λ's we then have that

$$R_\lambda = \lambda^{-1}(e - \lambda^{-1}x)^{-1} \to 0, \quad \text{as } \lambda \to \infty.$$

It follows that R is holomorphic at ∞. Therefore, R is holomorphic on $\mathbb{C}^* \setminus \overline{sp(x)}$. Thus, from Theorem 2.3.7(1) we have that $\sigma(x) \subset \overline{sp(x)}$.

In case $\mathcal{A}[\tau]$ is pseudo-complete, then by Corollary 2.3.8, $\sigma(x)$ is closed. Hence, $\overline{sp(x)} \subset \sigma(x)$, thus the desired equality follows. $\qquad\square$

Corollary 2.3.14 *Let $\mathcal{A}[\tau]$ be a pseudo-complete locally convex algebra with continuous inversion. Then, $x \in \mathcal{A}_0$, if and only if, $sp(x)$ is bounded.*

Proof By Theorem 2.3.13, $\overline{sp(x)} = \sigma(x)$. Let $x \in \mathcal{A}_0$ and assume that $sp(x)$ is not bounded. Then, there exists a sequence $(\lambda_n)_{n \in \mathbb{N}}$ in $sp(x)$, such that $\lambda_n \to \infty$. So, $\infty \in \overline{sp(x)} = \sigma(x)$, hence $x \notin \mathcal{A}_0$, a contradiction. For the reverse implication, assume that $sp(x)$ is bounded. Then, $\overline{sp(x)}$ is bounded [131, p. 25, 5.1]. So, $\infty \notin \overline{sp(x)} = \sigma(x)$ and thus $x \in \mathcal{A}_0$. $\qquad\square$

2.4 A Functional Calculus

In this section we develop a functional calculus for a pseudo-complete locally convex algebra. Let $\mathcal{A}$ be a pseudo-complete locally convex algebra and $x \in \mathcal{A}$. Let us denote with F_x *the set of all complex-valued functions, which are holomorphic on the spectrum* $\sigma(x)$ of x, hence on some neighbourhood of $\sigma(x)$. By F_x' we denote the *quotient set of F_x by the equivalence relation* $\sim$, which is given as follows: for $f, g \in F_x$, $f \sim g$, if and only if, f equals g on some neighbourhood of $\sigma(x)$. With the algebraic operations defined pointwise on suitable neighbourhoods of $\sigma(x)$, it follows that F_x' is an algebra.

We recall that in case $\mathcal{A}$ is a Banach algebra, it is known that for $f \in F_x$ and γ a closed rectifiable Jordan curve, such that its interior domain, say D, contains $\sigma(x)$ and f is holomorphic on D and continuous on $D \cup \gamma$, then the formula

$$f(x) = \frac{1}{2\pi i} \int_\gamma f(\lambda) R_\lambda(x) d\lambda$$

defines a homomorphism $f \mapsto f(x)$ of F_x into $\mathcal{A}$, which enjoys certain properties (see [115, Chapter III, §11.6, Theorem 7]).

An extension of this result can be established for the case where $\mathcal{A}[\tau]$ is a pseudo-complete locally convex algebra. Towards this direction we first show the following

Proposition 2.4.1 *Let $\mathcal{A}[\tau]$ be a pseudo-complete locally convex algebra and $x \in \mathcal{A}$ with $\rho(x) \neq \emptyset$. Let γ be any rectifiable Jordan arc in $\rho(x) \cap \mathbb{C}$ and f a complex-valued function, which is holomorphic on γ. Then, there is some $B \in \mathfrak{B}$, such that $R_\lambda \in \mathcal{A}[B]$, for all $\lambda \in \gamma$. Furthermore, the integral $\int_\gamma f(\lambda) R_\lambda(x) d\lambda$ exists, in the sense of norm convergence in $\mathcal{A}[B]$ and its value is an element of $\mathcal{A}[B]$.*

Proof Since the set γ is a compact subset of $\rho(x)$, by Lemma 2.3.10 we have that there is some $B \in \mathfrak{B}$, such that $R_\lambda \in \mathcal{A}[B]$, for all $\lambda \in \gamma$. Therefore, the function $\lambda \mapsto f(\lambda) R_\lambda$ is a holomorphic function on γ which takes values in the Banach algebra $\mathcal{A}[B]$. Then the result follows from the holomorphic functional calculus for Banach algebras (see [115, Chapter I, §4.7, Theorem I]). $\square$

The following definition is a slight variant of [145, Definition, p. 193].

Definition 2.4.2 A subset D of $\mathbb{C}^*$ is called a *Cauchy domain* if it fulfills the following conditions:

(1) D is open;
(2) D has a finite number of components, whose closures are pairwise disjoint;
(3) the boundary ∂D of D is a subset of $\mathbb{C}$, which consists of a finite number of closed rectifiable Jordan curves no two of which intersect.

The next result can be obtained with the use of Proposition 2.4.1 and by following very similar arguments to those developed in [145, Theorem 4.1] and so its proof is omitted.

Lemma 2.4.3 *Let $\mathcal{A}[\tau]$ be a pseudo-complete locally convex algebra and $x \in \mathcal{A}$, such that $\rho(x) \neq \emptyset$. Then for any $f \in F_x$ there is a Cauchy domain D, such that*

(i) $\sigma(x) \subset D$;

(ii) $\overline{D} \subset \Delta(f)$, where $\Delta(f)$ denotes the domain of f. Furthermore, the integral $\int_{\partial D} f(\lambda) R_\lambda(x) d\lambda$ defines an element of $\mathcal{A}_0$, which is independent of the choice of the Cauchy domain D satisfying (i) and (ii).

We are now in position to establish a functional calculus for a pseudo-compete locally convex algebra.

Theorem 2.4.4 *Let $\mathcal{A}[\tau]$ be a pseudo-complete locally convex algebra and $x \in \mathcal{A}$. Then, there is a homomorphism $f \mapsto f(x)$ from F'_x into $\mathcal{A}_0$, which is given by the following formulae:*

(1) *if $x \in \mathcal{A}_0$, then $f(x) = \frac{1}{2\pi i} \int_{\partial D} f(\lambda) R_\lambda(x) d\lambda$, where D is as in Lemma 2.4.3;*

(2) *if $x \notin \mathcal{A}_0$ and $\rho(x) \neq \emptyset$, then $f(x) = f(\infty)e + \frac{1}{2\pi i} \int_{\partial D} f(\lambda) R_\lambda(x) d\lambda$, where D is as before;*

(3) *if $\rho(x) = \emptyset$, then F_x contains only constant functions. If $f(\lambda) \equiv c$, then $f(x) = ce$.*

Furthermore, for all cases, if $\mathcal{C}$ is a maximal commutative subalgebra of $\mathcal{A}$, which contains x, then $f(x) \in \mathcal{A}_0 \cap \mathcal{C}$.

If u_0, u_1 denote the complex functions $u_0(\lambda) \equiv 1$, $u_1(\lambda) \equiv \lambda$, then in case (1), $u_1 \in F_x$ and $u_1(x) = x$ and in all cases (1)–(3), $u_0 \in F_x$ and $u_0(x) = e$.

Proof For the proof of (1) and (2) are used standard arguments that follow from similar arguments to those in the proof of [145, Theorem 4.3].

For (3), we have that $\sigma(x) = \mathbb{C}^*$. Then, any $f \in F_x$ is holomorphic on the whole complex plane and at ∞. Therefore, by Liouville's theorem as is applied to vector-valued functions (see [115, Chapter I, §3.12]), we have that F_x consists only of constant functions.

Moreover, with respect to the fact that $f(x) \in \mathcal{C} \cap \mathcal{A}_0$, where $\mathcal{C}$ is a maximal commutative subalgebra of $\mathcal{A}$ containing x, this follows directly from the fact that $e \in \mathcal{C}$ and that $(\lambda e - x)^{-1}$ commutes with all elements of $\mathcal{C}$, for every $\lambda \in \partial D$.

Now if $u_1(\lambda) = \lambda$ and x is in $\mathcal{A}_0$, then taking into account Proposition 2.3.12(1) we can choose D, such that its boundary ∂D is a circle of radius greater than $\|x\|_B$, where B is some element of $\mathfrak{B}$ with $x \in \mathcal{A}[B]$. Then, for $\lambda \in \partial D$ we have that

$$(\lambda e - x)^{-1} = \left(e - \lambda^{-1}x\right)^{-1}\lambda^{-1} = \lambda^{-1}e + \lambda^{-2}x + \cdots$$

Hence,

$$u_1(x) = \frac{1}{2\pi i} \int_{\partial D} \lambda(\lambda e - x)^{-1} d\lambda = \frac{1}{2\pi i} \int_{\partial D} \left(e + \lambda^{-1}x + \lambda^{-2}x^2 + \cdots\right) d\lambda = x.$$

The statement about u_0 follows from similar considerations. □

2.5   The Carrier Space

Consider a commutative and pseudo-complete locally convex algebra $\mathcal{A}[\tau]$ with an identity e. Recall that $\mathfrak{B}$ denotes the collection of subsets of $\mathcal{A}$ just as the one described in Proposition 2.2.8(3). It will be convenient for the purposes of this section to denote the elements of $\mathfrak{B}$ by $\{B_\alpha : \alpha \in \Delta\}$, where Δ is an index set. For $\alpha, \beta \in \Delta$, we use the notation, $\alpha \leq \beta$ if $B_\alpha \subset B_\beta$. Then, by Theorem 2.2.10 the index set Δ is outer-directed by the ordering $\leq$. To simplify the notation, for each $\alpha \in \Delta$, *we denote by* A_α *the Banach algebras* $\mathcal{A}[B_\alpha]$. Let $\mathfrak{M}_\alpha$ denote the set of all nonzero multiplicative linear functionals on A_α endowed with the weak* topology $\sigma(\mathfrak{M}_\alpha, A_\alpha)$.

The respective *set of all nonzero multiplicative linear functionals on* $\mathcal{A}_0$, denoted by $\mathfrak{M}_0$, and endowed with the weak* topology $\sigma(\mathfrak{M}_0, \mathcal{A}_0)$ is called the *carrier space* (or *Gelfand space, or maximal ideal space*) of $\mathcal{A}_0$.

Proposition 2.5.1 *There exists a natural homeomorphism, say j, of the carrier space $\mathfrak{M}_0$ with the projective limit $\varprojlim_{\alpha \in \Delta} \mathfrak{M}_\alpha$, where $j(\varphi) = (\varphi_\alpha)_{\alpha \in \Delta}$ with $\varphi_\alpha(x) =$*

$\varphi(x)$, $x \in A_\alpha$, $\varphi \in \mathfrak{M}_0$, $\alpha \in \Delta$.

Proof For $\alpha, \beta \in \Delta$ with $\alpha \leq \beta$ consider the map $\pi_{\alpha\beta} : \mathfrak{M}_\beta \to \mathfrak{M}_\alpha$, such that $\pi_{\alpha\beta}(\varphi_\beta) = \varphi_\beta|_{A_\alpha}$. The maps $\pi_{\alpha\beta}$ are well-defined, that is $\varphi_\beta|_{A_\alpha}$ is not zero, since $e \in A_\alpha$. It is also clear that $\pi_{\alpha\beta}$ are weak *-continuous maps, such that $\pi_{\alpha\alpha}$ is the identity map on $\mathfrak{M}_\alpha$, for all $\alpha \in \Delta$ and $\pi_{\alpha\beta} \circ \pi_{\beta\gamma} = \pi_{\alpha\gamma}$ for $\alpha \leq \beta \leq \gamma$ in Δ. Therefore, the spaces $\{\mathfrak{M}_\alpha : \alpha \in \Delta\}$ form a projective system. Since for each $\alpha \in \Delta$, A_α is a commutative Banach algebra with identity, $\mathfrak{M}_\alpha$ is a nonempty compact Hausdorff space. Hence, the projective limit $\varprojlim_{\alpha \in \Delta} \mathfrak{M}_\alpha$ is a non-empty compact Hausdorff space (see [35, Chapter I, §9, No. 6, Proposition 8]). The fact that the map j is a homeomorphism is then easily checked. $\square$

> ▶ A direct implication of the previous proposition is that *the carrier space $\mathfrak{M}_0$ of $\mathcal{A}_0$ is a non-empty compact Hausdorff space.*

Lemma 2.5.2 *Let $\mathcal{A}[\tau]$ be a commutative and pseudo-complete locally convex algebra and $x \in \mathcal{A}_0$. Then, x is invertible in $\mathcal{A}_0$, if and only if, $\varphi(x) \neq 0$, for every $\varphi \in \mathfrak{M}_0$.*

Proof The forward implication is immediate since $\varphi(e) = 1$, for all $\varphi \in \mathfrak{M}_0$. For the inverse implication let $x \in \mathcal{A}_0$ and suppose that $\varphi(x) \neq 0$, for every $\varphi \in \mathfrak{M}_0$. We show that there is $\alpha \in \Delta$, such that $x \in A_\alpha$ and $\varphi_\alpha(x) \neq 0$, for all $\varphi_\alpha \in \mathfrak{M}_\alpha$. From a well-known result in the theory of Banach algebras it will then follow that x is invertible in A_α and thus in $\mathcal{A}$. Let us suppose to the contrary that, for every $\alpha \in \Delta$, such that $x \in A_\alpha$, there is $\varphi_\alpha \in \mathfrak{M}_\alpha$ with $\varphi_\alpha(x) = 0$. Let us fix an index $\delta \in \Delta$, such that $x \in A_\delta$ and for every $\alpha \geq \delta$, let N_α denote the set $\{\varphi_\alpha \in \mathfrak{M}_\alpha : \varphi_\alpha(x) = 0\}$. From the assumption we have made, $N_\alpha \neq \varnothing$, for each $\alpha \geq \delta$. Also $N_\alpha, \alpha \geq \delta$, is closed, hence a compact subspace of $\mathfrak{M}_\alpha$. Since $\pi_{\alpha\beta}(N_\beta) \subset N_\alpha$, for all $\beta \geq \alpha \geq \delta$, we have that $\{N_\alpha\}_{\alpha \geq \delta}$ forms a projective system, whose projective limit $\varprojlim_{\alpha \geq \delta} N_\alpha$ is a non-empty compact Hausdorff space. Let $\{\psi_\alpha\}_{\alpha \geq \delta}$ be an element in $\varprojlim_{\alpha \geq \delta} N_\alpha$. Define $\{\varphi_\alpha\}_{\alpha \in \Delta} \in \prod_{\alpha \in \Delta} \mathfrak{M}_\alpha$ as follows

$$\varphi_\alpha = \begin{cases} \psi_\alpha, & \text{if } \alpha \geq \delta \\ \pi_{\alpha\beta}(\psi_\beta), & \text{otherwise, for some } \beta \geq \alpha, \ \beta \geq \delta. \end{cases}$$

It is clear that φ_α, $\alpha \in \Delta$, is well-defined since if $\alpha \ngeq \delta$ and $\beta, \beta' \in \Delta$, such that $\beta, \beta' \geq \alpha$, then there exists $\beta'' \in \Delta$ with $\beta'' \geq \beta$, β' and so

$$\pi_{\alpha\beta'}(\psi_{\beta'}) = \pi_{\alpha\beta'}(\pi_{\beta'\beta''}(\psi_{\beta''})) = \pi_{\alpha\beta''}(\psi_{\beta''}) = \pi_{\alpha\beta}(\pi_{\beta\beta''}(\psi_{\beta''})) = \pi_{\alpha\beta}(\psi_\beta).$$

Moreover, $\{\varphi_\alpha\}_{\alpha \in \Delta} \in \varprojlim_{\alpha \in \Delta} \mathfrak{M}_\alpha$ as can easily be verified. So, based on Proposition 2.5.1 there is $\varphi \in \mathfrak{M}_0$, such that $\varphi(x) = \varphi_\delta(x) = \psi_\delta(x) = 0$, a contradiction. Thus, the result follows. $\qquad\square$

Theorem 2.5.3 *Let $A[\tau]$ be a commutative and pseudo-complete locally convex algebra and x an element in A_0. Then, $\sigma(x) = \{\varphi(x) : \varphi \in \mathfrak{M}_0\}$.*

Proof Since $x \in A_0$ we have that $\infty \notin \sigma(x)$. Then, $\lambda \in \mathbb{C}$ belongs to $\sigma(x)$, if and only if, $\lambda e - x$ has no inverse in A_0. From Lemma 2.5.2 this is equivalent to $\varphi(\lambda e - x) = 0$, for some $\varphi \in \mathfrak{M}_0$, that is $\varphi(x) = \lambda$. Hence, the result follows. $\qquad\square$

The next result describes the unique extension of any functional $\varphi \in \mathfrak{M}_0$ to a bigger set, namely to $A_\rho := \{x \in A : \rho(x) \neq \varnothing\}$.

Proposition 2.5.4 *Let $A[\tau]$ be a commutative and pseudo-complete locally convex algebra. Then, to each functional $\varphi \in \mathfrak{M}_0$ corresponds a unique $\mathbb{C}^*$-valued function φ' on A_ρ, such that the following hold:*

(1) *φ' is an extension of φ.*

The $\mathbb{C}^*$-valued function φ' on A_ρ is a 'partial character' of A, in the following sense:

(2) *$\varphi'(\lambda x) = \lambda \varphi'(x)$, $\lambda \in \mathbb{C}$, $x \in A_\rho$, with the convention that $0 \cdot \infty = 0$;*
(3) *$\varphi'(x_1 + x_2) = \varphi'(x_1) + \varphi'(x_2)$, provided that $x_1, x_2, x_1 + x_2 \in A_\rho$ and $\varphi'(x_1)$, $\varphi'(x_2)$ are not both ∞;*
(4) *$\varphi'(x_1 x_2) = \varphi'(x_1)\varphi'(x_2)$, provided that $x_1, x_2, x_1 x_2 \in A_\rho$ and $\varphi'(x_1)$, $\varphi'(x_2)$ are not 0, ∞ in some order.*

Proof

(1) Let $x \in \mathcal{A}_\rho$ and $\mu \in \rho(x)$, such that $\mu \neq \infty$. Let $y = (\mu e - x)^{-1} \in \mathcal{A}_0$ and consider $\varphi \in \mathfrak{M}_0$. If an extension, say φ', of φ to $\mathcal{A}_\rho$ satisfying properties (2)–(4) is possible, then provided that $\varphi(y) \neq 0$, φ' must satisfy the relation

$$\varphi'(\mu e - x)\varphi(y) = \varphi(e) = 1 \;\Rightarrow\; \varphi'(x) = \mu - \varphi(y)^{-1}. \qquad (2.5.7)$$

If $\varphi(y) = 0$, then by (3),(4) we have that $\varphi'(x) = \infty$. So, the equality $\varphi'(x) = \mu - \varphi(y)^{-1}$ holds in any case and it is considered as the definition of φ'.

The definition of φ' is independent from the choice of $\mu \in \mathbb{C} \cap \rho(x)$. Indeed let $\mu_1, \mu_2 \in \mathbb{C} \cap \rho(x)$. If either of the elements $\varphi\big((\mu_1 e - x)^{-1}\big)$, $\varphi\big((\mu_2 e - x)^{-1}\big)$ is 0, then the other must be 0 also, as it follows from the relation $R_{\mu_1} - R_{\mu_2} = (\mu_2 - \mu_1) R_{\mu_1} R_{\mu_2}$ (see beginning of the proof of Lemma 2.3.5). If both elements are not 0, then

$$\frac{1}{\varphi\big((\mu_1 e - x)^{-1}\big)} - \frac{1}{\varphi\big((\mu_2 e - x)^{-1}\big)} = -\frac{\varphi(R_{\mu_1} - R_{\mu_2})}{\varphi(R_{\mu_1})\varphi(R_{\mu_2})}$$

$$= \frac{(\mu_1 - \mu_2)\varphi(R_{\mu_1} R_{\mu_2})}{\varphi(R_{\mu_1})\varphi(R_{\mu_2})}$$

$$= \mu_1 - \mu_2,$$

hence $\mu_1 - \frac{1}{\varphi((\mu_1 e - x)^{-1})} = \mu_2 - \frac{1}{\varphi((\mu_2 e - x)^{-1})}$. The previous argumentation establishes also the uniqueness of any possible extension of φ to $\mathcal{A}_\rho$, satisfying conditions (2)–(4).

Moreover, φ' is an extension of φ. Indeed if $x \in \mathcal{A}_0$, then from $\varphi(\mu e - x)\varphi\big((\mu e - x)^{-1}\big) = 1$ we have that $\varphi\big((\mu e - x)^{-1}\big) \neq 0$, therefore

$$\varphi'(x) = \mu - \frac{1}{\varphi\big((\mu e - x)^{-1}\big)} = \frac{\varphi\big(\mu(\mu e - x)^{-1} - e\big)}{\varphi((\mu e - x)^{-1})}$$

$$= \frac{\varphi\big(x(\mu e - x)^{-1}\big)}{\varphi((\mu e - x)^{-1})} = \varphi(x).$$

(2) Let $x \in \mathcal{A}_\rho$ and $\mu \in \rho(x) \cap \mathbb{C}$, such that $\varphi\big((\mu e - x)^{-1}\big) \neq 0$. If $\lambda \in \mathbb{C}$, $\lambda \neq 0$, then $\lambda\mu \in \rho(\lambda x)$, hence

$$\varphi'(\lambda x) = \lambda\mu - \frac{1}{\varphi\big((\lambda\mu e - \lambda x)^{-1}\big)} = \lambda\left(\mu - \frac{1}{\varphi\big((\mu e - x)^{-1}\big)}\right) = \lambda\varphi'(x).$$

If $\lambda = 0$, then clearly $\lambda\varphi'(x) = 0 = \varphi'(\lambda x)$. By using similar arguments for the case $\varphi\big((\mu e - x)^{-1}\big) = 0$, we derive the result.

(3) Let x_1, x_2, $x_1 + x_2 \in \mathcal{A}_\rho$, such that $\varphi'(x_1)$, $\varphi'(x_2)$ are not both ∞. Consider $\mu_1 \in \rho(x_1) \cap \mathbb{C}$, $\mu_2 \in \rho(x_2) \cap \mathbb{C}$, $\lambda \in \rho(x_1 + x_2) \cap \mathbb{C}$. We have that

$$
\begin{aligned}
&\varphi\big(R_\lambda(x_1 + x_2)\big)\big[(\lambda - \mu_1 - \mu_2)\varphi\big(R_{\mu_1}(x_1)\big)\varphi\big(R_{\mu_2}(x_2)\big) \\
&\quad + \varphi\big(R_{\mu_1}(x_1)\big) + \varphi\big(R_{\mu_2}(x_2)\big)\big] \\
&= \varphi\big(R_\lambda(x_1 + x_2)\big)\varphi\big(R_{\mu_1}(x_1)\big((\lambda - \mu_1 - \mu_2)e \\
&\quad + \mu_2 e - x_2 + \mu_1 e - x_1\big)R_{\mu_2}(x_2)\big) \\
&= \varphi\big(R_\lambda(x_1 + x_2)\big(\lambda e - (x_1 + x_2)\big)R_{\mu_1}(x_1)R_{\mu_2}(x_2)\big) \\
&= \varphi\big(R_{\mu_1}(x_1)\big)\varphi\big(R_{\mu_2}(x_2)\big),
\end{aligned}
\tag{2.5.8}
$$

where the last to one equality is derived due to commutativity of $\mathcal{A}$.

Therefore, if both $\varphi\big(R_{\mu_2}(x_2)\big)$, $\varphi\big(R_{\mu_1}(x_1)\big)$ are different from 0, then by (2.5.8) we have that $\varphi\big(R_\lambda(x_1 + x_2)\big) \neq 0$ and

$$
\frac{1}{\varphi\big(R_\lambda(x_1 + x_2)\big)} = \lambda - \mu_1 - \mu_2 + \frac{1}{\varphi\big(R_{\mu_1}(x_1)\big)} + \frac{1}{\varphi\big(R_{\mu_2}(x_2)\big)},
$$

so that (2.5.7) implies $\varphi'(x_1 + x_2) = \varphi'(x_1) + \varphi'(x_2)$.

Also if one of $\varphi\big(R_{\mu_1}(x_1)\big)$ or $\varphi\big(R_{\mu_2}(x_2)\big)$ is 0, then from (2.5.8) we have that $\varphi\big(R_\lambda(x_1 + x_2)\big) = 0$. Thus, $\varphi'(x_1 + x_2) = \infty = \varphi'(x_1) + \varphi'(x_2)$.

(4) Let x_1, x_2, $x_1 x_2 \in \mathcal{A}_\rho$. Consider $\mu_1 \in \rho(x_1) \cap \mathbb{C}$, $\mu_2 \in \rho(x_2) \cap \mathbb{C}$ and $\lambda \in \rho(x_1 x_2) \cap \mathbb{C}$. Then, we have

$$
\begin{aligned}
&\varphi\big(R_\lambda(x_1 x_2)\big)\Big((\lambda - \mu_1\mu_2)\varphi\big(R_{\mu_1}(x_1)\big)\varphi\big(R_{\mu_2}(x_2)\big) \\
&\quad + \mu_1\varphi\big(R_{\mu_1}(x_1)\big) + \mu_2\varphi\big(R_{\mu_2}(x_2)\big) - 1\Big) \\
&= \varphi\big(R_\lambda(x_1 x_2)\big)\Big(\lambda\varphi\big(R_{\mu_1}(x_1)R_{\mu_2}(x_2)\big) \\
&\quad - \varphi\big((\mu_1 R_{\mu_1}(x_1) - e)(\mu_2 R_{\mu_2}(x_2) - e)\big)\Big) \\
&= \varphi\big(R_\lambda(x_1 x_2)\big)\Big(\lambda\varphi\big(R_{\mu_1}(x_1)R_{\mu_2}(x_2)\big) - \varphi\big(R_{\mu_1}(x_1)x_1 x_2 R_{\mu_2}(x_2)\big)\Big) \\
&= \varphi\big(R_\lambda(x_1 x_2)\big)\Big(\varphi\big((\lambda e - x_1 x_2)R_{\mu_1}(x_1)R_{\mu_2}(x_2)\big)\Big) \\
&= \varphi\big(R_{\mu_1}(x_1)\big)\varphi\big(R_{\mu_2}(x_2)\big).
\end{aligned}
$$

$$
\tag{2.5.9}
$$

Suppose that $\varphi\big(R_{\mu_1}(x_1)\big) \neq 0$ and $\varphi\big(R_{\mu_2}(x_2)\big) \neq 0$. The previous two relations, due to the very definition of φ', result equivalently in that $\varphi'(x_1) \neq \infty$ and

$\varphi'(x_2) \neq \infty$. By (2.5.9), $\varphi(R_\lambda(x_1 x_2)) \neq 0 \ (\Leftrightarrow \varphi'(x_1 x_2) \neq \infty)$. Therefore,

$$\varphi'(x_1 x_2) = \lambda - \frac{1}{\varphi(R_\lambda(x_1 x_2))} = \frac{\varphi(R_\lambda(x_1 x_2) x_1 x_2)}{\varphi(R_\lambda(x_1 x_2))}$$

$$= \frac{\varphi(R_\lambda(x_1 x_2)(\mu_1 R_{\mu_1}(x_1) - e)(\mu_2 R_{\mu_2}(x_2) - e))}{\varphi(R_\lambda(x_1 x_2)) \varphi(R_{\mu_1}(x_1)) \varphi(R_{\mu_2}(x_2))}$$

$$= \left(\mu_1 - \frac{1}{\varphi(R_{\mu_1}(x_1))} \right) \left(\mu_2 - \frac{1}{\varphi(R_{\mu_2}(x_2))} \right)$$

$$= \varphi'(x_1) \varphi'(x_2).$$

Furthermore, it follows from (2.5.9) that $\varphi(R_\lambda(x_1 x_2)) = 0$ if one of the following cases holds:

  (i) $\varphi(R_{\mu_1}(x_1)) = 0$ and $\varphi(R_{\mu_2}(x_2)) = 0$,
  (ii) $\varphi(R_{\mu_1}(x_1)) = 0$ and $\varphi(R_{\mu_2}(x_2)) \neq \frac{1}{\mu_2}$,
  (iii) $\varphi(R_{\mu_1}(x_1)) \neq \frac{1}{\mu_1}$ and $\varphi(R_{\mu_2}(x_2)) = 0$.

For $x \in A_\rho$ and $\mu \in \rho(x) \cap \mathbb{C}$ we have that $\varphi(R_\mu(x)) = 0$, if and only if, $\varphi'(x) = \infty$, and $\varphi(R_\mu(x)) = \frac{1}{\mu}$, if and only if, $\varphi'(x) = 0$. Therefore, we conclude that $\varphi'(x_1 x_2) = \varphi'(x_1) \varphi'(x_2) = \infty$ in all cases (i)–(iii) above. This completes the proof of (4). $\qquad \square$

Proposition 2.5.5 *Let* $A[\tau]$ *be a commutative and pseudo-complete locally convex algebra and* $x \in A_\rho$. *Then,* $\sigma(x) = \{\varphi'(x) : \varphi \in \mathfrak{M}_0\}$.

Proof Let $\mu \in \rho(x) \cap \mathbb{C}$. Put $y \equiv (\mu e - x)^{-1} \in A_0$ and $z \equiv \mu e - x$. Consider $\lambda \in \mathbb{C}$ with $\lambda \neq 0$. Since $zy = e$, $\lambda e - z = -\lambda z(\lambda^{-1} e - y)$. Hence $\lambda e - z$ has an inverse in A, if and only if, $\lambda^{-1} e - y$ has an inverse in A. Thus, an easy calculation shows that

$$(\lambda e - z)^{-1} = \lambda^{-1} e + \lambda^{-2}(y - \lambda^{-1} e)^{-1}.$$

Therefore, $(\lambda e - z)^{-1} \in A_0$, if and only if, $(\lambda^{-1} e - y)^{-1} \in A_0$. Now $\lambda = 0 \in \sigma(y)$ is equivalent to $z = y^{-1} \notin A_0$, which occurs, if and only if, $\infty \in \sigma(z)$. So, in any case we have that $\sigma(z) = \{\lambda^{-1} : \lambda \in \sigma(y)\}$.

Now, $\rho \in \mathbb{C} \cap \sigma(x)$, if an only if, $\rho e - x$ has no inverse in A_0, if and only if, $(\mu - \rho)e - z$ has no inverse in A_0, if and only if, $(\mu - \rho)^{-1} \in \sigma(y)$. Note that $\mu - \rho \neq 0$ since $\rho \in \sigma(x)$ and $\mu \in \rho(x)$. Hence, there is a $\lambda \in \sigma(y)$, $\lambda \neq 0$, such that $\rho = \mu - \lambda^{-1}$. If $\infty \in \sigma(x)$, then $x \notin A_0$, so $z \notin A_0$ and thus $0 \in \sigma(y)$. Therefore, in any case we have that $\sigma(x) = \{\mu - \lambda^{-1} : \lambda \in \sigma(y)\}$. Based on

Theorem 2.5.3 and the proof of Proposition 2.5.4(1) we conclude that

$$\sigma(x) = \left\{ \mu - \frac{1}{\varphi((\mu e - x)^{-1})} : \varphi \in \mathfrak{M}_0 \right\} = \left\{ \varphi'(x) : \varphi \in \mathfrak{M}_0 \right\}.$$

□

Proposition 2.5.6 Let $\mathcal{A}[\tau]$ be a commutative and pseudo-complete locally convex algebra and $x \in \mathcal{A}_\rho$. Then, for $f \in F_x$ and $\varphi \in \mathfrak{M}_0$, we have $\varphi(f(x)) = f(\varphi'(x))$.

Proof Suppose first that $x \notin \mathcal{A}_0$. Then, by Theorem 2.4.4(2) we have that

$$f(x) = f(\infty)e + \int_{\partial D} f(\lambda) R_\lambda(x) d\lambda,$$

for a Cauchy domain D having the properties of Definition 2.4.2. By Proposition 2.4.1 we have that there is some $B \in \mathfrak{B}$, such that the integral appearing in the definition of $f(x)$ converges with respect to the norm of $\mathcal{A}[B]$. Since $\mathcal{A}[B]$ is a commutative Banach algebra, the restriction of φ to $\mathcal{A}[B]$ is continuous. Therefore, we have that

$$\varphi(f(x)) = f(\infty) + \int_{\partial D} f(\lambda) \varphi(R_\lambda(x)) d\lambda.$$

In case $\varphi'(x) = \infty$, then from the way φ' is defined (see (2.5.7)) we have $\varphi(R_\lambda) = 0$, for every $\lambda \in \rho(x)$. Thus, in this case, $\varphi(f(x)) = f(\infty) = f(\varphi'(x))$.

Suppose now that $\varphi'(x) \neq \infty$. Then, (for R_λ, see Definition 2.3.3) $\varphi(R_\lambda) = (\lambda - \varphi'(x))^{-1}$, hence

$$\varphi(f(x)) = f(\infty) + \int_{\partial D} f(\lambda)(\lambda - \varphi'(x))^{-1} d\lambda = f(\varphi'(x)),$$

where the last equality is derived by Cauchy's formula for an unbounded domain (see [145, p. 190]).

In case $x \in \mathcal{A}_0$, then similar considerations to the above (but without the term $f(\infty)$) gives us the result. □

The following result provides a *spectral mapping theorem* for a pseudo-complete locally convex algebra.

Proposition 2.5.7 Let $\mathcal{A}[\tau]$ be a pseudo-complete locally convex algebra and $x \in \mathcal{A}$. Then, for any $f \in F_x$, $\sigma(f(x)) = f(\sigma(x))$.

Proof Let us suppose first that $x \in \mathcal{A}_\rho$. If C is a maximal commutative subalgebra containing x, then by Theorem 2.4.4, $f(x) \in \mathcal{A}_0 \cap C$. Let $(\mathfrak{M}_0)_C$ denote the carrier space of $\mathcal{A}_0 \cap C$. Then, by Proposition 2.3.2, Theorem 2.5.3, Proposition 2.5.5 and

Proposition 2.5.6 we have the following:

$$\sigma_{\mathcal{A}}\big(f(x)\big) = \sigma_{\mathcal{C}}\big(f(x)\big) = \big\{\varphi\big(f(x)\big) : \varphi \in (\mathfrak{M}_0)_{\mathcal{C}}\big\}$$
$$= \big\{f\big(\varphi'(x)\big) : \varphi \in (\mathfrak{M}_0)_{\mathcal{C}}\big\} = f\big(\sigma_{\mathcal{C}}(x)\big)$$
$$= f\big(\sigma_{\mathcal{A}}(x)\big).$$

In case $x \notin \mathcal{A}_\rho$, that is $\rho(x) = \emptyset$, then by Theorem 2.4.4(3), F_x contains only constant functions. So, if $f(\lambda) \equiv c \in F_x$, then it is clear that $\sigma\big(f(x)\big) = \sigma(ce) = \{c\} = f\big(\sigma(x)\big)$. $\qquad\square$

Notes All the results presented in this chapter are due to G.R. Allan and can be found in [4].

Chapter 3
Generalized B*-Algebras: Functional Representation Theory

Having the background we need from Chap. 2, we introduce in the present chapter GB^*-algebras (abbreviation for generalized B^*-algebras (Sect. 3.3)), originated by G.R Allan, in 1967. These algebras generalize the celebrated C^*-algebras and they are often met in Analysis. A typical example of a GB^*-algebra is the Arens algebra $L^\omega[0, 1]$; for this and other examples, see 3.3.16.

In this chapter we are mainly concerned with commutative GB^*-algebras. In Sect. 3.4, Theorem 3.4.9 is proved, which is an "algebraic" commutative Gelfand–Naimark type theorem for GB^*-algebras; namely, it is shown that *every commutative GB^*-algebra $\mathcal{A}[\tau]$ with identity is algebraically *-isomorphic to the *-algebra $\widehat{\mathcal{A}}$ of all continuous $\mathbb{C}^*(:= \mathbb{C} \cup \{\infty\})$-valued functions on the Gelfand space $\mathfrak{M}_0$ of the commutative C^*-subalgebra $\mathcal{A}_0 = \mathcal{A}[B_0]$ of $\mathcal{A}[\tau]$.*

In the final Sect. 3.5, we exhibit the definition and basic properties of C^*-like locally convex *-algebras, introduced by A. Inoue and K.-D. Kürsten, in 2002. In Theorem 3.5.3, we prove that these topological *-algebras are GB^*-algebras. Moreover, we compare in the case of a GB^*-algebra the C^*-algebra $\mathcal{A}[B_0]$ with the Banach *-algebra $\mathcal{D}(p_\Lambda)$; see (3.3.9), (3.5.24), as well as Proposition 3.5.6 and the discussion that follows.

The first two Sects. 3.1, 3.2 of this chapter constitute a preparatory stage for the third Sect. 3.3. In particular, Sect. 3.2 shows how G.R. Allan was led to the definition of GB^*-algebras.

3.1 Hermitian and Symmetric Locally Convex *-Algebras

Section 3.1, together with Chap. 2, provide us with all ingredients needed for the definition and the study of the main theory of GB^*-algebras that are discussed in this book.

As we shall see in Sect. 3.3, a GB^*-algebra is by definition a locally convex *-algebra with some extra conditions, among them being symmetry, a property closely related with hermiticity. In the present section, some general comments relating the preceding (algebraic) concepts are presented and then we specialize in the properties and relationship of these concepts within the context of pseudo-complete (see discussion before Proposition 2.2.5) locally convex *-algebras.

We start with some standard algebraic notions that subsequently are combined with topological ones.

If $\mathcal{A}$ is an algebra, we shall call *involution*, a map $* : \mathcal{A} \to \mathcal{A} : x \mapsto x^*$, with the following properties:

$$(\lambda x + y)^* = \bar{\lambda} x^* + y^*, \quad (xy)^* = y^* x^*, \quad (x^*)^* = x,$$

for all $x, y \in \mathcal{A}$ and $\lambda, \in \mathbb{C}$.

An algebra $\mathcal{A}$ endowed with an involution $*$ will be called a *-algebra*. A subalgebra $\mathcal{B}$ of $\mathcal{A}$ invariant under involution is called a *-subalgebra*.

An element $x \in \mathcal{A}$ is called *self-adjoint*, respectively *normal*, if $x^* = x$, respectively $x^*x = xx^*$. The notation $H(\mathcal{A})$, $N(\mathcal{A})$ will stand for the *sets of self-adjoint*, respectively *normal elements* of $\mathcal{A}$. It is clear that $H(\mathcal{A})$ is a real vector subspace of $\mathcal{A}$, such that

$$\mathcal{A} = H(\mathcal{A}) \oplus i H(\mathcal{A}), \tag{3.1.1}$$

where i is the imaginary unit.

If $\mathcal{A}$ is a *-algebra without identity, *its unitization* $\mathcal{A}_1$ (see beginning of Sect. 2.1) *becomes a *-algebra*, by defining involution as follows

$$(x, \lambda)^* := (x^*, \bar{\lambda}), \ \forall\, x \in \mathcal{A} \text{ and } \lambda \in \mathbb{C}. \tag{3.1.2}$$

If $\mathcal{A}[\tau]$ *is a locally convex algebra* (see Definition 2.1.1) *and $*$ an involution on $\mathcal{A}$, this always will be considered continuous*. In this case, the real vector space $H(\mathcal{A})$ is closed.

Definition 3.1.1 A topological algebra $\mathcal{A}[\tau]$ (ibid.) endowed with a continuous involution will be called a *topological *-algebra*. When the underlying topological vector space of a topological *-algebra $\mathcal{A}[\tau]$ is a locally convex space, then we shall speak of a *locally convex *-algebra*. A metrizable and complete locally convex *-algebra is called a *Fréchet *-algebra*. A normed (resp. Banach) algebra, with continuous involution, is said to be a *normed *-(resp. Banach *-) algebra*. An algebra norm on a normed *-algebra, which preserves involution is called an m^*-*norm*.

It is known that if E, F, G are topological vector spaces, where E, F are metrizable with E also barrelled and G locally convex, then *every separately continuous bilinear map from $E \times F$ into G is continuous* [74, p. 357, Theorem 1]. Hence, every Fréchet locally convex (*-)algebra $\mathcal{A}[\tau]$, being barrelled, has continuous multiplication. In this case, one easily shows that the topology τ is defined by a sequence $\Gamma \equiv \{p_n\}_{n \in \mathbb{N}}$ of $(*-)$seminorms, such that

$$p_n(x) \leq p_{n+1}(x) \text{ and } p_n(xy) \leq p_{n+1}(x)p_{n+1}(y), \tag{3.1.3}$$

for all $x, y \in \mathcal{A}$, $n \in \mathbb{N}$. For completeness sake, we give a short proof of the second inequality in (3.1.3). Let $p_1 \in \Gamma$. By the joint continuity of multiplication in $\mathcal{A}$, there exist $n_1, m_1 \in \mathbb{N}$, such that

$$p_1(xy) \leq p_{n_1}(x)p_{m_1}(y), \ \forall \, x, y \in \mathcal{A}.$$

Hence, we have

$$p_1'(xy) \leq p_2'(x)p_2'(y), \ \forall \, x, y \in \mathcal{A}, \text{ with } p_1' := p_1 \text{ and } p_2' := p_{max\{n_1,m_1\}}.$$

Applying the preceding argument to p_2', we obtain

$$p_2'(xy) \leq p_3'(x)p_3'(y), \ \forall \, x, y \in \mathcal{A}, \text{ with } p_3' := p_{max\{n_2,m_2\}}.$$

Continuing in this way, we find sequences of natural numbers $(n_k)_{k \in \mathbb{N}}$, $(m_k)_{k \in \mathbb{N}}$, such that

$$p_k'(xy) \leq p_{k+1}'(x)p_{k+1}'(y), \ \forall \, x, y \in \mathcal{A}, \text{ with } p_{k+1}' := p_{max\{n_k, m_k\}}.$$

Notice now that the family of seminorms $\Gamma' \equiv \{p_n'\}$, $n \in \mathbb{N}$, satisfies (3.1.3) and besides, equivalently defines the given topology τ of $\mathcal{A}$. This completes the proof of (3.1.3).

Furthermore, observe that in a Banach algebra $\mathcal{A}[\|\cdot\|]$ with continuous involution $*$, we can always suppose, without loss of generality, that the involution is isometric. Indeed, the function

$$\|x\|' := \max\{\|x\|, \|x^*\|\}, \ x \in \mathcal{A},$$

is a $*$-preserving norm on $\mathcal{A}$, equivalent to $\|\cdot\|$, making $\mathcal{A}$ a Banach algebra with isometric involution.

If $\mathcal{A}[\tau]$ is a topological $*$-algebra without identity, its unitization $\mathcal{A}_1[\tau_1]$ (see discussion after Definition 2.1.1) endowed with the involution given by (3.1.2) is also a topological $*$-algebra.

By a topological $(*-)$isomorphism between two topological $(*-)$algebras, we mean an algebraic $(*-)$isomorphism (i.e., a bijective $(*-)$homomorphism), which is a homeomorphism. When we use the symbol "$\cong$" between two topological $(*-)$ algebras, this will always mean a topological $(*-)$isomorphism. By a topological $(*-)$embedding, we mean an algebraic $(*-)$monomorphism, which is bicontinuous on its image.

If $\mathcal{A}$ is a *-algebra and p is a seminorm (resp. norm) on $\mathcal{A}$, we say that p has the C^*-*property* or equivalently p is a C^*-*seminorm* (resp. a C^*-norm), if $p(x^*x) = p(x)^2$, for all $x \in \mathcal{A}$. Z. Sebestyén proved, in 1979, that *every C^*-seminorm on a *-algebra is automatically *-preserving and submultiplicative* (see, for instance, [52, p. 167, §38]). In this regard, we now set the following

Definition 3.1.2 A locally convex algebra $\mathcal{A}[\tau]$ with involution, whose topology τ is defined by an upwards directed family of C^*-seminorms is called a C^*-*convex algebra;* its involution is automatically continuous, as follows from the above comments. A complete C^*-convex algebra, is called a *pro-C^*-algebra* [124] (this is what Apostol [8] called a b^*-algebra, Inoue [75] a locally C^*-algebra and Schmüdgen [132] an LMC^*-algebra). A metrizable pro-C^*-algebra is said to be a σ-C^*-*algebra* [124]. A *-algebra endowed with a C^*-norm is called a *pre C^*-algebra.* A complete pre C^*-algebra is said to be a C^*-*algebra.* Clearly, every C^*-algebra is a pro-C^*-algebra.

It is obvious from the preceding comments that every pro-C^*-algebra is *a complete m^*-convex algebra,* in the sense that all C^*-seminorms defining its topology are m^*-seminorms, that is *-preserving m-seminorms (see discussion before Definition 3.1.2).

Let $\mathcal{A}$ be a *-algebra. The involution of $\mathcal{A}$ is said to be *hermitian,* whenever the spectrum $sp(h)$ (see beginning of Sect. 2.3) of any element $h \in H(\mathcal{A})$ is a subset of the real line. A *-algebra with a hermitian involution is called a *hermitian algebra;* cf., e.g., [52, p. 128, (32.2)].

Furthermore, a *-algebra $\mathcal{A}$ is called *symmetric* if for every $x \in \mathcal{A}$, the element x^*x is quasi-invertible.

Every symmetric algebra is hermitian [52, p. 129, (32.4) Corollary]; the converse is true for every Banach algebra with involution and this result is known as the Shirali–Ford theorem; see [52, p. 136, (33.2) Theorem] and [60, p. 297, Theorem 22.23]; in the latter, the reader will find a proof based on Pták's theory [126] for hermitian algebras. Thus, *symmetry and hermiticity coincide on Banach algebras with involution.* In the non-normed case, one has that *every pro-C^*-algebra* (and a fortiori every C^*-algebra) *is symmetric* [60, p. 268, (21.6)]. An analogue of the Shirali–Ford theorem, in the aforementioned setting is valid *for every hermitian spectral complete m-convex algebra* (see [60, Theorem 22.27]). Note that an algebra $\mathcal{A}$ is called *spectral* (Palmer, [120]), if it can be equipped with an m-seminorm p, such that $r_{\mathcal{A}}(x) \le p(x)$, for all $x \in \mathcal{A}$.

▶ An element $x \in H(\mathcal{A})$ with $sp(x) \subseteq [0, +\infty)$, respectively $sp(x) \subseteq (0, +\infty)$, is called *positive,* respectively *strictly positive* and we write $x \ge 0$, respectively $x > 0$. The set of all positive elements of $\mathcal{A}$ will be denoted by $\mathcal{A}^+$.

In this regard, a geometric characterization of symmetry, reads as follows: *a *-algebra A is symmetric, if and only if, $x^*x \geq 0$, for every $x \in A$* (see [52, (32.5) Proposition]).

In the case of a locally convex *-algebra $A[\tau]$, the definition of "hermiticity" and "symmetry" is further specialized by using the notion of (Allan-)bounded elements (see Definitions 2.2.1 and 2.3.1).

Definition 3.1.3 (Allan) The involution on a locally convex *-algebra $A[\tau]$ is said to be *hermitian* if $\sigma(h)$ is real, for every $h \in H(A)$; note that ∞ is counted real, in this case. When a locally convex *-algebra $A[\tau]$ has a hermitian involution, it will be called *hermitian*.

Recall that given a locally convex algebra $A[\tau]$, A_0 denotes the set of all bounded elements of $A[\tau]$; see Definition 2.2.1 and the comments that follow. It is evident that *if $A[\tau]$ is a locally convex *-algebra and x an element of A, then $x \in A_0$, if and only if, $x^* \in A_0$.*

Lemma 3.1.4 *Let $A[\tau]$ be a hermitian pseudo-complete locally convex *-algebra with identity e. Then, for every $h \in H(A)$, the following hold:*

(i) *the element $e + h^2$ has a bounded inverse;*
(ii) *the element $h(e + h^2)^{-1}$ is bounded.*

Proof

(i) Since the involution of A is hermitian, we have that $i \in \rho(h)$ and $i \in \rho(-h)$, for every $h \in H(A)$. Therefore, the elements $(ie - h)^{-1}$ and $(ie + h)^{-1}$ exist and belong to A_0. Since moreover, they commute and $A[\tau]$ is pseudo-complete, their product belongs also to A_0 (see e.g, Corollary 2.2.11). Hence, we obtain that the element

$$-(ie - h)(ie + h) = e + h^2, \tag{3.1.4}$$

has an inverse that belongs to A_0, for every $h \in H(A)$.

(ii) Let now $u \equiv h(e + h^2)^{-1}$, $h \in H(A)$. Then,

$$u^2 = h^2(e + h^2)^{-2} = (e + h^2)^{-1} - (e + h^2)^{-2}, \forall h \in H(A),$$

where since $(e + h^2)^{-1}$ and $(e + h^2)^{-2}$ commute and $A[\tau]$ is pseudo-complete, we conclude that $u^2 \in A_0$ (see, for instance, Corollary 2.2.11). This implies that the set

$$S = \left\{ (\lambda u)^{2n} : n \in \mathbb{N} \right\}, \quad \text{for some } \lambda \in \mathbb{C} \backslash \{0\},$$

is bounded. It is now clear that $\left\{ (\lambda u)^n : n \in \mathbb{N} \right\} \subset S \cup (\lambda u)S$, therefore it is bounded; i.e., $u \in A_0$ and this completes the proof. ☐

The next Proposition 3.1.5 provides a characterization of hermiticity.

Proposition 3.1.5 *Let* $\mathcal{A}[\tau]$ *be a pseudo-complete locally convex *-algebra with identity* e. *The following are equivalent:*

(i) $\mathcal{A}[\tau]$ *is hermitian;*
(ii) *the element* $e + h^2$ *has a bounded inverse, for every* $h \in H(\mathcal{A})$.

Proof (i) $\Rightarrow$ (ii) follows from Lemma 3.1.4(i).

(ii) $\Rightarrow$ (i) By the assumption (ii) and (3.1.4), we conclude that the element $ie - h, h \in H(\mathcal{A})$ is invertible with inverse given by

$$(ie - h)^{-1} = -(ie + h)(e + h^2)^{-1} = -h(e + h^2)^{-1} - i(e + h^2)^{-1}, \ \forall \, h \in H(\mathcal{A}).$$

By Lemma 3.1.4, this implies that $(ie - h)^{-1} \in \mathcal{A}_0$, for every $h \in H(\mathcal{A})$, which in its turn gives that $i \notin \sigma(h)$, for all $h \in H(\mathcal{A})$.

Suppose now that $\alpha, \beta \in \mathbb{R}$, with $\beta \neq 0$. Then, $\beta^{-1}(h - \alpha e) \in H(\mathcal{A})$. Therefore, $i \notin \sigma(\beta^{-1}(h - \alpha e))$, which equivalently means that $\alpha + i\beta \notin \sigma(h)$, for any $\alpha, \beta \in \mathbb{R}$, with $\beta \neq 0$, and $h \in H(\mathcal{A})$. This completes the proof. $\square$

Definition 3.1.6 (Allan) A locally convex *-algebra $\mathcal{A}[\tau]$ is called *symmetric*, if for every $x \in \mathcal{A}$, the element x^*x is quasi-invertible with bounded quasi-inverse, which in the case of an identity element e, equivalently means that, for every $x \in \mathcal{A}$, the element $(e + x^*x)^{-1}$ exists and belongs to $\mathcal{A}_0$.

In other words, every symmetric locally convex *-algebra $\mathcal{A}[\tau]$ is "algebraically" symmetric, so that for every $x \in \mathcal{A}$, $x^*x \in G_{\mathcal{A}}^q$ and moreover $(x^*x)^\circ \in \mathcal{A}_0$.

An immediate consequence of Definition 3.1.6 and Proposition 3.1.5 is the following

Corollary 3.1.7 *Every symmetric pseudo-complete locally convex *-algebra is hermitian.*

The following Proposition 3.1.8 says that a locally convex *-algebra is symmetric, whenever it is "algebraically" symmetric and has a continuous inversion.

Proposition 3.1.8 *Let* $\mathcal{A}[\tau]$ *be a locally convex *-algebra with identity* e *and continuous inversion* (take, for instance an m-convex algebra). *Suppose that for every* $x \in \mathcal{A}$, $(e + x^*x)^{-1}$ *exists in* $\mathcal{A}$. *Then* $\mathcal{A}[\tau]$ *is symmetric.*

Proof By the discussion before Definition 3.1.3, we have that since $\mathcal{A}$ is "algebraically" symmetric, then $x^*x \geq 0$, for every $x \in \mathcal{A}$. This yields that

$$sp(e + x^*x) \subset [1, \infty) \quad \text{and} \quad sp\big((e + x^*x)^{-1}\big) \subset (0, 1], \ \forall \, x \in \mathcal{A}.$$

Now, since $\mathcal{A}[\tau]$ has continuous inversion, by Theorem 2.3.13, we conclude that

$$\sigma\big((e + x^*x)^{-1}\big) \subset \overline{sp\big((e + x^*x)^{-1}\big)} \subset [0, 1], \ \forall \, x \in \mathcal{A},$$

where the closure of $sp\big((e + x^*x)^{-1}\big)$ is taken in $\mathbb{C}^*$. It is now clear that $\infty \notin \sigma\big((e + x^*x)^{-1}\big)$, for all $x \in \mathcal{A}$. This, by definition of the spectrum in a locally convex algebra (see Definition 2.3.1), equivalently means that $(e + x^*x)^{-1} \in \mathcal{A}_0$, for all $x \in \mathcal{A}$, so that the proof is complete. □

3.2 Some Results on C*-Algebras

G.R. Allan noticed in [3], that studying certain locally convex algebras, the following problem cropped up: to express the C^*-condition in a given normed $*$-algebra using not its norm, but its properties as a locally convex $*$-algebra. A solution to this problem was given in the previous reference and we present it in this section. An application of these lines of thought were essentially applied for the development of the theory of GB^*-algebras (see [3, 5], as well as Chap. 2 and Sect. 3.3).

Let $\mathcal{A}[\| \cdot \|]$ be a normed algebra with an involution $*$. For convenience, we assume throughout this section that our algebras have an identity element, although all the results can be easily proved without this assumption. *Denote by* $\mathfrak{B}^*_{\mathcal{A}}$ (see also Definition 3.3.1, in Sect. 3.3) *the collection of all subsets B in* $\mathcal{A}[\| \cdot \|]$, *such that*:

(1) B is absolutely convex;
(2) $e \in B$, $B^2 \subset B$ and $B^* = B$, where $B^* := \{x^* : x \in B\}$;
(3) B is bounded and closed.

▶ We shall *consider* $\mathfrak{B}^*_{\mathcal{A}}$ *endowed with the partial ordering given by inclusion.*

We are going to give a characterization of a pre C^*-algebra, respectively C^*-algebra (see Theorems 3.2.9 and 3.2.10) through the collection $\mathfrak{B}^*_{\mathcal{A}}$. This will be done by using the given topological algebra as a locally convex algebra, rather than as a normed algebra. For the proof of the mentioned results we need a series of lemmas that we first axhibit.

Lemma 3.2.1 *Let* $\mathcal{A}[\| \cdot \|]$ *be a normed $*$-algebra with identity e and $\mathcal{B}$ a $*$-subalgebra of $\mathcal{A}$ containing e. Suppose that, for every $x \in \mathcal{B}$, the element $(e + x^*x)^{-1}$ exists in $\mathcal{A}$. Then, the following hold*:

(i) *for every $h \in H(\mathcal{B})$, $sp_{\mathcal{A}}(h) \subset \mathbb{R}$ (i.e., the spectrum of every self-adjoint element in $\mathcal{B}$ is real)*;
(ii) *for every $x \in \mathcal{B}$, $sp_{\mathcal{A}}(x^*x) \subset [0, \infty)$ (i.e., the spectrum of any element x^*x, $x \in \mathcal{B}$, is real and non-negative)*.

Proof

(i) Let $h \in H(\mathcal{B})$. Suppose that $\alpha + i\beta \in sp_{\mathcal{A}}(h)$, where $\alpha, \beta \in \mathbb{R}$, with $\beta \neq 0$. Then, considering the polynomial $f(t) = \beta^{-1}(\alpha^2 + \beta^2)^{-1}(\alpha t^2 + (\beta^2 - \alpha^2)t)$, $t \in \mathbb{C}$, we notice that $f(\alpha + i\beta) = i$. Applying the spectral mapping theorem (see, e.g., [127, Theorem (1.6.10)]), we obtain that $i \in sp_{\mathcal{A}}(f(h))$; by the same reason $i^2 \in sp_{\mathcal{A}}(f(h)^2)$, where $f(h) \in H(\mathcal{B})$. Hence, $e - i^2 f(h)^2 = e + f(h)^2$ does not have an inverse in $\mathcal{A}$, which contradicts our hypothesis. Consequently, $\beta = 0$, therefore the spectrum of h is real.

(ii) Let $\lambda \in \mathbb{R}$, with $\lambda > 0$. Consider the element $z := \lambda^{\frac{1}{2}}x$, $x \in \mathcal{B}$. Then, $z \in \mathcal{B}$ and $z^*z = \lambda x^*x \in H(\mathcal{B})$. Therefore by (i) z^*z has a real spectrum. Moreover, $e - (-\lambda)x^*x = e + z^*z$ is invertible in $\mathcal{A}$ by our hypothesis. Hence, $-\lambda \notin sp_{\mathcal{A}}(x^*x)$, which completes the proof of the assertion (ii). $\qquad \square$

If the involution in the preceding normed algebra $\mathcal{A}$ is supposed to be continuous, then without loss of generality we may suppose that $\|x^*\| = \|x\|$, for every $x \in \mathcal{A}$. Thus, the closed unit ball $U(\mathcal{A}) := \{x \in \mathcal{A} : \|x\| \leq 1\}$ in $\mathcal{A}$ is contained in $\mathfrak{B}_{\mathcal{A}}^*$. In particular, if B_0 is the greatest member in $\mathfrak{B}_{\mathcal{A}}^*$, one obtains

$$U(\mathcal{A}) \subset B_0 \subset \varepsilon U(\mathcal{A}),$$

for some $\varepsilon > 0$, which implies that the gauge function (see discussion after Definition 2.2.2) $\| \cdot \|_{B_0}$ is a norm on $\mathcal{A}$ equivalent to the given one. This allows us to suppose, without loss of generality, that B_0 is the closed unit ball of $\mathcal{A}$.

Denote by $\widetilde{\mathcal{A}}$ the completion of $\mathcal{A}[\| \cdot \|]$. Then, clearly $\widetilde{\mathcal{A}}$ is a Banach *-algebra.

Lemma 3.2.2 *Let $\mathcal{A}[\| \cdot \|]$ be a normed algebra with identity e and continuous involution. Suppose also that the collection $\mathfrak{B}_{\mathcal{A}}^*$ has a greatest member, say B_0. Then, $\|h\| = r_{\widetilde{\mathcal{A}}}(h)$, for every $h \in H(\widetilde{\mathcal{A}})$.*

Proof First, we suppose that $h \in H(\mathcal{A})$. Assume that $h \in B_0$. Then, clearly $h^2 \in B_0$. Conversely, let $h^2 \in B_0$. Then, choosing $\varepsilon \in \mathbb{R}$ with $\varepsilon > 1$, such that $h \in \varepsilon B_0$, we obtain

$$h^{2n} \in B_0^n \subset B_0 \subset \varepsilon B_0 \text{ and } h^{2n+1} \in \varepsilon B_0 B_0 \subset \varepsilon B_0, \ n \in \mathbb{N}.$$

It is now evident that if $C = \{e, h^n, n \in \mathbb{N}\}$, then C is a bounded set and $C^2 \subset C$; therefore the closed absolutely convex hull of C belongs to $\mathfrak{B}_{\mathcal{A}}^*$ and $C \subset B_0$, which yields $h \in B_0$. So we have proved

$$h \in B_0 \Leftrightarrow h^2 \in B_0.$$

This implies $\|h^2\| = \|h\|^2$ (see the discussion before Lemma 3.2.2).

Suppose now that $h \in H(\widetilde{\mathcal{A}})$. Then, there is a sequence $(x_n)_{n \in \mathbb{N}}$ in $\mathcal{A}$ such that $x_n \to h$. Put $h_n = \frac{1}{2}(x_n + x_n^*)$. Clearly, $h_n^* = h_n$, for each $n \in \mathbb{N}$ and $h_n \to h$.

Thus,

$$\|h^2\| = \lim_n \|h_n^2\| = \lim_n \|h_n\|^2 = \|h\|^2.$$

Inductively, we obtain

$$\|h^{2^n}\| = \|h\|^{2^n}, \, n \in \mathbb{N}, \, \forall \, h \in H(\tilde{\mathcal{A}}).$$

Applying now the spectral radius formula, valid in a Banach algebra, we have completed the proof.   □

Lemma 3.2.3 *Let* $\mathcal{A}[\| \cdot \|]$ *be a normed algebra with identity* e *and continuous involution. Suppose that the following condition holds: for each* $x \in \mathcal{A}$, *there exist sequences* $(u_n)_{n \in \mathbb{N}}$, $(v_n)_{n \in \mathbb{N}}$ *in* $\mathcal{A}$, *such that*

$$u_n(e + x^*x) \rightarrow e \leftarrow (e + x^*x)v_n. \tag{3.2.5}$$

Then, for every $x \in \mathcal{A}$, *the element* $(e + x^*x)^{-1}$ *exists in* $\tilde{\mathcal{A}}$.

Proof Put $y = e + x^*x$, $x \in \mathcal{A}$. Then, by (3.2.5), there exists a sequence $(u_n)_{n \in \mathbb{N}}$ in $\mathcal{A}$, such that $u_n y \rightarrow e$. So, there is $m \in \mathbb{N}$, with $\|u_m y - e\| < 1$. This implies that $u_m y$ is invertible in $\tilde{\mathcal{A}}$; therefore there is $u \in \tilde{\mathcal{A}}$, such that $(uu_m)y = e$. Similarly, there exists $l \in \mathbb{N}$ and $v \in \tilde{\mathcal{A}}$, with $y(v_l v) = e$. It is now clear that y is invertible in $\tilde{\mathcal{A}}$ and this completes the proof.   □

Lemma 3.2.4 *Let* $\mathcal{A}[\| \cdot \|]$ *be as in Lemma 3.2.3. Suppose also that the collection* $\mathfrak{B}_\mathcal{A}^*$ *has a greatest member. Then,* $\tilde{\mathcal{A}}$ *is a symmetric Banach algebra.*

Proof We must show that, for every $x \in \tilde{\mathcal{A}}$, the element $e + x^*x$ is invertible in $\tilde{\mathcal{A}}$. Taking an arbitrary $x \in \tilde{\mathcal{A}}$, there is a sequence $(x_n)_{n \in \mathbb{N}}$ in $\mathcal{A}$, such that $x_n \rightarrow x$. We set $y = e + x^*x$ and $y_n = e + x_n^*x_n$, $n \in \mathbb{N}$. Clearly, the elements y, y_n, $n \in \mathbb{N}$ are self-adjoint and by Lemma 3.2.3, $y_n^{-1} \in \tilde{\mathcal{A}}$, for all $n \in \mathbb{N}$. Thus, applying Lemma 3.2.1, with $\tilde{\mathcal{A}}$ in the place of $\mathcal{A}$ and $\mathcal{A}$ in the place of $\mathcal{B}$, we have that $sp_{\tilde{\mathcal{A}}}(x_n^*x_n)$ is real and non-negative, for all $n \in \mathbb{N}$. Hence,

$$sp_{\tilde{\mathcal{A}}}(y_n) \subset [1, \infty), \quad \text{which yields} \quad sp_{\tilde{\mathcal{A}}}(y_n^{-1}) \subset (0, 1], \, \forall \, n \in \mathbb{N}.$$

Lemma 3.2.2 now gives us that

$$\|y_n^{-1}\| = r_{\tilde{\mathcal{A}}}(y_n^{-1}) \leq 1, \, \forall \, n \in \mathbb{N}.$$

Furthermore,

$$\|y_n^{-1} - y_m^{-1}\| = \|y_m^{-1}(y_m - y_n)y_n^{-1}\| \leq \|y_m^{-1}\| \|y_m - y_n\| \|y_n^{-1}\| \leq \|y_m - y_n\|, \, n, m \in \mathbb{N}.$$

But, $y_n \rightarrow y$, so that $(y_n^{-1})_{n \in \mathbb{N}}$ is a Cauchy sequence, therefore it converges to an element $z \in \tilde{A}$. It follows that $yz = e = zy$ and this completes the proof. □

Lemma 3.2.5 *Let* $A[\| \cdot \|]$ *be a normed algebra with identity* e *and continuous involution. Suppose that the following conditions hold:*

(i) *the collection* $\mathfrak{B}^*_A$ *has a greatest member;*
(ii) *for each* $x \in A$, *there exist sequences* $(u_n)_{n \in \mathbb{N}}$, $(v_n)_{n \in \mathbb{N}}$ *in* A, *such that*

$$u_n(e + x^*x) \rightarrow e \leftarrow (e + x^*x)v_n.$$

Then, any maximal commutative $*$*-subalgebra* $\mathcal{C}$ *of* $\tilde{A}$ *becomes a* C^**-algebra under a* C^**-norm equivalent to the norm* $\| \cdot \|_{\mathcal{C}}$, *the restriction of the norm of* $\tilde{A}$ *to* $\mathcal{C}$.

Proof Let $\mathcal{C}$ be a maximal commutative $*$-subalgebra of $\tilde{A}$. Then, $\mathcal{C}$ is a closed $*$-subalgebra of $\tilde{A}$ with $sp_{\mathcal{C}}(x) = sp_{\tilde{A}}(x)$, for every $x \in \mathcal{C}$ (see [127, Theorem (4.1.3)]). Moreover, since $\tilde{A}$ is symmetric by Lemma 3.2.4, $\mathcal{C}$ is symmetric too (ibid., Corollary (4.7.7)). Let

$$r(x) \equiv r_{\mathcal{C}}(x) = r_{\tilde{A}}(x), \quad x \in \mathcal{C}.$$

Take an arbitrary $x \in \mathcal{C}$. Then, $x = h + ik$ with $h, k \in H(\mathcal{C})$. Clearly, $hk = kh$. Thus, from Lemma 3.2.2 and properties of the spectral radius, we obtain

$$r(x) \leq \|x\|_{\mathcal{C}} \leq \|h\|_{\mathcal{C}} + \|k\|_{\mathcal{C}} = r_{\mathcal{C}}(h) + r_{\mathcal{C}}(k) \leq 2r(x). \qquad (3.2.6)$$

To obtain the last inequality, just calculate the spectral radii of h and k using the expressions: $h = \frac{1}{2}(x + x^*)$, $k = \frac{1}{2i}(x - x^*)$ and standard properties of the spectral radius. Then, (3.2.6) shows that r is a norm equivalent to $\| \cdot \|_{\mathcal{C}}$. But, $\mathcal{C}$ is a symmetric Banach algebra, hence hermitian [127, Theorem (4.7.6)], therefore r fulfills the C^*-property (ibid., Lemma (4.2.1)). Thus, $\mathcal{C}[r]$ is a C^*-algebra and this completes the proof. □

Definition 3.2.6 Given an algebra A, we call the intersection of the kernels of all irreducible representations of A, *the Jacobson radical* of A and we denote it by J_A (see [32, p. 124, Definition 13]). For other expressions of J_A we refer to (ibid., p. 124, Proposition 14 and p. 125, Proposition 16). When $J_A = \{0\}$, A is called *semisimple*; see, for instance [127, Definition (2.3.1)].

Lemma 3.2.7 *Let* $A[\| \cdot \|]$ *be as in Lemma 3.2.5. Then,* $\tilde{A}$ *is semisimple.*

Proof We show that the Jacobson radical $J_{\tilde{A}}$ of $\tilde{A}$ is trivial. Let $x \in J_{\tilde{A}}$ be arbitrary. Then, $x = h + ik$, for some $h, k \in H(\tilde{A})$. Since, $J_{\tilde{A}}$ is a $*$-ideal and h, k are linear combinations of x, x^*, we conclude that h, $k \in J_{\tilde{A}}$. But, every element in the Jacobson radical has trivial spectral radius (cf., e.g., [60, Proposition 4.24(1)] and/or [52, Theorem (B.5.17)(c)]), therefore $r_{\tilde{A}}(h) = 0 = r_{\tilde{A}}(k)$. Now, Lemma 3.2.2 gives $h = 0 = k$; hence $x = 0$ and this completes the proof. □

Definition 3.2.8 ([126]) Let $\mathcal{A}[\| \cdot \|]$ be a Banach algebra with involution $*$ and $r_{\mathcal{A}}$ its spectral radius. Then, the function

$$p_{\mathcal{A}}(x) := r_{\mathcal{A}}(x^*x)^{1/2}, \ x \in \mathcal{A},$$

is called *the Pták function* and plays an important role in the theory of hermitian algebras developed by V. Pták [126], in 1972.

Theorem 3.2.9 *Let $\mathcal{A}[\| \cdot \|]$ be a normed algebra with identity e and continuous involution. Then, $\mathcal{A}$ is a pre C^*-algebra under a C^*-norm equivalent to the given one, if and only if, the following conditions hold:*

(i) *the collection $\mathfrak{B}_{\mathcal{A}}^*$ has a greatest member;*
(ii) *for each $x \in \mathcal{A}$, there exist sequences $(u_n)_{n \in \mathbb{N}}$, $(v_n)_{n \in \mathbb{N}}$ in $\mathcal{A}$, such that*

$$u_n(e + x^*x) \to e \leftarrow (e + x^*x)v_n.$$

Proof "Only if" part. Suppose that $\mathcal{A}[\| \cdot \|]$ is a pre C^*-algebra. Then, by the properties of the closed unit ball $U(\mathcal{A})$ of $\mathcal{A}[\| \cdot \|]$, we have that $U(\mathcal{A})$ is a greatest member in $\mathfrak{B}_{\mathcal{A}}^*$. On the other hand, the completion $\widetilde{\mathcal{A}}$ of $\mathcal{A}[\| \cdot \|]$ is a C^*-algebra, therefore symmetric (see [127, Theorem (4.8.9)]), consequently $(e + x^*x)^{-1}$ exists in $\widetilde{\mathcal{A}}$, for every $x \in \mathcal{A}$. Hence, for every $x \in \mathcal{A}$, there is a sequence $(u_n)_{n \in \mathbb{N}}$ in $\mathcal{A}$, such that $u_n \to (e + x^*x)^{-1}$, which yields that $u_n(e + x^*x) \to e \leftarrow (e + x^*x)v_n$, with $v_n = u_n^*$, for every $n \in \mathbb{N}$.

"If " part. The completion $\widetilde{\mathcal{A}}$ of $\mathcal{A}[\| \cdot \|]$ is a semisimple symmetric Banach algebra by Lemmas 3.2.7 and 3.2.4. This, by [127, Corollary (4.7.16)], means that $\widetilde{\mathcal{A}}$ admits a faithful $*$-representation, say π, on a Hilbert space $\mathcal{H}$. From [52, p. 100, Theorem (26.13)(b)], we then have that

$$\|\pi(x)\| \leq p_{\mathcal{A}}(x) \leq \|x\|, \ \forall \, x \in \widetilde{\mathcal{A}}.$$

The induced by π and the operator norm of $\mathcal{B}(\mathcal{H})$, C^*-norm on $\widetilde{\mathcal{A}}$, will be denoted by $\| \cdot \|_\pi$; that is,

$$\|x\|_\pi := \|\pi(x)\|, \ \forall \, x \in \widetilde{\mathcal{A}}.$$

Let $h \in H(\widetilde{\mathcal{A}})$. Denote by $\mathcal{C}_h$ the maximal commutative $*$-subalgebra of $\widetilde{\mathcal{A}}$ containing h. By Lemma 3.2.5, $\mathcal{C}_h$ is a C^*-algebra, under the C^*-norm $r \equiv r_{\mathcal{C}_h}$ (spectral radius of $\mathcal{C}_h$ equivalent to the norm $\| \cdot \|_{\mathcal{C}_h}$, the restriction of the norm of $\widetilde{\mathcal{A}}$ to $\mathcal{C}_h$). Restricting the continuous, faithful $*$-representation π of $\widetilde{\mathcal{A}}$ to $\mathcal{C}_h[r]$, we obtain an isometric $*$-isomorphism of $\mathcal{C}_h[r]$ in $\mathcal{B}(\mathcal{H})$. Hence, (see also Lemma 3.2.2)

$$\|h\|_\pi = r(h) = \|h\|_{\mathcal{C}_h} = \|h\|, \ \forall \, h \in H(\widetilde{\mathcal{A}}).$$

Take now an arbitrary $x \in \widetilde{\mathcal{A}}$. Then, $x = h + ik$, $h, k \in H(\widetilde{\mathcal{A}})$ and (see also the arguments used for the proof of (3.2.6))

$$\|x\| \leq \|h\| + \|k\| = \|h\|_\pi + \|k\|_\pi \leq 2\|x\|_\pi$$

$$\leq 2(\|h\|_\pi + \|k\|_\pi) = 2(\|h\| + \|k\|) \leq 4\|x\|.$$

This shows that the $*$-norms $\|\cdot\|$, $\|\cdot\|_\pi$ are equivalent on $\widetilde{\mathcal{A}}$. In this way, $\mathcal{A}$ becomes a pre C^*-algebra with respect to the C^*-norm $\|\cdot\|_\pi$ (equivalent to $\|\cdot\|$) and so the proof is complete.   □

Theorem 3.2.10 *Let $\mathcal{A}[\|\cdot\|]$ be a Banach $*$-algebra with identity e. Then, $\mathcal{A}$ is a C^*-algebra under a C^*-norm equivalent to the given norm $\|\cdot\|$, if and only if, the following conditions hold:*

(i) *the collection $\mathfrak{B}^*_{\mathcal{A}}$ has a greatest member;*
(ii) *$\mathcal{A}$ is symmetric.*

Proof "Only if " part. The condition (i) is proved, in a similar way, as (i) in Theorem 3.2.9. The condition (ii) follows from the fact that every C^*-algebra is symmetric (cf. [127, Theorem (4.8.9)]).
  "if " part. Define $|x| := max\{\|x\|, \|x^*\|\}$, $x \in \mathcal{A}$. Then,

$$|x^*| = |x|, \quad \|x\| \leq |x|, \quad \|x^*\| \leq |x|, \quad \forall\, x \in \mathcal{A}.$$

It follows that if B is a subset of $\mathcal{A}$, such that $B^* = B$, then B is $\|\cdot\|$-bounded, if and only if, it is $|\cdot|$-bounded. Therefore, $\mathfrak{B}^*_{\mathcal{A}[\|\cdot\|]} = \mathfrak{B}^*_{\mathcal{A}[|\cdot|]}$. Thus, $\mathcal{A}[\|\cdot\|]$ satisfies the conditions (i) and (ii) of Theorem 3.2.9. We remark that condition (ii) of Theorem 3.2.9 follows from the symmetry of $\mathcal{A}$, which is a completely algebraic property (see beginning of Sect. 3.1 and proof of the "only if " part of Theorem 3.2.9). This implies that $\mathcal{A}$ is a pre C^*-algebra, under a C^*-norm equivalent to $|\cdot|$. In [127, p. 181], a Banach $*$-algebra $\mathcal{A}[\|\cdot\|]$ endowed with a C^*-norm is called an A^*-*algebra*. In this case, the involution on $\mathcal{A}$ is continuous with respect to both norms (ibid., Theorem (4.1.15)). Then, it turns out that the norms $\|\cdot\|$, $|\cdot|$ are equivalent, therefore $\mathcal{A}[\|\cdot\|]$ is a C^*-algebra and this completes the proof of the "if " part.   □

Remark 3.2.11 Note that the conditions (i), (ii) in Theorems 3.2.9 and 3.2.10 are necessary. Indeed, consider the Banach $*$-algebra $\mathcal{A} \equiv C^{(n)}[0, 1]$ of all n-continuously differentiable functions on the unit interval $[0, 1]$. Then, $\mathcal{A}$ fulfills the condition (ii) of Theorems 3.2.9 and 3.2.10, but not the condition (i). So, if the theorems are true without assuming condition (i), $\mathcal{A}$ would be a C^*-algebra, which is not true.
  On the other hand, the disc algebra $\mathcal{A}(\mathcal{D})$, $\mathcal{D} = \{z \in \mathcal{C} : |z| \leq 1\}$ is a Banach $*$-algebra that fulfills condition (i), but not condition (ii). So, if Theorems 3.2.9 and 3.2.10 are true by removing condition (ii), then $\mathcal{A}(\mathcal{D})$ would be a C^*-algebra, which is a contradiction.

Recall that Sect. 3.1, together with the ideas of Theorem 3.2.10, lead to the definition of *GB**-algebras (subject matter of this book), in the next section.

3.3 GB*-Algebras

The algebras we introduce in this section generalize the celebrated C^*-algebras. They were initiated by G.R. Allan, in 1967 (see [5]) and they were studied first by himself and later by his student P.G. Dixon in [48, 49]. Their significance is due to the fact that they are algebras of unbounded operators as P.G. Dixon showed in [48], therefore they play an important role, in mathematical physics.

The following definition stated for a locally convex ∗-algebra $\mathcal{A}[\tau]$ with identity, uses Definition 2.2.2 stated for arbitrary locally convex algebras, so naturally an extra condition related to the involution of $\mathcal{A}$ and a second one related to the identity element is added (see also comments before Lemma 3.2.1).

Definition 3.3.1 Given a locally convex ∗-algebra $\mathcal{A}[\tau]$ with identity e, denote by $\mathfrak{B}^*_{\mathcal{A}}$ the collection of all subsets B in $\mathcal{A}[\tau]$ that satisfy the following properties:

(i) B is absolutely convex, closed and bounded;
(ii) $e \in B$, $B^2 \subseteq B$ and $B^* = B$, where $B^* := \{x^* : x \in B\}$.

▶ *When $\mathcal{A}$ is endowed with two different locally convex ∗-algebra topologies* τ, τ'*, then for distinction, we shall use the symbols* $\mathfrak{B}^*_{\mathcal{A}[\tau]}$*,* $\mathfrak{B}^*_{\mathcal{A}[\tau']}$ *respectively, for the preceding collections of subsets of* $\mathcal{A}$*.*

For $B \in \mathfrak{B}^*_{\mathcal{A}}$, recall the algebra $\mathcal{A}[B] = \{\lambda x : \lambda \in \mathbb{C}, x \in B\}$, generated by B (cf. (2.2.1)). We note that in this case, $\mathcal{A}[B]$ is a normed algebra with continuous involution, as it easily follows from the fact that $B = B^*$. Hence, each $\mathcal{A}[B]$, $B \in \mathfrak{B}^*_{\mathcal{A}}$, is a normed ∗-algebra.

By the standard locally convex space theory, it is evident that the topology τ of $\mathcal{A}$ restricted to $\mathcal{A}[B]$, $B \in \mathfrak{B}^*_{\mathcal{A}}$, is coarser than the ∗-normed topology induced on $\mathcal{A}[B]$ by the gauge function $\| \cdot \|_B$, $B \in \mathfrak{B}^*_{\mathcal{A}}$ (see (2.2.2)); that is,

$$\tau \upharpoonright_{\mathcal{A}[B]} \prec \| \cdot \|_B, \quad \forall B \in \mathfrak{B}^*_{\mathcal{A}}. \tag{3.3.7}$$

According to the preceding, G.R. Allan defined in [5, (2.5) Definition] a *GB**-algebra (abbreviation of generalized *B**-algebra) as follows.

Definition 3.3.2 (Allan) A *GB*-algebra* is a locally convex *-algebra $\mathcal{A}[\tau]$ with an identity element, such that:

(i) $\mathfrak{B}^*_{\mathcal{A}}$ has a greatest ($\Leftrightarrow$ maximal) member (under the partial ordering of $\mathfrak{B}^*_{\mathcal{A}}$ by inclusion), denoted by B_0;
(ii) $\mathcal{A}[\tau]$ is symmetric (in the sense of Definition 3.1.6);
(iii) $\mathcal{A}[\tau]$ is pseudo-complete.

Occasionally, we shall also use the term *GB*-algebra over B_0*, instead of just *GB*-algebra*.

P.G. Dixon introduced in [48, (2.5) Definition] a more general than Allan's Definition 3.3.2, by which one also can have examples of *GB*-algebras* that are not locally convex. For this purpose, he used a slightly different collection than $\mathfrak{B}^*_{\mathcal{A}}$ of Allan.

More precisely, if $\mathcal{A}[\tau]$ is a topological *-algebra with identity e, let $\mathfrak{B}_{\mathcal{A}}$ (or $\mathfrak{B}_{\mathcal{A}[\tau]}$, for taking into account the given topology), denote the collection of subsets B of $\mathcal{A}$ with the properties:

(i) B is closed and bounded;
(ii) $e \in B$, $B^2 \subset B$, $B^* = B$.

According to Dixon [48, p. 694], a topological *-algebra is *symmetric* if fo every $x \in \mathcal{A}$ the element $e + x^*x$ has an inverse in $\mathcal{A}_0$ (see also Definition 3.1.6).

Then, the author sets the following

Definition 3.3.3 (Dixon) A *GB*-algebra* *is a topological *-algebra* $\mathcal{A}[\tau]$*, such that:*

(i) $\mathfrak{B}_{\mathcal{A}}$ has a greatest member B_0 that is absolutely convex;
(ii) $\mathcal{A}[\tau]$ is symmetric;
(iii) $\mathcal{A}[B_0]$ is complete.

Remark 3.3.4

(1) *For every locally convex *-algebra* $\mathcal{A}[\tau]$ *with identity, the condition* (i) *of Definition 3.3.2 is equivalent to the condition* (i) *of Definition 3.3.3. Since this holds, then* $\mathfrak{B}^*_{\mathcal{A}}$ *and* $\mathfrak{B}_{\mathcal{A}}$ *have the same greatest member.*

Indeed, let $B \in \mathfrak{B}_{\mathcal{A}}$. Denote by $\overline{\Gamma(B)}^\tau$ the closed, absolutely convex hull of B. Then, by the properties of B, the separate continuity of multiplication of $\mathcal{A}$ and the continuity of the involution, we conclude that $\overline{\Gamma(B)}^\tau \in \mathfrak{B}_{\mathcal{A}}$, but also $\overline{\Gamma(B)}^\tau \in \mathfrak{B}^*_{\mathcal{A}}$. Thus, if $\mathfrak{B}^*_{\mathcal{A}}$ has a greatest member B_0, we shall have that $B \subset \overline{\Gamma(B)}^\tau \subset B_0$. This happens for every $B \in \mathfrak{B}_{\mathcal{A}}$, so that B_0 is the greatest member in $\mathfrak{B}_{\mathcal{A}}$. If now $\mathfrak{B}_{\mathcal{A}}$ has a greatest member B_0, which is absolutely convex, then $B_0 = \overline{\Gamma(B_0)}^\tau \in \mathfrak{B}^*_{\mathcal{A}}$, and so B_0 is the greatest member in $\mathfrak{B}^*_{\mathcal{A}}$.

(2) *In the case of locally convex *-algebras, the notion of a GB*-algebra of Dixon (Definition 3.3.5) is weaker than that of a GB*-algebra of Allan, because the condition* (iii) *of Definition 3.3.3 does not imply the condition* (iii) *of Definition 3.3.2. Indeed, let* $\mathcal{A}[\tau]$ *be a locally convex *-algebra and let* B_0 *be*

the greatest member in $\mathfrak{B}_A^*$. Then, by (1) B_0 is also the greatest member in $\mathfrak{B}_A$. Suppose that $\mathcal{A}[B_0]$ is complete. Then, $\mathcal{A}[B]$ is complete, for every $B \in \mathfrak{B}_A^*$, but it is not necessarily complete, for every $B \in (\mathfrak{B}_0)_A$ (see Definition 2.2.2); that is, $\mathcal{A}[\tau]$ is not pseudo-complete, when we refer to the collection $(\mathfrak{B}_0)_A$ (cf. discussion before Proposition 2.2.5).

We show now that if B_0 is the greatest member in $\mathfrak{B}_A^*$ and $\mathcal{A}[B_0]$ is complete, then $\mathcal{A}[B]$ is complete, for every $B \in \mathfrak{B}_A^*$. Indeed, take an arbitrary element B in $\mathfrak{B}_A^*$. For a Cauchy sequence (x_n) in $\mathcal{A}[B]$, we have that, for each $\varepsilon > 0$, there exists $n_0 \in \mathbb{N}$, such that

$$\|x_m - x_n\|_B < \frac{\varepsilon}{2}, \ \forall \, m, n \geq n_0. \tag{3.3.8}$$

Since, $\mathcal{A}[B] \subset \mathcal{A}[B_0]$ and $\|\cdot\|_{B_0} \leq \|\cdot\|_B$, the sequence (x_n) is a Cauchy sequence in the Banach space $\mathcal{A}[B_0]$. Therefore, there exists $x \in \mathcal{A}[B_0]$, such that $\lim_{n\to\infty} \|x_n - x\|_{B_0} = 0$. Taking into account (3.3.8), we obtain

$$\|x_m - x_{n_0}\|_{B_0} \leq \|x_m - x_{n_0}\|_B < \frac{\varepsilon}{2}, \ \forall \, m \geq n_0.$$

Since, $\tau \prec \|\cdot\|_{B_0}$ on $\mathcal{A}[B_0]$, we conclude that $x_m - x_{n_0} \xrightarrow{\tau} x - x_{n_0}$, where $x_m - x_{n_0} \in \frac{\varepsilon}{2} B$, for all $m \geq n_0$ and B is τ-closed, hence $x - x_{n_0} \in \frac{\varepsilon}{2} B$. Summing up, we have that $x \in \mathcal{A}[B]$ and $\|x - x_{n_0}\|_B < \frac{\varepsilon}{2}$. It follows that

$$\|x_m - x\|_B \leq \|x_m - x_{n_0}\|_B + \|x_{n_0} - x\|_B < \frac{\varepsilon}{2} + \frac{\varepsilon}{2} = \varepsilon, \ \forall \, m \geq n_0.$$

This proves that $\mathcal{A}[B]$ is complete.

According to Remark 3.3.4 we may define the notion of a GB^*-algebra of Dixon in the case of a locally convex $*$-algebra, as follows.

Definition 3.3.5 A locally convex $*$-algebra $\mathcal{A}[\tau]$ with an identity element is called a (locally convex) GB^*-algebra of Dixon (see also Definition 3.3.3), if

(i) The collection $\mathfrak{B}_A^*$ has a greatest member B_0;
(ii) $\mathcal{A}[\tau]$ is symmetric;
(iii) $\mathcal{A}[B_0]$ is complete.

Then $\mathcal{A}[B]$ is complete, for all $B \in \mathfrak{B}_A^*$, according to Remark 3.3.4 (2), but $\mathcal{A}[\tau]$ is not necessarily pseudo-complete (see Definition 2.2.2 and discussion before Proposition 2.2.5), since $\mathcal{A}[B]$ is not complete, for every $B \in (\mathfrak{B}_0)_A$. This clearly shows that Definition 3.3.2 implies Definition 3.3.5. Hence, Definition 3.3.5 weakens Allan's definition of a GB^*-algebra.

▶ From now on *we shall think* of a GB^*-algebra *to be that of Allan* (Definition 3.3.2) *and call it a* GB^*-*algebra,* for short. In the opposite case, the reader will be always warned. For instance, in Chap. 5, *we shall treat with a* (locally convex) GB^*-*algebra of Dixon* (Definition 3.3.5), which is not necessarily a GB^*-algebra (of Allan). We recall that existence of an *identity* element *is always assumed.* Definition 3.3.6 below introduces the notion of a GB^*-algebra without identity. *Whenever this concept is used, it will be explicitly stated.*

Often we need to know that a proper closed $*$-ideal I of a GB^*-algebra is also a GB^*-algebra. Hence, we have to modify the definition of a GB^*-algebra, in the case of the absence of identity. Then, we shall be able to prove that every closed $*$-subalgebra of a GB^*-algebra, which does not necessarily have an identity element, is a GB^*-algebra too (see Proposition 3.3.19). Thus, we set the following

Definition 3.3.6 *A pseudo-complete locally convex* $*$*-algebra* $\mathcal{A}[\tau]$*, not necessarily having an identity element,* is called a GB^**-algebra,* (without identity) if the following conditions are satisfied:

(i) $\mathcal{A}[\tau]$ is symmetric, in the sense that for every $x \in \mathcal{A}$, the element x^*x is quasi-invertible, with bounded quasi-inverse (i.e., $(x^*x)^\circ \in \mathcal{A}_0$); and
(ii) the collection $\mathfrak{B}_{\mathcal{A}}^*$, as in Definition 3.3.1, but without the first claim of the property (ii) for the elements $B \in \mathfrak{B}_{\mathcal{A}}^*$, has a greatest member denoted by B_0.

We present now various results that concern the basic theory of GB^*-algebras.

Lemma 3.3.7 *Let* $\mathcal{A}[\tau]$ *be a* GB^**-algebra. Then, the following hold:*

(i) $\mathcal{A}[B_0]$ *is a Banach* $*$*-algebra containing all normal elements of* $\mathcal{A}_0$ *and, in particular,* $\mathcal{A}[B_0]$ *is contained in* $\mathcal{A}_0$*;*
(ii) $\mathcal{A}[B_0] = \mathcal{A}_0$*, whenever* $\mathcal{A}$ *is commutative.*

Proof

(i) $\mathcal{A}[B_0] = \{\lambda x \ : \ \lambda \in \mathbb{C}, x \in B_0\}$. According to the comments after Definition 3.3.1, $\mathcal{A}[B_0]$ is a normed $*$-algebra under the $*$-norm $\|\cdot\|_{B_0}$, therefore a Banach $*$-algebra, since $\mathcal{A}[\tau]$ is pseudo-complete.

Let now $x \in \mathcal{A}_0$ be normal. Then, from $x = h + ik$, with $h, k \in H(\mathcal{A})$ and the normality of x, we obtain that h commutes with k. Moreover, by Remark 2.2.12, h, k also belong to $\mathcal{A}_0$. Then, arguing as in the second part of the proof of Proposition 2.2.4, there is $B \in \mathfrak{B}$ and $\lambda \in \mathbb{R}\backslash\{0\}$, such that $\lambda h \in B$. Therefore, $\lambda h \in B \cap B^*$, where $B \cap B^* \in \mathfrak{B}_{\mathcal{A}}^*$, consequently $h \in \mathcal{A}[B_0]$. In the same way, we have that $k \in \mathcal{A}[B_0]$ and so $x \in \mathcal{A}[B_0]$. Now, since $\tau \upharpoonright_{\mathcal{A}[B_0]} \prec \|\cdot\|_{B_0}$, clearly $\mathcal{A}[B_0]$ is contained in $\mathcal{A}_0$.

(ii) Since $\mathcal{A}$ is commutative Corollary 2.2.11 implies that $\mathcal{A}_0$ is a subalgebra of $\mathcal{A}$. Since moreover, $x^* \in \mathcal{A}_0$, if and only if, $x \in \mathcal{A}_0$, $\mathcal{A}_0$ is a *-subalgebra of $\mathcal{A}$ containing $\mathcal{A}[B_0]$ by (i). On the other hand, if $x \in \mathcal{A}_0$ and $\lambda \in \mathbb{C}$, $\lambda \neq 0$, the set $S \equiv \{(\lambda x)^n : n \in \mathbb{N}\}$ is bounded in $\mathcal{A}[\tau]$. It is easy to show that the closure of the absolutely convex hull of S belongs to $\mathfrak{B}_0$ and since $\mathfrak{B}_{\mathcal{A}}^*$ is a basic subcollection of $\mathfrak{B}_0$, there is some $B \in \mathfrak{B}_{\mathcal{A}}^*$ (Proposition 2.2.4), such that $S \subset B$. Thus, $\lambda x \in B$, hence in B_0 too. It follows that $x \in \mathcal{A}[B_0]$ and this completes the proof of (ii).

□

Corollary 3.3.8 *Let $\mathcal{A}[\tau]$ be a GB*-algebra. Then, $\mathcal{A}[B_0]$ contains all self-adjoint elements of $\mathcal{A}_0$.*

Our next result shows that for every GB^*algebra $\mathcal{A}[\tau]$, the Banach algebra $\mathcal{A}[B_0]$ is a C^*-algebra. This is a particularly helpful result for the study of GB^*algebras, since on the one hand, it constitutes the bounded part of $\mathcal{A}[\tau]$ and on the other hand, being a C^*-algebra, has a very rich structure, through which the structure of $\mathcal{A}[\tau]$ is investigated.

Theorem 3.3.9 *Let $\mathcal{A}[\tau]$ be a GB*-algebra. The following hold:*

(i) *The identity e of $\mathcal{A}$, as well as the elements $(e + x^*x)^{-1}$, x in $\mathcal{A}$, belong to $\mathcal{A}[B_0]$;*
(ii) *$\mathcal{A}[B_0]$ is a C*-algebra.*

Proof

(i) Let x be an arbitrary element in $\mathcal{A}$. Then, since e and $(e+x^*x)^{-1}$ are self-adjoint and bounded elements of $\mathcal{A}$, they belong to $\mathcal{A}[B_0]$, according to Corollary 3.3.8. Thus, the *-algebra $\mathcal{A}[B_0]$ has an identity element and it is symmetric (see discussion before Definition 3.1.3).
(ii) From the properties of $\mathcal{A}[B_0]$ and (i), we have that $\mathcal{A}[B_0]$ is a symmetric Banach *-algebra. Let now $\mathfrak{B}_0^*$ be the corresponding to $\mathfrak{B}_{\mathcal{A}}^*$ collection for $\mathcal{A}[B_0]$. Then clearly, $B_0 \in \mathfrak{B}_0^*$. Moreover, every $B \in \mathfrak{B}_0^*$ is bounded in $\mathcal{A}[B_0]$, therefore bounded in $\mathcal{A}[\tau]$ (recall that $\tau \prec \|\cdot\|_{B_0}$). Hence, we have, $B \subset \overline{B}^\tau \subset B_0$, B_0 is greatest member in $\mathfrak{B}_{\mathcal{A}}^*[B_o]$. The result now follows from Theorem 3.2.10, and so the proof is complete.

□

Remark 3.3.10 Observe that Corollary 3.3.8 and Theorem 3.3.9 stated for GB^*-algebras of Allan (Definition 3.3.2) are also true for GB^*-algebras of Dixon (Definition 3.3.3). Namely, if $\mathcal{A}[\tau]$ is a GB^*-algebra as in Definition 3.3.3, one has the following (see [48, p. 695, (2.6) Lemma, (2.7) Corollary and comments after the preceding corollary]):

(a) *The subspace $\mathcal{A}[B_0]$ contains all self-adjoint elements of $\mathcal{A}_0$.*
(b) *The element $(e + x^*x)^{-1}$ belongs to $\mathcal{A}[B_0]$, for every $x \in \mathcal{A}$.*
(c) *The Banach *-algebra $\mathcal{A}[B_0]$ is a C*-algebra.*

To prove (a) take $h = h^*$ in $\mathcal{A}_0$. Then, for some $\lambda \in \mathbb{R}\backslash\{0\}$ the set $S \equiv \{(\lambda h)^n, n \in \mathbb{N}\}$ is bounded. Denote by B the closure of the set $S \cup \{e\}$. Then, $B \in \mathcal{B}_\mathcal{A}$, therefore $\lambda h \in B \subset B_0$ from which one has $h \in \mathcal{A}[B_0]$.

Clearly (b) follows from (a), while for the proof of (c) one relies on the steps of the proof of Theorem 3.3.9(ii).

Corollary 3.3.11 *The following hold:*

(i) *every GB^*-algebra $\mathcal{A}[\tau]$, which is a Banach algebra, is a C^*-algebra;*
(ii) *every C^*-algebra is a GB^*-algebra.*

Proof

(i) It is evident that in any Banach algebra every element is (Allan-)bounded. Therefore, $\mathcal{A} = \mathcal{A}_0$. On the other hand, if x is an arbitrary element in $\mathcal{A}$, then $x = h + ik$, with h, k in $H(\mathcal{A})$, so by the equality $\mathcal{A} = \mathcal{A}_0$ and Corollary 3.3.8, we obtain $\mathcal{A}_0 \subset \mathcal{A}[B_0]$, which leads to the equality $\mathcal{A} = \mathcal{A}_0 = \mathcal{A}[B_0]$.

Now, if $\| \cdot \|$ is the given norm on $\mathcal{A}$, such that $\mathcal{A}[\| \cdot \|]$ is a Banach algebra, then $\| \cdot \| = \tau \lceil_{\mathcal{A}[B_0]} \prec \| \cdot \|_{B_0}$ (see (3.3.7)), so by the open mapping theorem the (complete) norms $\| \cdot \|$, $\| \cdot \|_{B_0}$ are equivalent. Hence, $\mathcal{A}[\tau]$ is a C^*-algebra.
(ii) Let $\mathcal{A}[\| \cdot \|]$ be a C^*-algebra. It follows from Theorem 3.2.10 and Definition 3.3.2 that $\mathcal{A}[\| \cdot \|]$ is a GB^*-algebra.

$\square$

The previous corollary shows that the concept of a GB^*-algebra indeed generalizes the notion of a C^*-algebra.

Corollary 3.3.12 *Suppose that $\mathcal{A}[\tau]$ is a GB^*-algebra, all of whose elements are bounded (i.e., $\mathcal{A} = \mathcal{A}_0$). Suppose that $\mathcal{A}[\tau]$, as a locally convex space, is barrelled (see Definition 4.3.14). Then, $\mathcal{A}[\tau]$ is a C^*-algebra.*

Proof As in the proof of Corollary 3.3.11(i), we obtain that $\mathcal{A} = \mathcal{A}[B_0]$. Hence, the absolutely convex, closed set B_0 is also absorbing, so that by the barrelledness of $\mathcal{A}[\tau]$, we conclude that B_0 is a 0-neighbourhood. But B_0 is also bounded, therefore (cf. [74, p. 109, Proposition 1]), τ is defined by a single seminorm, hence a norm, since τ is Hausdorff. But, $\mathcal{A}[\tau]$ is pseudo-complete, so that we clearly have that it becomes a Banach algebra, consequently a C^*-algebra by Corollary 3.3.11(i). $\square$

For the corollary that follows, we need the definition of a Q-algebra that we first give. This concept goes back to 1947 and is due to I. Kaplansky [96]. For a survey on Q-algebras, including comments and examples, see [60, p. 69, Section 6].

Definition 3.3.13 A topological algebra with identity is called a *Q-algebra*, whenever the group of its invertible elements is open.

Clearly, every Banach algebra is a Q-algebra, but not every normed algebra is a Q-algebra (ibid.)

Corollary 3.3.14 *Let $\mathcal{A}[\tau]$ be a Fréchet GB^*-algebra, which is also a Q-algebra. Then, $\mathcal{A}[\tau]$ is a C^*-algebra.*

Proof In every Fréchet locally convex Q-algebra the inversion is continuous according to Corollary 7.8 in [158]. On the other hand, since $\mathcal{A}[\tau]$ is a Q-algebra, every element $x \in \mathcal{A}$ has compact spectrum [111, p. 60, Proposition 4.2]. That is, $sp_{\mathcal{A}}(x)$ is bounded, for every $x \in \mathcal{A}$. Thus, all assumptions of Corollary 2.3.14 are satisfied, therefore $\mathcal{A} = \mathcal{A}_0$. Now, $\mathcal{A}[\tau]$ being Fréchet is barrelled, consequently Corollary 3.3.12 yields that $\mathcal{A}[\tau]$ is a C^*-algebra.   □

Remark 3.3.15 It is known that in the setting of m-convex algebras the property Q, under certain conditions, acts as a catalyst. For instance, *if $\mathcal{A}[\tau]$ is an m-convex Q-algebra, such that τ is defined by a family of square-preserving seminorms, then $\mathcal{A}[\tau]$ is topologically isomorphic to a normed algebra with square-preserving norm* (Bhatt–Karia [27], see also [60, pp. 93, 94]).

Furthermore, *every pro-C^*-algebra, which is a Q-algebra is topologically ∗-isomorphic to a C^*-algebra;* see [60, p. 111, Corollary 8.2]. It is now evident that the previous Corollary 3.3.14 is exactly an analogue of the last result, in the case of GB^*-algebras.

Examples 3.3.16

(1) [5, p. 95, 2.] Let $\mathcal{A}[\tau]$ be a pro-C^*-algebra (see Definition 3.1.2) with identity e, where τ is determined by an upwards directed family $\Gamma = \{p_v\}_{v \in \Lambda}$ of C^*-seminorms. Recall that the seminorms in Γ are assumed mutually non-equivalent (see comments before Sect. 2.2). Thus, we may consider the subset of $\mathcal{A}$ given as follows

$$\mathcal{D}(p_{\Lambda}) = \{x \in \mathcal{A} : \sup_{v \in \Lambda} p_v(x) < \infty\}, \quad \text{with}$$
$$p_{\Lambda}(x) := \sup_{v \in \Lambda} p_v(x), \quad x \in \mathcal{D}(p_{\Lambda}). \tag{3.3.9}$$

The same set can be considered for a GB^*-algebra $\mathcal{A}[\tau]$ with jointly continuous multiplication and in Sect. 3.5, it will be compared with $\mathcal{A}[B_0]$; more precisely, see discussion after Proposition 3.5.6, together with Remark 3.5.7.

Since $\mathcal{A}$ has an identity element, we have $\mathcal{A}[B_0] = \mathcal{D}(p_{\Lambda})$ (cf. Proposition 3.5.6).

The function, p_{Λ} is clearly a C^*-norm on $\mathcal{D}(p_{\Lambda})$, making it a C^*-algebra with identity, dense in $\mathcal{A}[\tau]$ (Apostol [8], Schmüdgen [132]); see, e.g., [60, Theorem 10.23]. Considering the C^*-property of p_v's for e, it is clear that $p_v(e) = 1$, for every $v \in \Lambda$, therefore $p_{\Lambda}(e) = 1$ and so $e \in \mathcal{D}(p_{\Lambda})$.

Take now $B_0 \equiv \{x \in \mathcal{A} : p_{\Lambda}(x) \leq 1\}$. Since $\tau \prec p_{\Lambda}$, it is clear that $B_0 \in \mathfrak{B}^*_{\mathcal{A}}$. The set B_0 is a greatest member in $\mathfrak{B}^*_{\mathcal{A}}$. Indeed, let $B \in \mathfrak{B}^*_{\mathcal{A}}$ be arbitrary. Then, for $x \in B$, we have $y = x^*x \in B$ and $y^n \in B$, $n \in \mathbb{N}$. Suppose that $x \notin B_0$. Then, there is $v \in \Lambda$, such that $p_v(x) > 1$. Hence, $p_v(y) = p_v(x)^2 > 1$. This leads to a contradiction, since y is self-adjoint, lies in the bounded set B and $p_v(y^{2^n}) = p_v(y)^{2^n} \xrightarrow[n \to \infty]{} \infty$. Hence, $x \in B_0$. This proves our claim. It remains to show that $\mathcal{A}[\tau]$ is (Allan-)symmetric. It is known that every pro-C^*-

algebra is symmetric (see [60, (21.6)]); so, the element $(e+x^*x)^{-1}$ exists in the given pro-C^*-algebra $\mathcal{A}$, for every $x \in \mathcal{A}$. On the other hand, using standard properties of positive elements in a pro-C^*-algebra (ibid., Theorem 10.15 and Corollary 10.18), we have that, for every $x \in \mathcal{A}$

$$e+x^*x \geq e \Rightarrow (e+x^*x)^{-1} \leq e \Rightarrow p_v((e+x^*x)^{-1}) \leq p_v(e) = 1, \forall v \in \Lambda.$$

It is now clear that $(e+x^*x)^{-1}$ is bounded (and belongs to $\mathcal{D}(p_\Lambda)$). Thus, *every pro-C^*-algebra $\mathcal{A}[\tau]$ is a GB^*-algebra.*

(2) From Corollary 3.3.11(ii) or the example (1) above, it follows that *every C^*-algebra is a GB^*-algebra.*

   A topological space X is called *k-space* [97], if a subset F of X is closed (respectively, open), whenever $F \cap K$ is closed (respectively, open), for every compact subset K of X.

(3) Let X be a completely regular k-space and $\mathcal{C}_c(X)$ the algebra of all $\mathbb{C}$-valued continuous functions on X, under the topology "c" of uniform convergence on the compact subsets K of X. Then, under involution given by the complex conjugate, $\mathcal{C}_c(X)$ is a pro-C^*-algebra. The topology c is defined by the family of C^*-seminorms $\| \cdot \|_K$ (K as before), such that $\|f\|_K := \|f\upharpoonright_K \|_\infty$, $f \in \mathcal{C}_c(X)$; $\|\cdot\|_\infty$ is the supremum norm on the C^*-algebra $\mathcal{C}(K)$ of all $\mathbb{C}$-valued continuous functions on the compact space K. For more details, see [60, Example 3.10(4)]. It follows now from Example (1) above that $\mathcal{C}_c(X)$ *is a GB^*-algebra.*

(4) For $X = \mathbb{N}$, we conclude from (3) that *the algebra $\mathbb{C}^\mathbb{N} = \mathcal{C}_c(\mathbb{N})$ of all complex sequences is a Fréchet GB^*-algebra.*

   For further examples of GB^*-algebras, coming from pro-C^*-algebras, see [60, Examples 7.6]. A different one, which is not an m-convex algebra, is the following.

(5) [5, p. 95, 4.] Let $\mathcal{A}$ be the Arens algebra; i.e., $\mathcal{A} = L^\omega[0, 1] := \bigcap_{1 \leq p < \infty} L^p[0, 1]$ (see [9]). In other words, $\mathcal{A}$ consists of all $\mathbb{C}$-valued measurable functions on $[0, 1]$, equipped with the topology τ of the L^p-norms $\{\| \cdot \|_p\}$, $p = 1, 2, \ldots$. Under this topology, $\mathcal{A}$ is a Fréchet locally convex $*$-algebra (involution is given by the complex conjugate), with continuous multiplication, but not continuous inversion, so that it is not an m-convex algebra, not a Q-algebra and a fortiori not a pro-C^*-algebra (see, for instance, [111, p. 12, 4] and Corollary 3.3.14, in this section). For further details, see also [9, 10, 15].

   Let now $L^\infty[0, 1]$ denote the $*$-subalgebra of $L^\omega[0, 1]$, consisting of all (equivalence classes of) essentially bounded measurable functions on $[0, 1]$. Then, it is easily seen that $\mathcal{A}_0 = L^\infty[0, 1]$ and that a greatest member of the collection $\mathfrak{B}^*_{L^\omega[0,1]}$ is the closed unit ball

$$B_0 = \{f \in L^\omega[0, 1] : \|f\|_\infty \equiv esssup\{|f(t)| : t \in [0, 1]\} \leq 1\}.$$

On the other hand, if f in $L^{\omega}[0, 1]$ is arbitrary, then $(e + f^*f)^{-1} = (e + |f|^2)^{-1} \in \mathcal{A}_0$, where $|f|^2(t) := |f(t)|^2$, for every $t \in [0, 1]$. Thus, $\mathcal{A}$ is also symmetric, hence a GB^*-algebra.

Yet, note that the Banach algebra $L^{\infty}[0, 1]$, under the essential supremum, is a C^*-algebra dense in $L^{\omega}[0, 1]$.

For our next example, we require the following Theorem 3.3.17. For the terminology applied, see, e.g., [144] and [40].

Theorem 3.3.17 (Crowther) *Let $\mathcal{M}$ be a semifinite von Neumann algebra with a faithful semifinite normal trace ϕ. Suppose that the projection lattice $\mathcal{M}_p$ of $\mathcal{M}$ is atomic, i.e., for all nonzero projections p in $\mathcal{M}$, there exists a minimal projection q, such that $q \leq p$. Let $\widetilde{\mathcal{M}}$ denote the algebra of ϕ-measurable operators affiliated with $\mathcal{M}$. If*

$$\inf\{\phi(p) : p \in \mathcal{M}_p, \ \phi(p) \neq 0\} = 0 \ \text{ and } \ \exists \, k > 0 \ \text{ such that } \ \sum \phi(p) < \infty,$$

where the summation is taken over all atomic projections $p \in \mathcal{M}$ with $\phi(p) < k$, then the topology of convergence in measure on $\widetilde{\mathcal{M}}$ is locally convex.

(6) Let (X, Σ, μ) be a finite atomic measure space. Denote the algebra of all (equivalence classes of) μ-measurable functions on X by $L_0(X)$. The symbol $L^{\infty}(X)$ stands for the algebra of all (equivalence classes of) essentially bounded μ-measurable functions on X and we recall that $L^{\infty}(X)$ is a W^*-algebra with finite trace $f \mapsto \int f d\mu$. Assume also that the latter trace satisfies the hypothesis of Theorem 3.3.17. Since (X, Σ, μ) is an atomic measure space, the projection lattice of $L^{\infty}(X)$ is atomic, and therefore, by Theorem 3.3.17, $L_0(X)$ is locally convex with respect to the topology of convergence in measure. Now $\mathcal{M} := M_n(L^{\infty}(X))$ ($n \times n$-matrices with entries from $L^{\infty}(X)$) is a von Neumann algebra and $\mathcal{M} = L^{\infty}(X) \overline{\otimes} \mathfrak{B}(\mathcal{H})$, up to an isomorphism of von Neumann algebras, where $\mathfrak{B}(\mathcal{H})$ denotes the algebra of all bounded linear operators on an n-dimensional Hilbert space $\mathcal{H}$ and "$\overline{\otimes}$" means tensor product of W^*-algebras (see [144]). Since the projection lattice of $L^{\infty}(X)$ is atomic, we have from [151, Lemma 3.9] that the projection lattice of $\mathcal{M}$ is also atomic. Furthermore, assume that $\mathcal{M}$ is equipped with a finite trace satisfying the hypothesis of Theorem 3.3.17, so that by Theorem 3.3.17 itself, the $*$-algebra $\widetilde{\mathcal{M}}$ of all τ-measurable operators affiliated with $\mathcal{M}$ is locally convex with respect to the topology of convergence in measure. Yet, $\widetilde{\mathcal{M}}$ is a GB^*-algebra in the topology of convergence in measure [150, Theorem 1.5.29]. Now $\widetilde{\mathcal{M}} = M_n(L_0(X))$, up to an algebra isomorphism [16, Lemma 2], so that $\widetilde{\mathcal{M}} = L_0(X) \otimes \mathfrak{B}(\mathcal{H})$ (algebraically), therefore the latter tensor product, equipped with the topology of convergence in measure, is a GB^*-algebra.

This example is also relevant in Chap. 8, where we use it to give an example of a tensor product GB^*-algebra.

(7) [48, Example (3.3)] Let $\mathcal{H}$ be a Hilbert space and $\mathfrak{B}(\mathcal{H})$ the C^*-algebra of all bounded linear operators on $\mathcal{H}$, under the operator norm $\|\cdot\|$. Consider a

$\| \cdot \|$-closed $*$-subalgebra $\mathcal{A}$ of $\mathcal{B}(\mathcal{H})$, containing the identity operator. Endow $\mathcal{A}$ with the weak operator topology, given by the $*$-seminorms $p_{\xi,\eta}(T) := |\langle T\xi, \eta \rangle|$, $T \in \mathcal{A}$, where ξ, η run in $\mathcal{H}$. Then, $\mathcal{A}$ becomes a GB^*-algebra, with B_0 the unit ball of $\mathcal{A}[\| \cdot \|]$ and, of course, $\mathcal{A}[B_0] = \mathcal{A}[\| \cdot \|]$.

Using now the Definition 3.3.3 of P.G. Dixon, we shall give an example of a GB^*-algebra, which is not locally convex (see [48, Example (3.4)]).

(8) [48, p. 696, (3.4)] Let $\mathcal{A}$ be the algebra $\mathcal{M}[0, 1]$ of all measurable functions on $[0, 1]$ (modulo equality almost everywhere). Endow $\mathcal{A}$ with the topology of convergence in measure. Then, $\mathcal{A}$ becomes a Fréchet non locally convex GB^*-algebra with jointly continuous multiplication. It is easily checked that $\mathcal{A}[B_0] = L^\infty[0, 1]$ (see example 5, above), with

$$B_0 = \left\{ f \in \mathcal{M}[0, 1] : \operatorname*{esssup}_{t \in [0,1]} |f(t)| \leq 1 \right\}.$$

Symmetry of $\mathcal{A}$ results as in Example 5, above.

In [49, Sections 6, 7], P.G. Dixon constructs another example of a non locally convex GB^*-algebra, with continuous involution and jointly continuous multiplication. For this, he considers a W^*-algebra $\mathcal{A}[\| \cdot \|]$ on a separable Hilbert space $\mathcal{H}$ and the $*$-algebra $\mathfrak{Q}(\mathcal{A})$ of all "quasi-measurable" operators on $\mathcal{H}$, with respect to $\mathcal{A}[\| \cdot \|]$ (ibid., (6.4) Definition). He endows $\mathfrak{Q}(\mathcal{A})$ with a vector space topology resembling to the convergence in measure topology on a space of measurable functions and he proves that this topology connected with the algebraic structure of $\mathfrak{Q}(\mathcal{A})$, makes it into a (non locally convex) GB^*-algebra, with $\mathcal{A}[B_0] = \mathcal{A}[\| \cdot \|]$, where B_0 is clearly the closed unit ball of the W^*-algebra $\mathcal{A}[\| \cdot \|]$.

Other interesting examples of GB^*-algebras are given by the L^ω-algebras of operators associated with an unbounded Hilbert algebra; see [79, 81, 82], particularly, [80, Theorem 3.3, Corollary 3.4], [83, p. 32, Theorem 1].

Furthermore, S.J.L. van Eijndhoven and P. Kruszyński have considered in [56] a directed family $\mathcal{R}$ of commuting positive bounded operators on a Hilbert space $\mathcal{H}$ and constructed an inductive limit $\mathcal{S}_\mathcal{R} \subseteq \mathcal{H}$ of Hilbert spaces, that serves as the maximal common dense domain for the unbounded operator algebras $\mathcal{R}^c$, $\mathcal{R}^{cc}$, corresponding to the strong commutant respectively strong bicommutant of $\mathcal{R}$. It is proved (ibid., (3.10) Theorem) that both $\mathcal{R}^c$ and $\mathcal{R}^{cc}$ are GB^*-algebras.

Finally, note that there is a whole class of complete locally convex $*$-algebras, the so called C^*-*like locally convex $*$-algebras*, introduced in 2002, by A. Inoue and K.-D. Kürsten [88] (see Sect. 3.5 in this chapter), which generalize the classical C^*-algebras and, in particular, they are GB^*-algebras with jointly continuous multiplication.

We shall prove now the result mentioned before Definition 3.3.6. First, we need the following

Theorem 3.3.18 *Let $\mathcal{A}[\tau]$ be an arbitrary pseudo-complete locally convex algebra and $\mathcal{B}$ a closed subalgebra of $\mathcal{A}$. Then, for every $x \in \mathcal{B}$,*

(i) $\sigma_\mathcal{A}(x) \subset \sigma_\mathcal{B}(x)$;

(ii) $\partial \sigma_B(x) \subset \partial \sigma_A(x)$, where "$\partial$" means boundary;
(iii) $\infty \in \sigma_B(x)$, if and only if, $\infty \in \sigma_A(x)$.

Proof Note that $\sigma_A(x) := \sigma_{A_1}(x)$, for all $x \in A$ and that B_1 is also pseudo-complete (see Definition 2.3.1 and Proposition 2.2.8). It is clear then, that an element $x \in B$ is bounded, if and only if, it is bounded in A. Therefore, (i) and (iii) follow from Definition 2.3.1.

(ii) Without loss of generality, we suppose that A has an identity e, which is also contained in the subalgebra B. Let $x \in B$. The conclusion is trivial, if $\partial \sigma_B(x) = \emptyset$. So, let $\partial \sigma_B(x) \neq \emptyset$. Take an element $\lambda_0 \in \partial \sigma_B(x)$ and choose a sequence $(\lambda_n)_{n \in \mathbb{N}}$ in the resolvent set $\rho_B(x)$ of x, such that $\lambda_n \to \lambda_0$. Then, by (i), $(\lambda_n)_{n \in \mathbb{N}} \subset \rho_A(x)$. Suppose that $\lambda_0 \notin \sigma_A(x)$. By Theorem 2.3.7(3), we obtain that the resolvent map of x (Definition 2.3.3)

$$R : \rho_A(x) \to A[\tau] : \lambda \mapsto (\lambda e - x)^{-1},$$

is holomorphic at λ_0, so that $R(\lambda_n) \underset{n \to \infty}{\to} R(\lambda_0)$. But, $R(\lambda_n) \in B$, for all $n \in \mathbb{N}$ and B is closed in A; therefore,

$$R(\lambda_0) \in B \cap A_0, \quad \text{i.e.,} \quad \lambda_0 \in \rho_B(x),$$

which is a contradiction. Hence, $\lambda_0 \in \sigma_A(x)$ and since all λ_n belong to $\rho_A(x)$, we conclude that $\lambda_0 \in \partial \sigma_A(x)$. This completes the proof. □

Proposition 3.3.19 *Let $A[\tau]$ be a GB*-algebra, which does not necessarily have an identity element. Suppose that B is a closed *-subalgebra of $A[\tau]$. Then, $B[\tau \restriction_B]$ is a GB*-algebra.*

Proof It is readily verified that $\mathfrak{B}_B^* = \{B \cap B : B \in \mathfrak{B}_A^*\}$ is a collection in B, satisfying the properties of the collection in A of Definition 3.3.6, with greatest member $B \cap B_0$. Hence, it suffices to show that B is symmetric (see Definition 3.1.6). The given algebra $A[\tau]$ being symmetric is hermitian (Corollary 3.1.7), therefore (see Definition 3.1.3), we have that

$$\sigma_A(x^*x) \subset \mathbb{R}, \ \forall \, x \in B.$$

On the other hand, (see Definition 2.3.1)

$$\sigma_B(x) := \sigma_{B_1}(x), \ \forall \, x \in B.$$

Note now that B as a closed subalgebra of a pseudo-complete locally convex algebra is also pseudo-complete and subsequently the same is true for its unitization B_1 (Proposition 2.2.8). Therefore, the spectra $\sigma_B(x)$, $\sigma_A(x)$, with $x \in B$, are closed (see Corollary 2.3.8). Thus, applying Theorem 3.3.18, we have

$$\partial \sigma_B(x^*x) \subset \partial \sigma_A(x^*x) = \sigma_A(x^*x) \subset \mathbb{R}, \quad \text{and therefore}$$
$$\sigma_B(x^*x) = \partial \sigma_B(x^*x) \subset \sigma_A(x^*x) \subset \sigma_B(x^*x).$$

Hence,

$$\sigma_{\mathcal{B}}(x^*x) = \sigma_{\mathcal{A}}(x^*x), \ \forall \ x \in \mathcal{B}.$$

This shows that x^*x is quasi-invertible in $\mathcal{B}$, if and only if, it is quasi-invertible in $\mathcal{A}$. Thus, since $\mathcal{A}[\tau]$ is symmetric, x^*x is quasi-invertible in $\mathcal{B}$, for every $x \in \mathcal{B}$. Yet, $(x^*x)^\circ \in \mathcal{A}_0$, therefore $(x^*x)^\circ \in \mathcal{B} \cap \mathcal{A}_0$, for all $x \in \mathcal{B}$ and this completes the proof for the symmetry of $\mathcal{B}$. $\square$

It is evident that Proposition 3.3.19 remains true if $\mathcal{A}[\tau]$ has an identity, but $\mathcal{B}$ does not.

3.4 Commutative GB*-Algebras: Functional Representation Theory

In this section we deal with commutative pseudo-complete locally convex algebras with an identity element. If $\mathcal{A}[\tau]$ is such an algebra, which is also a GB^*-algebra, then from Lemma 3.3.7, we have that

$$\mathcal{A}[B_0] = \mathcal{A}_0, \tag{3.4.9}$$

where B_0 is the greatest member in both families $\mathfrak{B}^*_\mathcal{A}$ and $\mathfrak{B}_0$, *by imposing the extra condition $e \in B$, for all elements $B \in \mathfrak{B}_0$, since $\mathcal{A}$ carries the identity element e.* Indeed, let B_0 be the greatest member in $\mathfrak{B}^*_\mathcal{A}$. Take an arbitrary element $B \in \mathfrak{B}_0$ (in this regard, see also Proposition 2.2.8(iii)). By the continuity of the involution we have that B^* also belongs to $\mathfrak{B}_0$. Therefore, from Theorem 2.2.10 there exists $B_1 \in \mathfrak{B}_0$, such that $B \cup B^* \subset B_1 \cap B_1^*$, the later being an element in $\mathfrak{B}^*_\mathcal{A}$. Hence, $B \subset B \cup B^* \subset B_0$, so that B_0 is greatest member in $\mathfrak{B}_0$ too.

Consider the carrier space $\mathfrak{M}_0$ of $\mathcal{A}_0$ (see (3.4.9) and Sect. 2.5). Namely, $\mathfrak{M}_0$ consists of all characters (i.e., nonzero multiplicative linear functionals) of $\mathcal{A}_0$. As a consequence of Proposition 2.5.1, $\mathfrak{M}_0$ endowed with the weak $*$-topology $\sigma(\mathfrak{M}_0, \mathcal{A}_0)$ is a non-empty compact Hausdorff space, for every commutative pseudo-complete locally convex algebra $\mathcal{A}[\tau]$ with an identity element. In this respect, we give a variant of Proposition 2.5.4, under the extra assumption of symmetry for the topological $*$-algebra involved.

Let $\mathcal{A}[\tau]$ be a commutative symmetric pseudo-complete locally convex $*$-algebra with identity e. By Corollary 3.1.7, $\mathcal{A}[\tau]$ is hermitian, therefore $\sigma(h)$ is real, for every $h \in H(\mathcal{A})$ and so $H(\mathcal{A}) \subset \mathcal{A}_\rho := \{x \in \mathcal{A} : \rho(x) \neq \emptyset\}$. Hence, we can define a $\mathbb{C}^*$-valued function φ_1 on $H(\mathcal{A})$, as follows:

$$\varphi_1(h) := i - \frac{1}{\varphi((ie - h)^{-1})}, \quad h \in H(\mathcal{A}). \qquad (3.4.10)$$

This is the restriction to $H(\mathcal{A})$, of the function φ' defined on $\mathcal{A}_\rho$ in Proposition 2.5.4 (see also (2.5.7)). According to the latter proposition, we have

$$\varphi_1(h) = \varphi(h), \quad h \in \mathcal{A}_0 \cap H(\mathcal{A}), \qquad (3.4.11)$$

$$\varphi_1(\lambda h) = \lambda \varphi_1(h), \quad \lambda \in \mathbb{R}, \ h \in H(\mathcal{A}), \qquad (3.4.12)$$

$$\varphi_1(h + k) = \varphi_1(h) + \varphi_1(k), \quad h, k \in H(\mathcal{A}), \text{ when } \varphi_1(h), \varphi_1(k)$$
are not both ∞, $\qquad (3.4.13)$

$$\varphi_1(hk) = \varphi_1(h)\varphi_1(k), \quad h, k \in H(\mathcal{A}), \text{ when } \varphi_1(h) \text{ and } \varphi_1(k)$$
are not 0 or ∞, in any order. $\qquad (3.4.14)$

Proposition 3.4.1 *Let $\mathcal{A}[\tau]$ be a commutative pseudo-complete symmetric locally convex $*$-algebra. Then, for every $\varphi \in \mathfrak{M}_0$ there is a unique $\mathbb{C}^*$-valued function φ' on $\mathcal{A}$ with the following properties:*

(i) *φ' is an extension of φ;*
(ii) *φ' is a "partial homomorphism", in the sense that:*

> *(α) $\varphi'(\lambda x) = \lambda \varphi'(x), \ \forall \lambda \in \mathbb{C}, \ x \in \mathcal{A}$ (with the convention that $0 \cdot \infty = 0$);*
> *(β) $\varphi'(x + y) = \varphi'(x) + \varphi'(y), \ \forall x, y \in \mathcal{A}$, provided that only one of $\varphi'(x)$, $\varphi'(y)$ is ∞;*
> *(γ) $\varphi'(xy) = \varphi'(x)\varphi'(y), \ \forall x, y \in \mathcal{A}$, holds in the following cases, where $x = h_1 + ik_1$ and $y = h_2 + ik_2$, with, $h_i, k_i \in H(\mathcal{A}), i = 1, 2$:*
> *(γ_1) both $\varphi_1(h_1)$ and $\varphi_1(k_1)$ are not infinite and both $\varphi_1(h_2)$ and $\varphi_1(k_2)$ are not zero and vice versa;*
> *(γ_2) $\varphi_1(h_1) = \infty$, $\varphi_1(k_1) \neq \infty$ and $\varphi'(y) \neq 0$ and vice versa;*
> *(γ_3) $\varphi_1(h_1) = \infty$, $\varphi_1(k_1) \neq \infty$, $\varphi_1(h_2) \neq \infty \neq \varphi_1(k_1)$, but also $\left(\varphi_1(h_2), \varphi_1(k_2)\right) \neq (0, 0)$; similarly, when $\varphi_1(h_1) \neq \infty$, $\varphi_1(k_1) = \infty$, $\varphi_1(h_2) \neq \infty \neq \varphi_1(k_2)$, but also $(\varphi_1(h_2), \varphi_1(k_2)) \neq (0, 0)$.*

(iii) *$\varphi'(x^*) = \overline{\varphi'(x)}, \ x \in \mathcal{A}$ (with the convention that $\overline{\infty} = \infty$).*

Proof If there is an extension φ' of φ with the properties (i)–(iii), it will have the form

$$\varphi'(x) = \varphi_1(h) + i\varphi_1(k), \quad x \in \mathcal{A}, \text{ with } x = h + ik, \ h, k \in H(\mathcal{A}), \qquad (3.4.15)$$

where φ_1 is as in (3.4.10).

The statements (i) and (ii)(α) follow from (3.4.11), (3.4.12) and (3.4.15), respectively. We show (ii)(β). Suppose that $\varphi'(x) \neq \infty$ and let $x = h_1 + ik_1$, $y = h_2 + ik_2$ in $\mathcal{A}$, with $h_1, k_1, h_2, k_2 \in H(\mathcal{A})$. Then,

$$\varphi'(x + y) = \big(\varphi_1(h_1) + \varphi_1(h_2)\big) + i\big(\varphi_1(k_1) + \varphi_1(k_2)\big) = \varphi'(x) + \varphi'(y).$$

In the case when $\varphi'(y) \neq \infty$, we can similarly show that

$$\varphi'(x + y) = \varphi'(x) + \varphi'(y).$$

We now prove (ii)(γ). Since $xy = (h_1 h_2 - k_1 k_2) + i(h_1 k_2 + k_1 h_2)$, we have

$$\varphi'(xy) = \varphi_1(h_1 h_2 - k_1 k_2) + i\varphi_1(h_1 k_2 + k_1 h_2),$$
$$\varphi'(x)\varphi'(y) = \big(\varphi_1(h_1) + i\varphi_1(k_1)\big)\big(\varphi_1(h_2) + i\varphi_1(k_2)\big).$$

- (γ_1) Suppose that $\varphi'(h_1) \neq \infty \neq \varphi'(k_1)$ and $\varphi_1(h_2) \neq 0 \neq \varphi_1(k_2)$. Then, from (3.4.10) and (3.4.14) we obtain

$$\varphi'(xy) = \varphi'(x)\varphi'(y).$$

- (γ_2) Suppose now that $\varphi_1(h_1) = \infty$, $\varphi_1(k_1) \neq \infty$ and $\varphi'(y) \neq 0$. From the latter we shall have that either $\varphi_1(h_2) \neq 0$, or $\varphi_1(k_2) \neq 0$, or $\varphi_1(h_2) \neq 0 \neq \varphi_1(k_2)$. Suppose that $\varphi_1(k_2) \neq 0$. Then, from (3.4.13) and (3.4.14) we have

$$\varphi'(x)\varphi'(y) = \big(\varphi_1(h_1)\varphi_1(h_2) - \varphi_1(k_1)\varphi_1(k_2)\big) + i\big(\varphi_1(h_1)\varphi_1(k_2) + \varphi_1(h_2)\varphi_1(k_1)\big)$$

$$= \infty$$

$$= \big(\varphi_1(h_1 h_2) - \varphi_1(k_1 k_2)\big) + i\varphi_1(h_1 k_2 + h_2 k_1)$$

$$= \varphi'(xy).$$

In the same way, considering $\varphi_1(h_2) = \infty$, $\varphi_1(k_2) \neq \infty$ and $\varphi'(x) \neq 0$ we obtain

$$\varphi'(xy) = \infty = \varphi'(x)\varphi'(y).$$

- (γ_3) Suppose that $\varphi_1(h_1) = \infty$, $\varphi_1(k_1) \neq \infty$, $\varphi_1(h_2) \neq \infty \neq \varphi_1(k_2)$, $\varphi_1(h_2) \neq 0$, or $\varphi_1(k_2) \neq 0$. Then, according to (2.4.11), (2.4.12) we obtain

$$\varphi'(xy) = \infty = \varphi'(x)\varphi'(y).$$

(iii) It follows easily from (3.4.15). $\square$

With $\mathfrak{M}_0$ as in the discussion after (3.4.9), consider the algebra $C(\mathfrak{M}_0)$ of all $\mathbb{C}$-valued continuous functions on $\mathfrak{M}_0$ under the topology of the supremum norm $\|\cdot\|_\infty$. Suppose now that $\mathcal{A}[\tau]$ is a commutative GB^*-algebra with identity. Then, (see (3.4.9) and Theorem 3.3.9(ii)) $\mathcal{A}_0 = \mathcal{A}[B_0]$ is a commutative C^*-algebra with identity, therefore by the Gelfand–Naimark theorem

$$\mathcal{A}_0 \cong C(\mathfrak{M}_0) : x \mapsto \widehat{x}, \ \text{ with } \ \widehat{x}(\varphi) = \varphi(x), \ \forall \, \varphi \in \mathfrak{M}_0, \tag{3.4.16}$$

up to an isometric $*$-isomorphism given as before, where $\widehat{x}$ stands for the *Gelfand transform* of x. The preceding isometric $*$-isomorphism is called the *Gelfand representation* of $\mathcal{A}_0$. In what follows, we shall discuss an extension of the Gelfand representation of $\mathcal{A}_0$, which we shall name the *functional representation* of $\mathcal{A}[\tau]$ (see Theorem 3.4.9). For this we need a series of lemmas.

Lemma 3.4.2 *Let $\mathcal{A}[\tau]$ be a commutative GB^*-algebra. Then, the following hold:*

(i) *the elements $x(e + x^*x)^{-1}$ belong to $\mathcal{A}_0$, for every $x \in \mathcal{A}$;*
(ii) *an element x belongs to $\mathcal{A}_0$, if and only if, x^*x belongs to $\mathcal{A}_0$.*

Proof

(i) Take an element $x \in \mathcal{A}$. Then, $x = h + ik$, with $h, k \in H(\mathcal{A})$. Let $u := h(e+h^2)^{-1}$ and $v := (e+h^2)^{-1}$. Note that $\mathcal{A}[\tau]$, being symmetric, is hermitian (Corollary 3.1.7), hence Lemma 3.1.4 implies that $u, v \in \mathcal{A}_0$; therefore $u, v \in \mathcal{A}_0 \cap H(\mathcal{A})$. Furthermore, $v = u^2 + v^2$, so that v is a positive element in $\mathcal{A}_0$. From the functional calculus in C^*-algebras, we obtain now that there is an element $\omega \in \mathcal{A}_0 \cap H(\mathcal{A})$, such that $v = \omega^2$.

On the other hand, $x^*x = h^2 + k^2$ and $(e + h^2)(e + (k\omega)^2) = e + h^2 + k^2$, so that, taking also into account Theorem 3.3.9(i), we obtain

$$h(e + x^*x)^{-1} = h(e + h^2 + k^2)^{-1} = h(e + h^2)^{-1}(e + (k\omega)^2)^{-1} \in \mathcal{A}_0.$$

In the same way, we have $k(e + x^*x)^{-1} \in \mathcal{A}_0$, consequently

$$x(e + x^*x)^{-1} = (h + ik)(e + x^*x)^{-1} \in \mathcal{A}_0, \ \forall \, x \in \mathcal{A}.$$

(ii) Let $x \in \mathcal{A}_0$. Then, clearly $x^*x \in \mathcal{A}_0$. Conversely, if $x^*x \in \mathcal{A}_0$, then by (i) and the fact that $e \in \mathcal{A}_0$, we have

$$x = x(e + x^*x)^{-1}(e + x^*x) \in \mathcal{A}_0.$$

This completes the proof.

$\square$

Notice that *a noncommutative version of* Lemma 3.4.2(1), *the reader can find in* Lemma 7.3.2. Compare also with Lemma 7.1.1.

Remark 3.4.3

(1) If $A[\tau]$ is a GB^*-algebra, in the sense of Dixon, Definition 3.3.3, not necessarily commutative, then Lemma 3.4.2(i) is valid in the following form:

For every $h = h^* \in A$, one has that $h(e + h^2)^{-1} \in A[B_0]$;

see [48, (4.4) Lemma]. The proof may easily be built from that of Lemma 3.4.2(i).

With $A[\tau]$ as in (1), Dixon defines the *spectrum* $\sigma^D(x)$ (or $\sigma_A^D(x)$, for distinction, in case more than one GB^*-algebra is involved) of an element $x \in A$, by replacing A_0 in Allan's Definition 2.3.1 with the C^*-algebra $A[B_0]$ [48, (4.2) Definition], i.e.,

$$\sigma^D(x) = \big\{\lambda \in \mathbb{C} : \lambda e - x \text{ has no inverse in } A[B_0]\big\} \cup \big\{\infty \Leftrightarrow x \notin A[B_0]\big\}.$$

We know that $A[B_0] \subseteq A_0$ and if $A[\tau]$ is a GB^*-algebra in the sense of Allan (Definition 3.3.2), then it is readily seen that

$$\sigma_A(x) \subseteq \sigma^D(x), \ x \in A, \tag{3.4.17}$$

where on the left hand side sits the Allan spectrum of x.

(2) Considering again $A[\tau]$ as in (1) then for every $h = h^*$ in A, $\sigma^D(h)$ is real [48, (4.5) Lemma]. Indeed: a topological $*$-algebra $A[\tau]$ is said to be hermitian if for every $h = h^*$ in A the spectrum $\sigma^D(h)$ is real (this is in accordance with the definition of symmetry in the comments, before Definition 3.3.3). But, by Definition 3.3.3, $A[\tau]$ is symmetric, therefore hermitian, as one easily obtains by arguing as in Proposition 3.1.5, (ii) $\Rightarrow$ (i).

(3) Observe that Theorem 3.4.9 also holds for a GB^*-algebra, in the sense of Definition 3.3.3 (Dixon), following exactly the same proof steps; see [48, (4.6) Theorem] and the discussion before it.

Lemma 3.4.4 *Let $A[\tau]$ be a commutative GB^*-algebra and let $x \in A$. Put $y := (e + x^*x)^{-1}$ and $z := xy$. Let $\varphi \in \mathfrak{M}_0$ and φ' its extension to A, according to Proposition 3.4.1. Then, the following hold:*

(i) $\varphi(y) \neq 0$ *implies* $\varphi'(x) = \frac{\varphi(z)}{\varphi(y)}$;

(ii) $\varphi(y) = 0$, *if and only if,* $\varphi'(x) = \infty$;

(iii) $|\varphi'(x)|^2 = -1 + \frac{1}{\varphi(y)}$.

Proof (i) Let $x = h + ik$, $h, k \in H(A)$. Since $y \in H(A) \cap A_0$ and $\varphi(y) \neq 0$, it follows from (3.4.14) and (3.4.15) that

$$\varphi(z) = \varphi_1(hy) + i\varphi_1(ky) = \varphi_1(h)\varphi(y) + i\varphi_1(k)\varphi(y)$$

$$= \big(\varphi_1(h) + i\varphi_1(k)\big)\varphi(y) = \varphi'(x)\varphi(y).$$

Hence $\varphi'(x) = \frac{\varphi(z)}{\varphi(y)}$.

(ii), (iii) Since $(e + x^*x)^{-1} \in \mathcal{A}_0$, that is $-1 \in \rho(x^*x)$, follows from the proof of Proposition 2.5.4 that

$$\varphi'(x^*x) = -1 + \frac{1}{\varphi(y)}. \qquad (3.4.18)$$

We shall show now that

$$\varphi'(x^*x) = |\varphi'(x)|^2. \qquad (3.4.19)$$

Indeed, if $\varphi'(x) \neq \infty$, then $\varphi'(x^*) = \overline{\varphi'(x)} \neq \infty$, therefore by Proposition 2.5.4(4) we obtain

$$\varphi'(x^*x) = \varphi'(x^*)\varphi'(x) = |\varphi'(x)|^2.$$

Therefore, $\varphi'(x^*x) \neq 0$. Suppose that $\varphi'(x^*x) \neq \infty$. Then, by (3.4.18), we have $\varphi(y) \neq 0$. Hence, it follows from (i) that $\varphi'(x) = \frac{\varphi(z)}{\varphi(y)} \neq \infty$. This implies that $\varphi'(x^*x) = \infty$, if and only if, $\varphi'(x) = \infty$. Thus, (3.4.19) is shown. Now, by (3.4.18) and (3.4.19), the statement (iii) holds and clearly (iii) implies (ii) and this completes the proof. $\qquad \square$

Lemma 3.4.5 *Let $\mathcal{A}[\tau]$ be a commutative GB*-algebra and let $x \in \mathcal{A}$ with $z := xy$, where $y := (e + x^*x)^{-1}$. Then, the following hold:*

(i) $z^*z = y - y^2$;
(ii) *the Gelfand transform $\widehat{y}$ of y is a real non-negative function on $\mathfrak{M}_0$;*
(iii) $\|y\|_{B_0} \leq 1 \geq \|z\|_{B_0}$.

Proof (i) is straightforward.

(ii) Observe that $y \in \mathcal{A}_0$. From Lemma 3.4.4(iii) we have

$$\widehat{y}(\varphi) = \varphi(y) = \left(1 + |\varphi'(x)|^2\right)^{-1} \leq 1, \; \forall \, \varphi \in \mathfrak{M}_0, \qquad (3.4.20)$$

which implies (ii).

(iii) Since $\mathcal{A}_0[\|\cdot\|_{B_0}]$ is a commutative C^*-algebra with identity, (3.4.20) implies that

$$\|y\|_{B_0} = \|\widehat{y}\|_{\infty} \leq 1.$$

Furthermore, since $y \in \mathcal{A}_0$, from (i) follows that $z^*z \in \mathcal{A}_0$. Hence,

$$\widehat{z^*z}(\varphi) = \varphi(z^*z) = \varphi(y) - \varphi(y)^2 \geq 0, \; \forall \, \varphi \in \mathfrak{M}_0.$$

This implies that $0 \leq \widehat{z}(\varphi) \leq 1$, for all $\varphi \in \mathfrak{M}_0$. Consequently, $\|z\|_{B_0} \leq 1$. $\qquad \square$

Lemma 3.4.6 *Let $A[\tau]$ be a commutative GB^*-algebra and let $x \in A$. Then, the set*

$$N = \{\varphi \in \mathfrak{M}_0 : \varphi'(x) = \infty\}$$

is a closed, nowhere-dense subset of $\mathfrak{M}_0$.

Proof By Lemma 3.4.4(ii) we conclude that

$$N = \{\varphi \in \mathfrak{M}_0 : \varphi(y) = 0\},$$

where $y := (e+x^*x)^{-1} \in A_0$. Clearly, N is closed. Assume now that there is a non-empty open subset U of $\mathfrak{M}_0$, such that $U \subset N$. Let $\varphi_0 \in U$ and recall that $A_0 \cong C(\mathfrak{M}_0)$. Then, by Urysohn's lemma, there is a continuous function $\widehat{u} : \mathfrak{M}_0 \mapsto [0, 1]$ (or equivalently $u \in A_0$), such that $\widehat{u}(\varphi) = 0$, for all $\varphi \in \mathfrak{M}_0 \backslash U$ and $\widehat{u}(\varphi_0) = 1$. Then, it is clear that

$$\widehat{yu}(\varphi) = \varphi(yu) = \varphi(y)\varphi(u) = 0, \ \forall \, \varphi \in \mathfrak{M}_0.$$

This yields $\widehat{yu} = 0$ and so $yu = 0$ (from the injectivity of the Gelfand representation). But y is invertible in A_0, hence $u = 0$, which is a contradiction. Consequently, the set N is nowhere dense in $\mathfrak{M}_0$ and this completes the proof. □

Definition 3.4.7 (Allan) Let $A[\tau]$ be a commutative GB^*-algebra. The map $x \mapsto \widehat{x}$ of $A[\tau]$ into the set of all $\mathbb{C}^*$-valued functions $\widehat{x}$, $x \in A$, on $\mathfrak{M}_0$, such that

$$\widehat{x}(\varphi) := \varphi'(x), \ \forall \, \varphi \in \mathfrak{M}_0,$$

is called *the functional representation of $A[\tau]$*, where φ' is the extension of φ on A defined in Proposition 3.4.1.

It is clear that the functional representation of $A[\tau]$ extends the Gelfand representation of the C^*-algebra A_0.

Definition 3.4.8 (Allan) Let X be a topological space and $\mathcal{F}$ a collection of continuous $\mathbb{C}^*$-valued functions on X. We call $\mathcal{F}$ a $*$-algebra of functions on X if the following conditions hold:

(i) Each function f in $\mathcal{F}$ takes the value ∞ on at most a nowhere-dense subset of X.

(ii) For any functions f, g in $\mathcal{F}$ and $\lambda \in \mathbb{C}^*$, the functions λf, $f + g$, fg, $f^* := \overline{f}$, defined pointwise on the dense subset of X, where f and g are both finite, are uniquely extendable to continuous $\mathbb{C}^*$-valued functions on X, which also belong to $\mathcal{F}$. We keep exactly the same symbols for the corresponding extensions.

Condition (ii) of the previous definition claims that it is possible to extend some continuous functions from certain dense domains to the whole of X. Note that such extensions are possible if, for instance, X is a Stonean space (i.e., extremely disconnected compact space) or, more generally, X is an F-space (cf. [67, p. 208, §14.25]).

The collection of the functions $\widehat{x}$, $x \in \mathcal{A}$, considered in Definition 3.4.7 on $\mathfrak{M}_0$ *will be denoted by* $\widehat{\mathcal{A}}$. We are now ready to prove the aforementioned theorem concerning the functional representation of $\mathcal{A}[\tau]$, which is an "algebraic" commutative Gelfand–Naimark type theorem for GB*-algebras.

Theorem 3.4.9 (Allan) *Let* $\mathcal{A}[\tau]$ *be a commutative GB*-algebra. The functional representation of* $\mathcal{A}[\tau]$ *is an algebraic $*$-isomorphism of* $\mathcal{A}$ *onto a $*$-algebra* $\widehat{\mathcal{A}}$ *of continuous* $\mathbb{C}^*$*-valued functions on* $\mathfrak{M}_0$. *The map* $\mathcal{A} \to \widehat{\mathcal{A}} : x \mapsto \widehat{x}$ *extends the Gelfand representation of the C^*-algebra* $\mathcal{A}_0$, *and so* $\widehat{\mathcal{A}}$ *contains all continuous* $\mathbb{C}$*-valued functions on* $\mathfrak{M}_0$.

Proof Let $x \in \mathcal{A}$. Then, by Lemma 3.4.6, $\widehat{x}$ takes the value ∞ at most on a nowhere-dense closed subset of $\mathfrak{M}_0$. We shall show that $\widehat{x}$ is continuous on $\mathfrak{M}_0$. So, let $\varphi_0 \in \mathfrak{M}_0$. By Theorem 3.3.9(i) and Lemma 3.4.2(i), we have that the elements $y \equiv (e + x^*x)^{-1}$ and $z \equiv xy$ belong to the C^*-algebra $\mathcal{A}_0$. Therefore, the functions $\widehat{y}$, $\widehat{z}$ are continuous on $\mathfrak{M}_0$.

We first suppose that $\widehat{x}(\varphi_0) \neq \infty$, which by Lemma 3.4.4(ii) means that $\widehat{y}(\varphi_0) \neq 0$. Consequently, there is a neighbourhood V of φ_0, on which $\widehat{y}$ does not vanish. Hence (see Lemma 3.4.4(i)),

$$\widehat{x}(\varphi) = \frac{\widehat{z}(\varphi)}{\widehat{y}(\varphi)}, \ \forall \, \varphi \in V. \tag{3.4.21}$$

Suppose now that $\widehat{x}(\varphi_0) = \infty$. Then (see Lemma 3.4.4(ii)), $\widehat{y}(\varphi_0) = 0$. Take a net (φ_α) in $\mathfrak{M}_0$, such that $\varphi_\alpha \to \varphi_0$. Then, $\widehat{y}(\varphi_\alpha) \to \widehat{y}(\varphi_0) = 0$, so that from Lemma 3.4.4(iii), $|\widehat{x}(\varphi_\alpha)| \to \infty$. This together with (3.4.21) completes the continuity of $\widehat{x}$ at φ_0.

It remains to show that $\widehat{\mathcal{A}}$ is a $*$-algebra of functions and that the map $\mathcal{A} \to \widehat{\mathcal{A}} : x \mapsto \widehat{x}$ is a $*$-homomorphism.

Take x, y in $\mathcal{A}$. Then $\widehat{x}$, $\widehat{y}$, $\widehat{x+y}$ are continuous $\mathbb{C}^*$-valued functions, each one of which takes the value ∞ on at most a nowhere-dense subset of $\mathfrak{M}_0$. In particular, if φ belongs to that dense subset, say D, of $\mathfrak{M}_0$ on which both functions $\widehat{x}$ and $\widehat{y}$ take finite values, then by Proposition 3.4.1(ii)(β), we conclude that

$$(\widehat{x+y})(\varphi) = \widehat{x}(\varphi) + \widehat{y}(\varphi), \ \forall \, \varphi \in D;$$

therefore $\widehat{x+y}$ is that extension of $\widehat{x} + \widehat{y}$ to $\mathfrak{M}_0$, which is required by Definition 3.4.8(ii). Hence,

$$\widehat{x+y} = \widehat{x} + \widehat{y}.$$

The rest of the other algebraic operations, as in Definition 3.4.8(ii), are obtained in a similar way. What remains to be proved is that the $*$-homomorphism $\mathcal{A} \to \widehat{\mathcal{A}} :$

$x \mapsto \widehat{x}$ is injective. So, let us take $x \in \mathcal{A}$, such that $\widehat{x} = 0$. Then, Lemma 3.4.4 implies that $\widehat{y} \neq 0$ and $\widehat{z} = 0$, where y and z are as in the aforementioned lemma. On the other hand, by Lemma 3.4.2(i), $z \in \mathcal{A}_0$ and since $\mathcal{A}_0$ is a C^*-algebra, its Gelfand representation is an isometric $*$-isomorphism, therefore $z = 0$. But, $z = xy$ and y is invertible (see Lemma 3.4.4), consequently $x = zy^{-1} = 0$, so that the proof of theorem is complete.   □

If x is an element of an algebra $\mathcal{A}[\tau]$ as in Theorem 3.4.9, then its spectrum (see Definition 2.3.1) is given by the values of the Gelfand transform $\widehat{x}$ of x on the elements of $\mathfrak{M}_0$. This follows easily from Theorem 3.4.9 (see also Definition 3.4.7) and the fact that $\mathcal{A}_0 \cong \mathcal{C}(\mathfrak{M}_0)$, by the commutative Gelfand–Naimark theorem for the C^*-algebra $\mathcal{A}_0$. Namely, we have the following, which should be compared with Proposition 2.5.5.

Corollary 3.4.10 *Let $\mathcal{A}[\tau]$ be a commutative GB^*-algebra. Let $x \in \mathcal{A}$. Then,*

$$\sigma(x) = \{\widehat{x}(\varphi) := \varphi'(x) : \varphi \in \mathfrak{M}_0\}.$$

3.5 C*-Like Locally Convex *-Algebras as GB*-Algebras

In this section, we introduce a class of complete locally convex $*$-algebras, called 'C^*-like locally convex $*$-algebras', that on the one hand, generalize the C^*-algebras and on the other hand, they are GB^*-algebras, whose bounded part (cf. Definition 3.3.9) coincides with the C^*-algebra generated by the greatest member B_0 of the corresponding collection $\mathfrak{B}^*_{\mathcal{A}}$ that describes the structure of a GB^*-algebra $\mathcal{A}[\tau]$ (see discussion after Definition 3.3.1). *All algebras throughout this section have an identity element e.*

It is known that the study of a *general* locally convex $*$-algebra, which is neither m-convex nor a C^*-convex algebra, is more difficult than that of the latter topological $*$-algebras, even if they are equipped with jointly continuous multiplication. So, in what follows, we define and study the so called M^*-like, respectively C^*-like locally convex $*$-algebras and we prove that the C^*-like ones are GB^*-algebras (see Theorem 3.5.3).

Let $\mathcal{A}[\tau]$ be a locally convex $*$-algebra with an identity element. An upwards directed family $\Gamma = \{p_\nu\}_{\nu \in \Lambda}$ of seminorms determining the topology τ is said to be M^*-like, if for any $\nu \in \Lambda$ there is a $\nu' \in \Lambda$, such that

$$p_\nu(xy) \leq p_{\nu'}(x)p_{\nu'}(y) \text{ and } p_\nu(x^*) \leq p_{\nu'}(x), \ \forall \, x, y \in \mathcal{A} \text{ and } \nu \in \Lambda. \tag{3.5.22}$$

Furthermore, if

$$p_\nu(x)^2 \leq p_{\nu'}(x^*x), \ \forall \, x \in \mathcal{A} \text{ and } \nu \in \Lambda, \tag{3.5.23}$$

then Γ is said to be *C*-like. It is obvious that if Γ is M*-like, respectively C*-like, then p_v's are not necessarily m*-seminorms, respectively C*-seminorms, but the involution is always continuous and the multiplication is jointly continuous.* Moreover, if

$$\mathcal{D}(p_\Lambda) = \{x \in \mathcal{A} : \sup_{v \in \Lambda} p_v(x) < \infty\} \text{ and } p_\Lambda(x) := \sup_{v \in \Lambda} p_v(x), \tag{3.5.24}$$

for all $x \in \mathcal{D}(p_\Lambda)$, then clearly p_Λ is an m^*-norm, respectively C^*-norm on $\mathcal{D}(p_\Lambda)$. Such a norm is said to be an unbounded m^*-norm, respectively unbounded C^*-norm on $\mathcal{A}$. In general, if $\mathcal{A}$ is a *-algebra and p an m^*-seminorm, respectively C^*-seminorm defined on a *-subalgebra $\mathcal{D}(p)$ of $\mathcal{A}$, then p is called *unbounded m^*-seminorm*, respectively *unbounded C^*-seminorm* of $\mathcal{A}$. For the basic theory and applications of the unbounded *-seminorms, the reader is referred to [22, 24, 25, 61]. It is worth mentioning that unbounded C^*-seminorms have appeared in various mathematical and physical subjects, like for instance, in locally convex *-algebras, the moment problem, the quantum field theory, etc.; see, e.g., [6, 54, 137].

Definition 3.5.1 (Inoue–Kürsten)   A locally convex *-algebra $\mathcal{A}[\tau]$ is said to be an *M*-like*, respectively *C*-like locally convex *-algebra* if it is complete and there exists an M^*-like, respectively a C^*-like family $\Gamma = \{p_v\}_{v \in \Lambda}$ of seminorms determining the topology τ, such that $\mathcal{D}(p_\Lambda)$ is τ-dense in $\mathcal{A}$.

Evidently, *every pro-C*-algebra is a C*-like locally convex *-algebra and every Fréchet *-algebra is an M*-like locally convex *-algebra* (see discussion after Definition 3.1.1). Also *the well-known Arens algebra $L^\omega[0, 1] := \bigcap_{1 \leq p < \infty} L^p[0, 1]$* (see (5) in Examples 3.3.16) *is a C*-like locally convex *-algebra, but not a pro-C*-algebra.*

After discussing the structure of C^*-like locally convex *-algebras, we shall show that each such a topological *-algebra $\mathcal{A}[\tau]$ is a GB^*-algebra (see Theorem 3.5.3).

Let $\mathcal{A}[\tau]$ be a complete locally convex *-algebra with a C^*-like family $\Gamma = \{p_v\}_{v \in \Lambda}$ of seminorms. From what was said above p_Λ is a C^*-norm on $\mathcal{D}(p_\Lambda)$, such that $\tau \restriction_{\mathcal{D}(p_\Lambda)} \prec p_\Lambda$. Denote by $U(p_\Lambda)$ the closed unit ball of the C^*-algebra $\mathcal{D}(p_\Lambda)[p_\Lambda]$. Since, p_Λ is a C^*-norm on $\mathcal{D}(p_\Lambda)$, it follows that

$$e \in U(p_\Lambda), \text{ if } e \in \mathcal{D}(p_\Lambda). \tag{3.5.25}$$

If $e \notin \mathcal{D}(p_\Lambda)$, we consider the unitization $\mathcal{D}(p_\Lambda)_1$ of $\mathcal{D}(p_\Lambda)$ (see beginning of Sect. 2.1 and before Sect. 2.2), with

$$\mathcal{D}(p_\Lambda)_1 \equiv \{(\mu, x) \equiv \mu e + x : \mu \in \mathbb{C}, \ x \in \mathcal{D}(p_\Lambda)\}, \tag{3.5.26}$$

$$p_\Lambda^1(\mu, x) = \sup\{p_\Lambda((\mu, x)y) : y \in \mathcal{D}(p_\Lambda), \ p_\Lambda(y) \leq 1\}, \ x \in \mathcal{D}(p_\Lambda).$$

According to our preceding notation $U(p_\Lambda^1)$ denotes the closed unit ball of the C^*-algebra $\mathcal{D}(p_\Lambda)_1[p_\Lambda^1]$. Then, $U(p_\Lambda^1)$ belongs to the collection $\mathfrak{B}_{\mathcal{A}}^*$ and the C^*-algebra $\mathcal{D}(p_\Lambda)_1[p_\Lambda^1]$ equals the C^*-algebra $\mathcal{A}[U(p_\Lambda^1)]$.

Now $\mathfrak{B}_{\mathcal{A}}^*$ does not have, in general, a greatest member. But, if $\mathcal{A}[\tau]$ is a GB^*-algebra, then $\mathfrak{B}_{\mathcal{A}}^*$ has a greatest member denoted by B_0 (see Definition 3.3.2). From the properties of B_0 and the uniqueness of the C^*-norm p_Λ^1 on the C^*-algebra $\mathcal{D}(p_\Lambda)_1[p_\Lambda^1]$, it follows that

$$\mathcal{A}[U(p_\Lambda^1)] = \mathcal{D}(p_\Lambda)_1 \subseteq \mathcal{A}[B_0] \text{ and } p_\Lambda^1(z) = \|z\|_{B_0}, \qquad (3.5.27)$$

for every $z \in \mathcal{D}(p_\Lambda)_1$; hence, we also have $\mathcal{D}(p_\Lambda) \subseteq \mathcal{A}[B_0]$ with $p_\Lambda = \| \cdot \|_{B_0}$ on $\mathcal{D}(p_\Lambda)$. But, as we shall see later in this section, in general, $\mathcal{D}(p_\Lambda) \neq \mathcal{A}[B_0]$; see discussion after Proposition 3.5.6, together with Remark 3.5.7.

Before we state the main result of this section we prove the following technical result.

Lemma 3.5.2 *Let $\mathcal{A}[\tau]$ be a C^*-like locally convex $*$-algebra, with $\Gamma = \{p_\nu\}_{\nu \in \Lambda}$ a C^*-like family of seminorms. Then, for each $x \in \mathcal{A}$, the elements $(e + x^*x)^{-1}$, $x(e + x^*x)^{-1}$ and $(e + x^*x)^{-1}x$, exist in $\mathcal{A}$ and belong to $U(p_\Lambda)$.*

Proof Let $x \in \mathcal{A}$. Since $\mathcal{D}(p_\Lambda)$ is τ-dense in $\mathcal{A}$, there is a net (x_α) in $\mathcal{D}(p_\Lambda)$, such that $x = \tau - \lim_\alpha x_\alpha$. Then, clearly we shall also have that $x^*x = \tau - \lim_\alpha x_\alpha^*x_\alpha$, where from the functional calculus for C^*-algebras with identity we conclude that the elements $(e + x_\alpha^*x_\alpha)^{-1}$ belong to $U(p_\Lambda^1)$, for each index α. We show that $\left((e + x_\alpha^*x_\alpha)^{-1}\right)$ is a Cauchy net in $\mathcal{A}[\tau]$. Let $p_\nu \in \Gamma$ be arbitrary. By (3.5.22) and the fact that $\tau \prec p_\Lambda^1$, there exist $p_{\nu'}, p_{\nu''} \in \Gamma$, such that

$$
\begin{aligned}
p_\nu\left((e + x_\alpha^*x_\alpha)^{-1} - (e + x_\beta^*x_\beta)^{-1}\right) &= p_\nu\left((e + x_\alpha^*x_\alpha)^{-1}(x_\beta^*x_\beta - x_\alpha^*x_\alpha)(e + x_\beta^*x_\beta)^{-1}\right) \\
&\leq p_{\nu'}\left((e + x_\alpha^*x_\alpha)^{-1}(x_\beta^*x_\beta - x_\alpha^*x_\alpha)\right)p_{\nu'}\left((e + x_\beta^*x_\beta)^{-1}\right) \\
&\leq p_{\nu''}\left((e + x_\alpha^*x_\alpha)^{-1}\right)p_{\nu''}(x_\beta^*x_\beta - x_\alpha^*x_\alpha)p_{\nu'}\left((e + x_\beta^*x_\beta)^{-1}\right) \\
&\leq p_\Lambda^1\left((e + x_\alpha^*x_\alpha)^{-1}\right)p_\Lambda^1\left((e + x_\beta^*x_\beta)^{-1}\right)p_{\nu''}(x_\beta^*x_\beta - x_\alpha^*x_\alpha) \\
&\leq p_{\nu''}(x_\beta^*x_\beta - x_\alpha^*x_\alpha).
\end{aligned}
$$

So our claim follows and the element $y \equiv \tau - \lim_\alpha (e \mid x_\alpha^*x_\alpha)^{-1}$ belongs to $\mathcal{A}[\tau]$. Taking now τ-limits in the relation

$$(e + x_\alpha^*x_\alpha)(e + x_\alpha^*x_\alpha)^{-1} = e = (e + x_\alpha^*x_\alpha)^{-1}(e + x_\alpha^*x_\alpha),$$

we conclude that $y = (e + x^*x)^{-1}$. We prove that $y \in U(p_\Lambda)$. Indeed,

$$p_\nu(y) = p_\nu\big((e + x^*x)^{-1}\big) = \lim_\alpha p_\nu\big((e + x_\alpha^*x_\alpha)^{-1}\big)$$

$$\leq \lim_\alpha p_\Lambda\big((e + x_\alpha^*x_\alpha)^{-1}\big)$$

$$\leq \lim_\alpha p_\Lambda^1\big((e + x_\alpha^*x_\alpha)^{-1}\big) \leq 1, \ \forall \ \nu \in \Lambda.$$

It is now clear from the definition of p_Λ that $y = (e + x^*x)^{-1} \in U(p_\Lambda)$.

In the same way, we can show that $x^*x(e + x^*x)^{-1}$ belongs to $U(p_\Lambda)$, which for any $\nu \in \Lambda$, implies that (see (3.5.23))

$$p_\nu\big(x(e + x^*x)^{-1}\big)^2 \leq p_{\nu'}\big((e + x^*x)^{-1}x^*x(e + x^*x)^{-1}\big)$$

$$\leq p_\Lambda\big((e + x^*x)^{-1}\big)p_\Lambda\big(x^*x(e + x^*x)^{-1}\big)$$

$$\leq 1,$$

for some $\nu' \in \Lambda$. Hence, $x(e + x^*x)^{-1} \in U(p_\Lambda)$. Similarly, $(e + x^*x)^{-1}x \in U(p_\Lambda)$. □

The following theorem provides a characterization of C^*-like locally convex *-algebras.

Theorem 3.5.3 (Inoue–Kürsten) *Let $\mathcal{A}[\tau]$ be a complete locally convex *-algebra with a C^*-like family $\Gamma = \{p_\nu\}_{\nu \in \Lambda}$ of seminorms. The following statements are equivalent:*

(i) *$\mathcal{A}[\tau]$ is a C^*-like locally convex *-algebra.*
(ii) *For every $h^* = h \in \mathcal{A}$, the elements $(e + h^2)^{-1}$ and $h(e + h^2)^{-1}$, as well as the identity element e of $\mathcal{A}$ belong to $U(p_\Lambda)$.*
(iii) *$\mathcal{A}[\tau]$ is a GB^*-algebra, with $B_0 = U(p_\Lambda)$.*

Proof (i) $\Rightarrow$ (ii) Let $h^* = h \in \mathcal{A}$. It is an immediate consequence of Lemma 3.5.2 that the elements $(e + h^2)^{-1}$, $h(e + h^2)^{-1}$ belong to $U(p_\Lambda)$, which implies that $\frac{1}{2}e$ is an element of $U(p_\Lambda)$ and so of $\mathcal{D}(p_\Lambda)$ too. Hence, from (3.5.25) we have that $e \in U(p_\Lambda)$.

(ii) $\Rightarrow$ (iii) Take an arbitrary $B \in \mathfrak{B}_\mathcal{A}^*$ and $h^* = h \in B$. Let $\mathcal{C}$ be a maximal commutative *-subalgebra of $\mathcal{A}$ containing h. Then, $\mathcal{C}$ is a complete commutative locally convex *-algebra, with $\Gamma_\mathcal{C} = \{p_\nu \restriction \mathcal{C}\}_{\nu \in \Lambda}$ as a C^*-like family of seminorms. We denote by $\mathfrak{B}_\mathcal{C}^*$ the collection of subsets of $\mathcal{C}$ as in Definition 3.3.1. Then, $\mathfrak{B}_\mathcal{C}^* = \{B \cap \mathcal{C} : B \in \mathfrak{B}_\mathcal{A}^*\}$ and $U(p_\Lambda^\mathcal{C}) \equiv U(p_\Lambda) \cap \mathcal{C} \in \mathfrak{B}_\mathcal{C}^*$, where $U(p_\Lambda) \in \mathfrak{B}_\mathcal{A}^*$ (see comments after (3.5.26)) and $p_\Lambda^\mathcal{C}$ is the corresponding C^*-norm on $\mathcal{D}(p_\Lambda^\mathcal{C})$ as in (3.5.24).

We show that $B \cap \mathcal{C} \subset U(p_\Lambda^\mathcal{C})$. Since $\mathcal{C}$ is commutative, it follows from Theorem 2.2.10 that $\mathfrak{B}_\mathcal{C}^*$ is directed, so that there exists $B' \in \mathfrak{B}_\mathcal{C}^*$ such that

$(B \cap \mathcal{C}) \cup U(p_\Lambda^\mathcal{C}) \subset B'$. Then, since the C^*-algebra $\mathcal{D}(p_\Lambda^\mathcal{C}) \equiv \mathcal{D}(p_\Lambda) \cap \mathcal{C}$ is contained in the Banach $*$-algebra $\mathcal{C}[B']$, it follows from [144, I, p. 22, Proposition 5.3] that $p_\Lambda^\mathcal{C}(x) \leq \|x\|_{B'}$, for each $x \in \mathcal{D}(p_\Lambda^\mathcal{C})$. On the other hand, since $U(p_\Lambda^\mathcal{C}) \subset B'$, it follows from the definition of the norms $p_\Lambda^\mathcal{C}(\cdot)$ and $\| \cdot \|_{B'}$ that $\|x\|_{B'} \leq p_\Lambda(x)$, for each $x \in \mathcal{D}(p_\Lambda^\mathcal{C})$. Hence, we have

$$p_\Lambda^\mathcal{C}(x) = \|x\|_{B'}, \ \forall \, x \in \mathcal{D}(p_\Lambda^\mathcal{C}). \tag{3.5.28}$$

Take an arbitrary $b^* = b \in \mathcal{C}[B']$. By the assumption (ii) $(e+b^2)^{-1}$ and $b(e+b^2)^{-1}$ belong to $U(p_\Lambda^\mathcal{C})$, and so $b(e + \frac{1}{n}b^2)^{-1} \in \mathcal{C} \cap \mathcal{D}(p_\Lambda)$, for each $n \in \mathbb{N}$. Furthermore, it follows from (3.5.28) that

$$
\begin{aligned}
\left\| b(e + \frac{1}{n}b^2)^{-1} - b \right\|_{B'} &= \frac{1}{n} \left\| b^3(e + \frac{1}{n}b^2)^{-1} \right\|_{B'} \\
&\leq \frac{1}{n} \|b^3\|_{B'} \left\| (e + \frac{1}{n}b^2)^{-1} \right\|_{B'} \\
&= \frac{1}{n} \|b^3\|_{B'} \, p_\Lambda^\mathcal{C}\big((e + \frac{1}{n}b^2)^{-1}\big) \\
&\leq \frac{1}{n} \|b^3\|_{B'}, \forall \, n \in \mathbb{N}.
\end{aligned}
$$

This implies that $\mathcal{D}(p_\Lambda^\mathcal{C})$ is dense in $\mathcal{C}[B']$. Hence, $\mathcal{D}(p_\Lambda^\mathcal{C}) = \mathcal{C}[B']$ and so $U(p_\Lambda^\mathcal{C}) = B'$. Therefore, $B \cap \mathcal{C} \subset U(p_\Lambda) \cap \mathcal{C}$, so that $h \in U(p_\Lambda)$. Thus, we have $B \subset U(p_\Lambda)$, which implies that $U(p_\Lambda)$ is the greatest element in $\mathfrak{B}_\mathcal{A}^*$. Consequently, $\mathcal{A}[\tau]$ is a GB^*-algebra over $B_0 = U(p_\Lambda)$.

(ii) $\Rightarrow$ (i) We have that $\mathcal{A} = H(\mathcal{A}) + i H(\mathcal{A})$ and for each $h^* = h \in \mathcal{A}$ (i.e., $h \in H(\mathcal{A})$) we have by (ii) that $h(e + \varepsilon h^2)^{-1} \in \mathcal{D}(p_\Lambda)$, for each $\varepsilon > 0$. Moreover, again by (ii) $\mathcal{D}(p_\Lambda)$ has an identity element, therefore

$$p_\nu\big(h(e+\varepsilon h^2)^{-1} - h\big) = \varepsilon p_\nu\big(h^3(e+\varepsilon h^2)^{-1}\big) \leq \varepsilon p_{\nu'}(h^3) p_{\nu'}\big((e+\varepsilon h^2)^{-1}\big) \leq \varepsilon p_{\nu'}(h^3),$$

for each $\varepsilon > 0$ and some $\nu' \in \Lambda$. This yields that $\mathcal{D}(p_\Lambda)$ is τ-dense in $\mathcal{A}$, therefore $\mathcal{A}[\tau]$ is a C^*-like locally convex $*$-algebra.

(iii) $\Rightarrow$ (ii) This is immediate from the fact that $e \in B_0$ and $\mathcal{A}[B_0]$ is a C^*-algebra containing the elements $(e + h^2)^{-1}$, $h^* = h \in \mathcal{A}$ (see Theorem 3.3.9). Thus, the proof is complete. $\qquad \square$

The following corollary is an immediate consequence of Theorem 3.5.3.

Corollary 3.5.4 *For every C^*-like locally convex $*$-algebra $\mathcal{A}[\tau]$, one has that $\mathcal{A}[B_0] = \mathcal{D}(p_\Lambda)$.*

Remark 3.5.5

(1) Theorem 3.5.3 implies that a C^*-like locally convex $*$-algebra is independent of the method of taking the C^*-like families.

(2) If $\mathcal{A}[\tau]$ is a locally convex $*$-algebra with a C^*-like family $\Gamma = \{p_\nu\}_{\nu \in \Lambda}$ of seminorms, such that $\mathcal{D}(p_\Lambda)$ is τ-dense in $\mathcal{A}$, then the completion $\widetilde{\mathcal{A}}[\tau]$ of $\mathcal{A}[\tau]$ is a C^*-like locally convex $*$-algebra. Hence, the assumption of completeness of the locally convex $*$-algebra in Theorem 3.5.3 is not essential.

(3) Another proof of (i) $\Rightarrow$ (iii) in Theorem 3.5.3 has been given by S. J. Bhatt in [17, Theorem 2] using the extended Vidav–Palmer theorem of Wood [156] (see Theorem 7.5.49, in Sect. 7.5 of this book).

(4) Let $\mathcal{A}[\tau]$ be a locally convex $*$-algebra and $(\mathcal{A}_0)_h := \{x \in \mathcal{A}_0 : x^* = x\}$. Denote by $B(\mathcal{A})$ the $*$-subalgebra of $\mathcal{A}$ generated by $(\mathcal{A}_0)_h$.

If $\mathcal{A}[\tau]$ is a GB^*-algebra, the C^*-algebra $\mathcal{A}[B_0]$ contains the set $(\mathcal{A}_0)_h$ (see Corollary 3.3.8), so that we obtain $B(\mathcal{A}) \subseteq \mathcal{A}[B_0]$. For the inverse inclusion use the fact that every element x in $\mathcal{A}[B_0]$ is bounded and that $x = h + ik$, with h, k self-adjoint elements in $\mathcal{A}[B_0]$. Hence finally,

$$\mathcal{A}[B_0] = B(\mathcal{A}).$$

Suppose now that $\mathcal{A}[\tau]$ is a C^*-like locally convex $*$-algebra; then by Corollary 3.5.4, $\mathcal{D}(p_\Lambda) = \mathcal{A}[B_0]$.

We can now state the following

Proposition 3.5.6 *For every C^*-like locally convex $*$-algebra, one has the equality*

$$B(\mathcal{A}) = \mathcal{A}[B_0] = \mathcal{D}(p_\Lambda).$$

▶ *We now compare the C^*-algebra $\mathcal{A}[B_0]$ with the Banach $*$-algebra $\mathcal{D}(p_\Lambda)$ in the case of a GB^*-algebra $\mathcal{A}[\tau]$.*

Let $\mathcal{A}[\tau]$ *be a complete GB^*-algebra with jointly continuous multiplication*, where τ is given by a directed family $\Gamma = \{p_\nu\}_{\nu \in \Lambda}$ of $*$-seminorms. Following the same strategy as in the second part of the proof in [60, Theorem 10.23, p. 133], we show that $\mathcal{D}(p_\Lambda)$ is a Banach $*$-algebra, under the norm p_Λ. In fact, recalling that $\mathcal{A}[\tau]$ has an identity and following the comments before and after (3.5.27), we conclude that $\mathcal{D}(p_\Lambda) = \mathcal{A}[B]$, with $B \equiv U(p_\Lambda) \in \mathcal{B}_\mathcal{A}^*$. Since $B \subset B_0$, one obtains that the Banach $*$-algebra $\mathcal{D}(p_\Lambda)$, under the $*$-norm p_Λ, is contained in the C^*-algebra $\mathcal{A}[B_0]$, under the C^*-norm $\| \cdot \|_{B_0}$. Summing up, we attain

$$\mathcal{D}(p_\Lambda) \subseteq \mathcal{A}[B_0] \text{ and } \|x\|_{B_0} = p_\Lambda(x), \ \forall \, x \in \mathcal{D}(p_\Lambda). \tag{3.5.29}$$

We consider now when $\mathcal{D}(p_\Lambda) = \mathcal{A}[B_0]$. It is easily seen that the following statements are equivalent:

(i) $\mathcal{A}[B_0] = \mathcal{D}(p_\Lambda)$;
(ii) $\|\cdot\|_{B_0} = p_\Lambda$;
(iii) p_Λ is a C^*-norm on $\mathcal{D}(p_\Lambda)$;
(iv) there exists a constant $\delta > 0$, such that $p_\nu(x) \leq \delta$, for all $\nu \in \Lambda$ and $x \in B_0$.

Indeed, the equivalence of (i), (ii), (iii) follows easily from C^*-algebra theory (see, e.g., [144, p. 22, Proposition 5.3]). The implication (ii) $\Rightarrow$ (iv) is trivial. Now (iv) implies that $\mathcal{A}[B_0] \subseteq \mathcal{D}(p_\Lambda)$, while (3.5.29) gives the inverse inclusion. Thus, (iv) $\Rightarrow$ (i).

Remark 3.5.7

(1) The set B_0 in $\mathfrak{B}^*_\mathcal{A}$ is τ-bounded, therefore for each $\nu \in \Lambda$, there exists a constant $\delta_\nu > 0$, depending on ν, such that $p_\nu(x) \leq \delta_\nu$, for all $x \in B_0$. The statement (iv) above, asserts a constant $\delta > 0$, but this does not depend on ν.

(2) Note that $\mathcal{D}(p_\Lambda)$ depends on the family of $*$-seminorms $\Gamma = \{p_\nu\}_{\nu \in \Lambda}$ defining the topology τ of $\mathcal{A}$. Indeed, suppose that $\mathcal{A}[\tau]$ is a GB^*-algebra, where τ is defined by an increasing sequence $\{p_n\}_{n \in \mathbb{N}}$ of $*$-seminorms. Put

$$p'_n := np_n, \quad n \in \mathbb{N}.$$

Then, the family $\{p'_n\}_{n \in \mathbb{N}}$ defines equivalently the given topology τ on $\mathcal{A}$. But, if $\mathcal{D}(p'_\mathbb{N})$ is the Banach $*$-algebra corresponding to $(\mathcal{A}, \{p'_n\}_{n \in \mathbb{N}})$, then clearly,

$$\mathcal{D}(p'_\mathbb{N}) \underset{\neq}{\subseteq} \mathcal{D}(p_\mathbb{N}).$$

(3) Note that when $\mathcal{A}[\tau]$ is a pro-C^*-algebra, we have $B_0 = \{x \in \mathcal{A} : p_\Lambda(x) \leq 1\} \equiv U(p_\Lambda)$ (see (1) in Examples 3.3.16), for any family of C^*-seminorms defining the topology τ on $\mathcal{A}$. Hence, in this case, we denote

$$\mathcal{D}(p_\Lambda) \text{ and } p_\Lambda \text{ by } \mathcal{A}_b \text{ and } \|\cdot\|_b, \text{ resp.}$$

The C^*-algebra $\mathcal{A}_b[\|\cdot\|_b]$ is called *bounded part* of $\mathcal{A}[\tau]$.

▶ *The same happens when* $\mathcal{A}[\tau]$ *is a* C^*-*like locally convex* $*$-*algebra, as* follows from Corollary 3.5.4. Thus, *from now on, we shall also use the notation* $\mathcal{A}_b$, $\|\cdot\|_b$, respectively, *instead of* $\mathcal{D}(p_\Lambda)$, p_Λ, *for a* C^*-*like locally convex* $*$-*algebra* $\mathcal{A}[\tau]$.

(4) In case of a general GB^*-algebra $\mathcal{A}[\tau]$ with jointly continuous multiplication, there does not necessarily exist a family $\Gamma = \{p_\nu\}_{\nu \in \Lambda}$ of $*$-seminorms defining

the topology τ, such that $\mathcal{D}(p_\Lambda) = \mathcal{A}[B_0]$. Even if there exists a family Γ' with subindex, say Λ', the equality $\mathcal{D}(p'_{\Lambda'}) = \mathcal{A}[B_0]$ does not necessarily hold for Γ' (see (2) above).

Thus, a family $\Gamma = \{p_\nu\}_{\nu \in \Lambda}$ of *-seminorms defining the topology τ, such that $\mathcal{D}(p_\Lambda) = \mathcal{A}[B_0]$ will be called *natural* and the corresponding algebra $\mathcal{A}[\tau]$ will be called a *GB*-algebra with a natural family of *-seminorms*.

(5) Suppose that the multiplication on a GB^*-algebra $\mathcal{A}[\tau]$ is separately continuous. Let $\{p_\nu\}_{\nu \in \Lambda}$ be an upwards directed family of *-seminorms defining the GB^*- topology τ on $\mathcal{A}$. Then, $\mathcal{A}[B_0]$ is a C^*-algebra, but $\mathcal{D}(p_\Lambda)$ is not even an algebra, in general.

Combining Proposition 3.5.6 with Remark 3.5.7(3), we have the following

Proposition 3.5.8 *In every C^*-like locally convex *-algebra $\mathcal{A}[\tau]$, we have that*

$$B(\mathcal{A}) = \mathcal{A}[B_0] = \mathcal{A}_b.$$

We shall finish this section with a result on commutative C^*-like locally convex *-algebras and some examples.

A *-algebra of functions* (see Definition 3.4.8) *consisting of $\mathbb{C}^*$-valued continuous functions on a compact space*, is said to be C^*-*like* if it is a C^*-like locally convex *-algebra. By Theorem 3.5.3 and Theorem 3.4.9 we have the following

Theorem 3.5.9 *Every commutative C^*-like locally convex *-algebra is isomorphic to a C^*-like algebra of $\mathbb{C}^*$-valued continuous functions on a compact space.*

For other results concerning C^*-like locally convex *-algebras, see end of Sect. 5.3 and Chap. 8.

We now present some examples of C^*-like locally convex *-algebras.

Examples 3.5.10

(1) Every pro$-C^*$-algebra $\mathcal{A}[\tau]$ is a C^*-like locally convex *-algebra with $B_0 = U(\|\cdot\|_b)$.
(2) In particular, the function algebra $\mathcal{C}(\mathbb{R})$ of all $\mathbb{C}$-valued continuous functions on $\mathbb{R}$ equipped with the family $\Gamma = \{p_n\}_{n \in \mathbb{N}}$ of C^*-seminorms, such that

$$p_n(f) = \sup_{t \in [-n,n]} |f(t)|, \quad f \in \mathcal{C}(\mathbb{R})$$

is a pro-C^*-algebra with $\mathcal{C}(\mathbb{R})_b[\|\cdot\|_b] = \mathcal{C}_b(\mathbb{R})$, the C^*-algebra of all continuous bounded functions on $\mathbb{R}$ and $\|f\|_b = \sup_{t \in \mathbb{R}} |f(t)|, f \in \mathcal{C}_b(\mathbb{R})$.
(3) Let (h_n) be a sequence of continuous $\mathbb{C}^*$-valued functions on a compact space X, such that $\mathbf{1} \leq h_1 \leq h_2 \leq \cdots$ and for all $n \in \mathbb{N}$, there exists $n_1 \in \mathbb{N}$ in such a way that $h_{n+1} \leq h_n^2 \leq h_{n_1}$, where $\mathbf{1}$ is the constant function 1. We denote by $\mathcal{F}_{h_n}$ the set of all continuous $\mathbb{C}^*$-valued functions f on X, such that $\sup\{|f(x)|/h_n(x) : x \in X\} < \infty$. Since $\mathcal{F}_{h_n} \subset \mathcal{F}_{h_{n+1}}$ and $h_n^2 \leq h_{n_1}$, for all

$n \in \mathbb{N}$, it follows that $\mathcal{F}(\{h_n\}) \equiv \bigcup_{n=1}^{\infty} \mathcal{F}_{h_n}$ is a $*$-algebra containing $\mathcal{C}(X)$. Since $h_{n+1} \leq h_n^2$, for all $n \in \mathbb{N}$, we can define a sequence $\Gamma = \{p_n\}_{n \in \mathbb{N}}$ of seminorms on $\mathcal{F}(\{h_n\})$ by $p_n(f) = \sup \left\{ e^{-\frac{h_1(x)}{n}} |f(x)| : x \in X \right\}, n \in \mathbb{N}$. Then, we have

$$p_n(fg) \leq p_{2n}(f) p_{2n}(g), \quad p_n(f)^2 \leq p_n(f^*f), \quad p_n(\mathbf{1}) \leq 1,$$

for all $n \in \mathbb{N}$ and $f, g \in \mathcal{F}(\{h_n\})$. Further,

$$\mathcal{F}(\{h_n\})_b[\| \cdot \|_b] = C(X) \quad \text{and} \quad \|f\|_b = \sup_{x \in X} |f(x)|, \ f \in \mathcal{F}(\{h_n\})_b[\| \cdot \|_b].$$

Hence, the completion $\widetilde{\mathcal{F}}(\{h_n\})$ of $\mathcal{F}(\{h_n\})$ is a C^*-like locally convex $*$-algebra.

(4) Let $(X, \mathcal{B}, \mu)$ be a finite measure space. Then, the function algebra

$$L^{\omega}(X, \mathcal{B}, \mu) \equiv \bigcap_{1 \leq p < \infty} L^p(X, \mathcal{B}, \mu)$$

is a C^*-like locally convex $*$-algebra endowed with a C^*-like family consisting of the norms $\{\mu(X)^{-1/p} \| \cdot \|_p\}_{1 \leq p < \infty}$. In particular, the Arens algebra $L^{\omega}[0, 1] \equiv \bigcap_{1 \leq p < \infty} L^p[0, 1]$ is a C^*-like locally convex $*$-algebra, with $L^{\omega}[0, 1]_b = L^{\infty}[0, 1]$.

(5) We consider now the noncommutative case. Let $\mathcal{M}$ be a von Neumann algebra with a faithful finite trace τ and $L^p(\tau)$ the Segal L^p-space with respect to τ [140]. Then,

$$L^{\omega}(\tau) \equiv \bigcap_{1 \leq p < \infty} L^p(\tau)$$

is a C^*-like locally convex $*$-algebra equipped with a C^*-like family provided by the norms $\left\{\tau(I)^{-1/p} \| \cdot \|_p\right\}_{1 \leq p < \infty}$ [81], where I is the identity operator in $\mathcal{M}$. See the references [140] and [81] for more details.

For an example of a C^*-like locally convex $*$-algebra that is also an $\mathcal{O}^*$-algebra, see [88, Example 3.4]. For $\mathcal{O}^*$-algebras, see Sect. 5.1.

Notes The majority of results presented in the Sects. 3.1–3.4 of Chap. 3 are due to G.R. Allan and can be found in [3, 5]. The results of Sect. 3.5 belong jointly to A. Inoue and K.-D. Kürsten and come from [88]. The Example 3.3.16(6) is due to C.L. Crowther [40], while 3.3.16(7) is due to P.G. Dixon [48]. For Theorem 3.3.17, see [40, Theorem 1.5.3]. Proposition 3.3.19 is a joint result of the first three authors of this book and it is Lemma 6.5 in [64].

Chapter 4
Commutative Generalized B*-Algebras: Functional Calculus and Equivalent Topologies

In this chapter we are going to introduce (see [5, Section 3]) a functional calculus for commutative GB^*-algebras, analogue to that of C^*-algebras. It will be often used in various results in this book, but also similar line of thoughts can be applied in further generalizations of C^*-algebras (see, e.g., [14, Sections 5 and 6]). For a functional calculus, in the case of a not necessarily commutative GB^*-algebra, see Sect. 6.2.

Furthermore, using the fact that in a commutative GB^*-algebra there are sufficiently many positive linear functionals to separate its elements (see Corollary 4.2.6), it is proved (cf. Corollary 4.3.10) that for a commutative $*$-algebra $\mathcal{A}$, any two distinct topologies making it a GB^*-algebra are "equivalent", in a sense involving the corresponding $\mathfrak{B}^*$-collections defining the respective GB^*-structures on $\mathcal{A}$ (see Definition 4.3.1). Moreover, there exists a finest locally convex $*$-algebra topology on a commutative GB^*-algebra $\mathcal{A}[\tau]$ that is equivalent to the given topology τ (Theorem 4.3.13).

4.1 Functional Calculus

Recall that all GB^*-algebras we deal with have an identity element, unless mention is made to the contrary. Let $\mathcal{A}[\tau]$ be a commutative GB^*-algebra. *For an element $x \in \mathcal{A}$ we set $S \equiv \sigma(x)$ and denote by $\mathcal{C}_0(S)$ the algebra of all continuous $\mathbb{C}$-valued functions on S, vanishing at ∞.*

© The Author(s), under exclusive license to Springer Nature Switzerland AG 2022
M. Fragoulopoulou et al., *Generalized B*-Algebras and Applications*, Lecture Notes in Mathematics 2298, https://doi.org/10.1007/978-3-030-96433-7_4

First we need the following

Proposition 4.1.1 *Let $A[\tau]$ be a commutative GB*-algebra. Then, for every $x \in A$, there exists a unique *-homomorphism Φ from $C_0(S)$ into $A[\tau]$ taking f from $C_0(S)$ to $f(x)$ in $A[\tau]$, satisfying the properties:*

(i) *if $\lambda \in S$ and $f(\lambda) := (1 + |\lambda|^2)^{-1}$, then $f(x) = (e + x^*x)^{-1}$ and if $g(\lambda) :=$*
    $\lambda f(\lambda)$, then $g(x) = xf(x)$;
(ii) *$\widehat{f(x)}(\varphi) = f(\widehat{x}(\varphi))$, for any $f \in C_0(S)$ and $\varphi \in \mathfrak{M}_0$.*

Furthermore,

$$f(x) \in A_0, \quad for\ all\ f \in C_0(S)$$

and when $C_0(S)$ carries the supremum norm and A_0 its usual norm, then the map

$$C_0(S) \to A[\tau] : f \mapsto f(x)$$

is an isometric *-isomorphism onto the closed *-subalgebra of A_0 generated by the element $x(e + x^*x)^{-1}$.

Proof Let $f \in C_0(S)$. From Theorem 3.4.9 and Corollary 3.4.10, we have that $f \circ \widehat{x}$ is an element of $C(\mathfrak{M}_0)$. Therefore, by the commutative Gelfand–Naimark theorem it coincides with the Gelfand transform of a unique element, say y, of the C^*-algebra A_0. Following the usual notation, we denote y by $f(x)$. Thus, we put

$$f \circ \widehat{x} \equiv \widehat{f(x)}, \ f \in C_0(S). \tag{4.1.1}$$

It is then easy to verify that the map $C_0(S) \to A[\tau] : f \mapsto f(x)$ is a *-homomorphism. For instance, let $f, g \in C_0(S)$. Then, applying Theorem 3.4.9, we conclude that

$$\widehat{(f + g)}(x) = (f + g) \circ \widehat{x} = f \circ \widehat{x} + g \circ \widehat{x}$$
$$= \widehat{f(x)} + \widehat{g(x)} = \widehat{f(x) + g(x)}.$$

Hence, by uniqueness, we obtain $(f + g)(x) = f(x) + g(x)$. Now, (i) follows from Theorem 3.4.9 , while (ii) is an immediate consequence of (4.1.1).

Furthermore, taking into account Corollary 3.4.10 and the fact that $A_0 = A[B_0]$ is a commutative C^*-algebra, we obtain

$$\|f\|_\infty = sup\{|f(\lambda)| : \lambda \in S\} = sup\{|f(\varphi(x))| : \varphi \in \mathfrak{M}_0\}$$
$$= sup\{|\widehat{f(x)}(\varphi)| : \varphi \in \mathfrak{M}_0\} = \|f(x)\|_{B_0}, \ \forall\ f \in C_0(S).$$

Therefore, our map $f \mapsto f(x)$ is an isometry.

Now, take the function $g \in \mathcal{C}_0(S)$ given in (i). Then, the closed $*$-subalgebra generated by g is nothing other than $\mathcal{C}_0(S)$, according to the Stone–Weierstass theorem. But $g(x) = x(e + x^*x)^{-1}$ and by Lemma 3.4.2(1), the element on the right hand side of the previous equality belongs to $\mathcal{A}_0$. Hence, the proof is complete. $\square$

In what follows, we extend the previous functional calculus to a more general form (see Theorem 4.1.2). It is clear, that if in $\hat{\mathcal{A}}$ were sitting all continuous $\mathbb{C}^*$-valued functions on $\mathfrak{M}_0$ that take the value ∞ on at most a nowhere dense set, then the result would be trivial. On the contrary, the usefulness of Theorem 4.1.2 comes exactly from the fact that $\hat{\mathcal{A}}$ does not include all these functions.

We adopt the notation $\mathcal{C}_1(S)$ for *the set of all continuous $\mathbb{C}^*$-valued functions f on $S \cap \mathbb{C}$, such that the function*

$$\lambda \mapsto \frac{f(\lambda)}{(1 + |\lambda|^2)^n} \text{ is contained in } \mathcal{C}_0(S),$$

for some non-negative integer n, which may vary for different functions f. It is evident that the set $\mathcal{C}_1(S)$ is an algebra in which the algebra $\mathcal{C}_0(S)$ is included.

Theorem 4.1.2 *Let $\mathcal{A}[\tau]$ be a commutative GB*-algebra and $x \in \mathcal{A}$. Then, the map $\mathcal{C}_0(S) \to \mathcal{A}[\tau] : f \mapsto f(x)$ of Proposition 4.1.1 may be extended uniquely to a $*$-isomorphism from $\mathcal{C}_1(S)$ into $\mathcal{A}[\tau]$ (that we also denote by $f \mapsto f(x)$), such that*

(i) *if $u_0(\lambda) = 1$, then $u_0(x) = e$;*
(ii) *if $u_1(\lambda) = \lambda$, then $u_1(x) = x$;*
(iii) *for any $f \in \mathcal{C}_1(S)$ and $\varphi \in \mathfrak{M}_0$, one has that $\widehat{f(x)}(\varphi) = f(\widehat{x}(\varphi))$.*

Proof Let $f \in \mathcal{C}_1(S)$. Then, we can find $n \in \mathbb{N} \cup \{0\}$, such that the function

$$g_n(\lambda) := \frac{f(\lambda)}{(1 + |\lambda|^2)^n}, \ \lambda \in \sigma(x) \cap \mathbb{C},$$

sits in $\mathcal{C}_0(S)$. According to Proposition 4.1.1, we have that

$$g_n(x) = f(x)(e + x^*x)^{-n}, \ n \in \mathbb{N} \cup \{0\}.$$

Then, if a map $f \in \mathcal{C}_1(S) \mapsto f(x) \in \mathcal{A}[\tau]$ extending the homomorphism of Proposition 4.1.1 is to be defined, the only possible choice for $f \in \mathcal{C}_1(S)$ is

$$f(x) := g_n(x)(e + x^*x)^n.$$

We prove that the function f does not depend on the selected n, such that $g_n \in \mathcal{C}_0(S)$. Indeed, let us take $k \in \mathbb{N} \cup \{0\}$ with $k \le n$ and $\lambda \in \sigma(x) \cap \mathbb{C}$. Then, we have

$$g_n(\lambda) = \frac{f(\lambda)}{(1 + |\lambda|^2)^n} = \frac{f(\lambda)}{(1 + |\lambda|^2)^k} \frac{1}{(1 + |\lambda|^2)^{n-k}} = g_k(\lambda) \frac{1}{(1 + |\lambda|^2)^{n-k}}.$$

Hence, $g_n(x) = g_k(x)(e + x^*x)^{-(n-k)} \in \mathcal{A}[\tau]$ and

$$g_n(x)(e + x^*x)^n = g_k(x)(e + x^*x)^k; \tag{4.1.2}$$

Let us now consider $k \in \mathbb{N}$ with $k > n$. Then, we work in a similar way, as before, with the positive integers $k,\ k - n$, so that we are led again to (4.1.2). This proves our claim.

Now, taking into account the proof of Proposition 4.1.1, one verifies easily that the map $\mathcal{C}_1(S) \to \mathcal{A} : f \mapsto f(x)$ is a *-monomorphism that satisfies the required properties (i)–(iii).   □

In what follows we need the concept of a positive element in a GB^*-algebra. For the general definition and basic properties, see Sect. 6.2. Here the definition is given in the commutative case.

Definition 4.1.3 Let $\mathcal{A}[\tau]$ be a commutative GB^*-algebra and $h \in H(\mathcal{A})$. We say that h is *positive* and we write $h \geq 0$, if the Gelfand transform $\widehat{h}$ of h is a non-negative function in $C(\mathfrak{M}_0)$. Note that we consider ∞ as non-negative, in this case. Denote by $\mathcal{A}^+$ the set of positive elements in $\mathcal{A}[\tau]$.

We have the following

Theorem 4.1.4 *Let $\mathcal{A}[\tau]$ be a commutative GB^*-algebra and $h \in H(\mathcal{A})$. Then, the following statements hold:*

(i) *The element h is positive, if and only if, there is $k \in H(\mathcal{A})$, such that $h = k^2$.*
(ii) *There are positive elements h^+, h^- in $\mathcal{A}$, such that $h = h^+ - h^-$ and $h^+ h^- = 0$.*

Proof

(i) Suppose that $h = k^2$, for some $k \in H(\mathcal{A})$. Observe that $\mathcal{A}[\tau]$ as a GB^*-algebra is symmetric, hence hermitian, therefore k as a self-adjoint element has real spectrum. Taking also into account Corollary 3.4.10, we conclude that the Gelfand transform $\widehat{k}$ of k is a real, or infinite function on $\mathfrak{M}_0$. Thus, $\widehat{h}(= \widehat{k}^2)$ is a non-negative function on $\mathfrak{M}_0$, therefore h is a positive element in $\mathcal{A}$ from the preceding definition.

Conversely, suppose that h is a positive element in $\mathcal{A}$. Then, Proposition 3.4.1(iii) together with Corollary 3.4.10 imply that $S \equiv \sigma_{\mathcal{A}}(h) \subset \mathbb{R}_+$ and the function

$$f(\lambda) := +\sqrt{\lambda},\ \lambda \in S \cap \mathbb{C},$$

belongs to $\mathcal{C}_1(S)$ and satisfies the relation $f^2 = u_1$ (Theorem 4.1.2(ii)). Hence, $f(h)^2 = h$, with $f(h) \in \mathcal{A}$. From the definition of $\sigma_{\mathcal{A}}(h)$ and the fact that the function $\widehat{h}$ is positive, we conclude that $f(h)$ has a real spectrum. Since, as we said before, $\mathcal{A}$ is hermitian, we obtain that $k \equiv f(h)$ is a self-adjoint element in $\mathcal{A}$; thus, $h = k^2$.

(ii) Since $\mathcal{A}[\tau]$ is symmetric (hence hermitian by Corollary 3.1.7), we have that $S \equiv \sigma(h) \subset \mathbb{R} \cup \{\infty\}$. We consider the functions

$$f^+(\lambda) := \begin{cases} 0, & \text{if } \lambda \leq 0 \\ \lambda, & \text{if } \lambda > 0 \end{cases} \quad \text{and} \quad f^-(\lambda) := \begin{cases} -\lambda, & \text{if } \lambda \leq 0 \\ 0, & \text{if } \lambda > 0 \end{cases}, \quad \lambda \in S \cap \mathbb{C}.$$

It is easily seen that the functions f^+, f^- have values in $\mathbb{R}_+$, belong to $\mathcal{C}_1(S)$ and moreover

$$f^+(\lambda) - f^-(\lambda) = \lambda, \ \lambda \in S \cap \mathbb{C}.$$

Take now $h^+ := f^+(h)$, $h^- := f^-(h)$ and $u_1 := f^+ - f^-$. Then, simple calculations show that h^+, h^- are positive elements of $\mathcal{A}$, such that $h^+ - h^- = h$ and $h^+h^- = 0$. This completes the proof of (ii). $\qquad \square$

4.2 Positive Linear Functionals

This section treats positive linear functionals on commutative GB^*-algebras. Among others it is proved that in such an algebra there are plenty of positive linear functionals to separate its points. For the shown results, commutativity plays an important role. The methods of constructing positive linear functionals in the case of C^*-algebras seems not to apply in the present situation, in the sense that new techniques must be employed in many cases.

It is well known that positive linear functionals play an essential role in the representation theory of Banach $*$-algebras and C^*-algebras, as well as that of topological $*$-algebras and pro-C^*-algebras (see, for instance, [7, 45, 52, 60, 85]).

We start with the definition of a positive linear functional and some of its basic properties, which can be found, for instance, in [51, 52, 60].

Definition 4.2.1 A linear functional f on a $*$-algebra $\mathcal{A}$ is said to be *self-adjoint* if $f(x^*) = \overline{f(x)}$, for all $x \in \mathcal{A}$, while f is called *positive* if $f(x^*x) \geq 0$, for all $x \in \mathcal{A}$.

The following two propositions give some standard properties of positive linear functionals. The behaviour of positive linear functionals depends on whether the involved $*$-algebra has an identity element or not. In this regard, we have

Proposition 4.2.2 *Let $\mathcal{A}$ be a $*$-algebra and f a positive linear functional on $\mathcal{A}$. Then,*

(i) $f(y^*x) = \overline{f(x^*y)}$, *for all x, $y \in \mathcal{A}$;*
(ii) (Cauchy–Schwarz inequality) $|f(y^*x)|^2 \leq f(y^*y)f(x^*x)$, *for all $x, y \in \mathcal{A}$;*

Suppose that A has an identity e. Then,

(i)′ f *is ∗-preserving, i.e.,* $f(x^*) = \overline{f(x)}$, *for all* $x \in A$;
(ii)′ $|f(x)|^2 \le f(e)f(x^*x)$, *for all* $x \in A$.

It is known that a positive linear functional f on a ∗-algebra A without identity is *extendable*; that is f *can be extended to a positive linear functional on the unitization* A_1 *of* A, *if and only if,* f *is ∗-preserving and there exists a constant* $\gamma > 0$, *such that* (see, for instance, [52, Proposition 21.7] and/or [115, p. 186, IV])

$$|f(x)|^2 \le \gamma\, f(x^*x), \ \forall\, x \in A.$$

For the continuity of positive linear functionals on locally convex ∗-algebras we have the following result (see, for instance, [60, Section 12 and, in particular, Theorem 15.5 and Theorem 11.5]).

Proposition 4.2.3

(i) *If A is a Banach algebra with identity and isometric involution, then f is continuous, such that* $\|f\| = f(e)$.
(ii) (Dixon) *Every positive linear functional on a Fréchet locally convex ∗-algebra with a uniformly bounded left approximate identity is continuous.*
(iii) *Every positive linear functional on a σ-C^*-algebra is continuous.*

Examples 4.2.4

(1) Let $A[\tau]$ be a commutative GB^*-algebra. Then, *any multiplicative linear functional f on A is a positive linear functional*. Indeed, let $x \in A$ be arbitrary and $y \equiv (e + x^*x)^{-1}$. Put $z \equiv xy$. By commutativity we have that $A_0 = A[B_0]$ (Lemma 3.3.7(ii)). Hence, $y \in A_0$ (Theorem 3.3.9(i)). Observe that also $z \in A_0$, since $y \in A_0$ and $A[\tau]$ has separately continuous multiplication. Furthermore, since A_0 is a commutative C^*-algebra, the restriction of f to A_0 is a positive linear functional, so that the equality $z^*z = x^*xy^2$ implies that $f(x^*x) = f(z^*z)/f(y^2) \ge 0$.
(2) Although a commutative C^*-algebra with an identity element always has non-trivial multiplicative linear functionals, in the case of a commutative GB^*-algebra (with an identity element), it may happen that there are no non-trivial multiplicative linear functionals. Indeed, consider the Arens algebra $L^\omega[0, 1]$, which as we have shown in Example 3.3.16(5) is a commutative GB^*-algebra. Suppose φ is a non-trivial multiplicative linear functional on $L^\omega[0, 1]$. Consider the subalgebra $C[0, 1]$ of all continuous $\mathbb{C}$-valued functions on $[0, 1]$. Then, the restriction of φ to $C[0, 1]$ is non-trivial, therefore it is a character of the C^*-algebra $C[0, 1]$. Hence, it coincides with a point evaluation δ_{t_0}, for some $t_0 \in [0, 1]$. That is, $\varphi(g) = g(t_0)$, for all $g \in C[0, 1]$. But, one clearly can find a continuous $\mathbb{C}^*$-valued function $h \in L^\omega[0, 1]$, such that $h(t_0) = \infty$ and $h(t) \ge 0$, for all $t \in [0, 1]$. Furthermore, one can construct a function $f_n \in C[0, 1]$, $n \in \mathbb{N}$, such that $f_n(t_0) = 1$ and $nf_n(t) \le h(t)$, $t \in [0, 1]$. Now, using the fact that φ, as a positive linear functional, preserves order, we

conclude that $\varphi(h) \geq n$, $n \in \mathbb{N}$, which evidently leads to a contradiction. The last example has the following effect: Regarding the functional representation of a GB^*-algebra $\mathcal{A}[\tau]$, one generally is obliged to allow the functions to take the value ∞ too, otherwise one can have a non-trivial multiplicative linear functional on $\mathcal{A}$. Consequently, some of the more elegant results obtained for m-convex algebras by Arens and Michael, for instance, [11, Theorem 3.1, Corollary 3.4], respectively [112, Theorem 10.1, Corollary 10.4], cannot be extended in the case of GB^*-algebras.

The theorem that follows implies that a commutative GB^*-algebra has a plethora of positive linear functionals (see Corollary 4.2.6). It is well known that any continuous self-adjoint linear functional on a C^*-algebra can be expressed as a difference of positive linear functionals on this algebra (see, for instance, [58, p. 365, Theorem VIII.1] and/or [122, Theorem 3.2.5]). An extension of this result to commutative GB^*-algebras is given by Theorem 4.2.5, below, where commutativity plays a key role. It is not known if this result also holds for noncommutative GB^*-algebras. The techniques applied for constructing positive linear functionals on an arbitrary C^*-algebra, seems not to work in the general case of GB^*-algebras.

Theorem 4.2.5 *If $\mathcal{A}[\tau]$ is a commutative GB^*-algebra, then for any continuous self-adjoint linear functional f on $\mathcal{A}$, there exist* (not necessarily continuous) *positive linear functionals f^+ and f^- on $\mathcal{A}$, such that $f = f^+ - f^-$.*

Proof Since $\mathcal{A}$ is commutative, $\mathcal{A}$ is algebraically $*$-isomorphic to $\widehat{\mathcal{A}} = \{\widehat{x} : x \in \mathcal{A}\}$, where $\widehat{x}$ stands for the Gelfand transform of $x \in \mathcal{A}$ (see Definition 3.4.7 and Theorem 3.4.9). Therefore, from here on, we may consider $\widehat{\mathcal{A}}$ instead of $\mathcal{A}$. If $h, k \in \mathcal{A}^+$ and $0 \leq \alpha \in \mathbb{C}$, then it is easily seen that $h + k \in \mathcal{A}^+$ and $\alpha h \in \mathcal{A}^+$.

Since f is self-adjoint, a function $g : \mathcal{A}^+ \to \mathbb{R}$ can be defined by

$$g(h) = \sup\left\{f(k) : 0 \leq k \leq h, \ k \in \mathcal{A}^+\right\}, \ h \in \mathcal{A}^+. \tag{4.2.3}$$

It is clear that $0 \leq g(h)$ and $f(h) \leq g(h)$, $h \in \mathcal{A}^+$.

We show that $g(h) < +\infty$. Since the restriction of τ to $\mathcal{A}[B_0]$ is weaker than $\|\cdot\|_{B_0}$ on $\mathcal{A}[B_0]$, and since $x \mapsto x(e + h^2)$ is $\tau - \tau$ continuous from $\mathcal{A}_0 = \mathcal{A}[B_0]$ into $\mathcal{A}[\tau]$, we obtain that this map is also $\|\cdot\|_{B_0} - \tau$ continuous. Hence,

$$\exists \ \varepsilon > 0 : \ \text{if} \ \|x\|_{B_0} < \varepsilon, \ \text{then} \ |f(x(e + h^2))| \leq 1. \tag{4.2.4}$$

If $0 \leq k \leq h$, then $k(e + h^2)^{-1} \in \mathcal{A}_0$ (see Lemma 3.3.7(ii), Theorem 3.3.9(i) recalling that the multiplication of $\mathcal{A}[\tau]$ is separately continuous) and

$$\|k(e + h^2)^{-1}\|_{B_0} \leq \|h(e + h^2)^{-1}\|_{B_0} \leq 1,$$

where the last inequality follows from [127, Lemma (4.8.13)]. Therefore, since $\|\frac{\varepsilon}{2}k(e + h^2)^{-1}\|_{B_0} < \varepsilon$, from (4.2.4) we conclude that

$$\left| f\left(\frac{\varepsilon}{2}k\right) \right| = \left| f\left(\frac{\varepsilon}{2}k(e + h^2)^{-1}(e + h^2)\right) \right| \leq 1.$$

This is true for every $k \in \mathcal{A}^+$, with $0 \leq k \leq h$. Consequently, $f(k) \leq \frac{2}{\varepsilon}$, for all $0 \leq k \leq h$, and thus $g(h) \leq \frac{2}{\varepsilon}$, implying that $g(h) < +\infty$.

It is clear that $g(\alpha k) = \alpha g(k)$, for all $\alpha \geq 0$ and $k \in \mathcal{A}^+$.

We now show that g is additive. Let $h = h_1 + h_2$, where h_1 and h_2 are in $\mathcal{A}^+$. It is easy to verify that $g(h_1) + g(h_2) \leq g(h)$. It remains to prove that $g(h) \leq g(h_1) + g(h_2)$. Suppose first that h is invertible, and $0 \leq k \leq h$. Then, since $\mathcal{A}$ is commutative, $0 \leq h^{-1}h_1 k \leq h_1$ and $0 \leq h^{-1}h_2 k \leq h_2$ (this follows easily from the identification of $\mathcal{A}$ with $\widehat{\mathcal{A}}$), and so

$$\begin{aligned}
f(k) = f(h^{-1}hk) &= f\left(h^{-1}(h_1 + h_2)k\right) \\
&= f(h^{-1}h_1 k) + f(h^{-1}h_2 k) \\
&\leq g(h_1) + g(h_2), \ \forall\, 0 \leq k \leq h.
\end{aligned}$$

Therefore, $g(h) \leq g(h_1) + g(h_2)$, so that $g(h) = g(h_1) + g(h_2)$ if h is invertible.

Suppose now that $h = h_1 + h_2$, with h_1 and h_2 in $\mathcal{A}^+$, is not invertible. Using again the functional representation of $\mathcal{A}[\tau]$ (Theorem 3.4.9) we conclude that the positive elements $h + 2e$, $h_1 + e$ and $h_2 + e$ of $\mathcal{A}$ (Definition 4.1.3) are invertible in $\mathcal{A}$ and their inverses belong to $\mathcal{A}_0$. Thus, from the previous paragraph we have that

$$\begin{aligned}
g(h) + g(2e) = g(h + 2e) &= g(h_1 + e) + g(h_2 + e) \\
&= g(h_1) + g(h_2) + 2g(e) \\
&= g(h_1) + g(h_2) + g(2e).
\end{aligned}$$

Hence, $g(h) = g(h_1) + g(h_2)$ and this holds for all $h \in \mathcal{A}^+$.

Any $h \in H(\mathcal{A})$ can be expressed as a difference of positive elements, i.e. $h = h^+ - h^-$, where h^+ and h^- are in $\mathcal{A}^+$ (Theorem 4.1.4). Define

$$g_1 : H(\mathcal{A}) \to \mathbb{R} : g_1(h) = g(h^+) - g(h^-), \ \forall\, h \in H(\mathcal{A}).$$

We show that g_1 is well defined. Suppose that

$$h = h^+ - h^- = h_1^+ - h_1^-. \ \text{Then,} \ h^+ + h_1^- = h_1^+ + h^-.$$

Since g is additive,

$$g(h^+ + h_1^-) = g(h^+) + g(h_1^-) \text{ and } g(h_1^+ + h^-) = g(h_1^+) + g(h^-).$$

Furthermore, since

$$g(h^+ + h_1^-) = g(h_1^+ + h^-),$$

it follows that

$$g(h^+) + g(h_1^-) = g(h_1^+) + g(h^-).$$

Therefore,

$$g(h^+) - g(h^-) = g(h_1^+) - g(h_1^-),$$

which proves that g_1 is well defined.

The map g_1 is clearly a real valued linear functional, which extends g. If $x \in A$, then $x = h + ik$, where $h, k \in H(A)$. We define now a function

$$f^+ : A \to \mathbb{C} : f^+(x) = g_1(h) + ig_1(k).$$

Clearly, f^+ is a linear functional on A extending g_1. Moreover, f^+ and $f^- := f^+ - f$ are positive linear functionals by using (4.2.3) and properties of $g(h)$. This completes the proof. $\qquad\square$

For a noncommutative version of Theorem 4.2.5, see Corollary 6.3.16.

Corollary 4.2.6 *Let $A[\tau]$ be a commutative GB*-algebra. Then, there are sufficient positive linear functionals on A to separate its points.*

Proof $A[\tau]$ is a locally convex space, therefore by Hahn–Banach theorem, there are enough continuous linear functionals on $A[\tau]$ to separate its points. On the other hand, every continuous linear functional on $A[\tau]$ is a linear combination of self-adjoint continuous linear functionals on $A[\tau]$. Hence, the result follows by the preceding Theorem 4.2.5. $\qquad\square$

Corollary 4.2.7 *Let $A[\tau]$ be a commutative GB*-algebra. Then, every continuous linear functional f on A, is a linear combination of* (not necessarily continuous) *positive linear functionals.*

Proof Every continuous linear functional is a linear combination of self-adjoint continuous linear functionals. Hence, the result follows from Theorem 4.2.5. $\qquad\square$

Remark 4.2.8 According to the comments before Theorem 4.2.5, there is no analogue of Corollary 4.2.6 in the noncommutative case. Nevertheless, we may say that a partial answer to this problem is given by Theorem 6.3.4 (Dixon), that plays an essential role in the proof of the noncommutative Gelfand–Naimark type theorem for GB^*-algebras (see Theorem 6.3.5).

In this regard, see also [155, Proposition 3.16], where Theorem 4.2.5 is proven for a particular case of a non necessarily commutative Fréchet GB^*-algebra.

Lemma 4.2.9 *Let $A[\tau]$ be a commutative GB^*-algebra and $h \in A^+$. Then, there is an increasing sequence $(h_n)_{n \in \mathbb{N}}$ of positive elements in $A^+ \cap A_0$, such that for every positive linear functional f on A, one has $f(h_n) \to f(h)$, $n \in \mathbb{N}$.*

Proof Let $h \in A^+$. The following functions are then defined in $C(\mathfrak{M}_0)$:

$$\psi_n(\varphi) := \begin{cases} \widehat{h}(\varphi), & \text{when } \widehat{h}(\varphi) \leq n \\ n, & \text{when } \widehat{h}(\varphi) > n \end{cases}, \quad n \in \mathbb{N}, \ \varphi \in \mathfrak{M}_0.$$

It is then clear that

$$\psi_n(\varphi) \leq \psi_{n+1}(\varphi) \leq \widehat{h}(\varphi), \ \forall \varphi \in \mathfrak{M}_0, \ n \in \mathbb{N}.$$

In particular, each one of ψ_n is continuous, such that $\psi_n(\varphi) \in [0, n]$, $\varphi \in \mathfrak{M}_0$, $n \in \mathbb{N}$. Now, from Definition 4.1.3 and relation (3.4.16), it follows that ψ_n is the Gelfand transform of a unique element h_n in $A^+ \cap A_0$, $n \in \mathbb{N}$. From the preceding, it is evident that

$$0 \leq h_1 \leq h_2 \leq \cdots \leq h_n \leq \cdots \leq h.$$

Thus, considering a positive linear functional f on A, we obtain that

$$0 \leq f(h_1) \leq f(h_2) \leq \cdots \leq f(h_n) \leq \cdots \leq f(h).$$

Hence, $\sup_n f(h_n) < +\infty$, so that setting $g(h) := \sup_n f(h_n)$, we have

$$0 \leq g(h) \leq f(h). \tag{4.2.5}$$

We shall show that

$$g(h) = \sup \{ f(u) : u \in A^+ \cap A_0, \ 0 \leq u \leq h \} < \infty. \tag{4.2.6}$$

Indeed, let $\alpha \equiv \sup\{ f(u) : u \in A^+ \cap A_0, \ 0 \leq u \leq h \}$. Then, $\alpha < +\infty$ and $\alpha \geq g(h)$. On the other hand, since $u \in A^+ \cap A_0$ with $0 \leq u \leq h$, there will be $m \in \mathbb{N}$, such that

$$u \leq h_n, \ \text{ with } n \geq m.$$

This implies that $\alpha \le g(h)$, so that (4.2.6) is proved. It is then clear that

$$g(h) = f(h), \text{ when } h \in \mathcal{A}^+ \cap \mathcal{A}_0. \tag{4.2.7}$$

Now arguing as in the proof of Theorem 4.2.5, it can be shown that g is additive and positively homogeneous on $\mathcal{A}^+$, so that it is extended to a real linear functional on the real vector space $H(\mathcal{A})$. Because of (3.1.1), we further obtain an extension of the latter to a positive (see (4.2.5)) linear functional on $\mathcal{A}$, which we also denote by g.

Thus, defining

$$f_0(x) := f(x) - g(x), \ x \in \mathcal{A},$$

we obtain a positive linear functional f_0 on $\mathcal{A}$, as easily follows from (3.1.1) and (4.2.5), while f_0 vanishes on $\mathcal{A}^+ \cap \mathcal{A}_0$ by (4.2.7). It is also evident that $f_0(e) = 0$, so by the Cauchy–Schwarz inequality (Proposition 4.2.2(ii)′), we obtain that $f_0(x) = 0$, for all $x \in \mathcal{A}$, therefore

$$f(x) = g(x), \ \forall \, x \in \mathcal{A}.$$

It it clear now that

$$f(h) = g(h) = \sup_n f(h_n) = \lim_n f(h_n), \ h \in \mathcal{A}^+;$$

this completes the proof.   □

Corollary 4.2.10 *Let $\mathcal{A}[\tau]$ be a commutative GB^*-algebra and $x \in \mathcal{A}$. Then, there is a sequence $(x_n)_{n \in \mathbb{N}}$ in $\mathcal{A}_0$, such that $f(x_n) \to f(x)$, for every positive linear functional f on $\mathcal{A}$.*

Proof Let $x \in \mathcal{A}$. Then, $x = u + iv$ with $u, \, v \in H(\mathcal{A})$. Applying Theorem 4.1.4(ii), we have that $u = u^+ - u^-$ and $v = v^+ - v^-$, with u^+, u^- and v^+, v^- elements from $\mathcal{A}^+$. Then, Lemma 4.2.9 provides us with increasing sequences $(u_n^\iota)_{n \in \mathbb{N}}$ and $(v_n^\iota)_{n \in \mathbb{N}}$, $\iota = +, -$ from $\mathcal{A}^+ \cap \mathcal{A}_0$, such that $f(u_n^\iota) \to f(u^\iota)$ and $f(v_n^\iota) \to f(v^\iota)$, $\iota = +, -$, for any positive linear functional f on $\mathcal{A}$. The assertion now follows by setting $x_n := u_n^+ - u_n^- + i(v_n^+ - v_n^-)$, $n \in \mathbb{N}$.   □

Theorem 4.2.11 (Bhatt) *Let $\mathcal{A}[\tau]$ be a GB^*-algebra. Then, $\mathcal{A}[B_0]$ is sequentially dense in $\mathcal{A}[\tau]$.*

Proof Let $h \in H(\mathcal{A})$ and $\mathcal{C}$ be the closed $*$-subalgebra of $\mathcal{A}[\tau]$ generated by h and the identity element $e \in \mathcal{A}$. Then, $\mathcal{C}$ is a commutative GB^*-algebra (Proposition 3.3.19). Since h is an arbitrary self-adjoint element of $\mathcal{A}$, we may suppose, without loss of generality, that $\mathcal{A}[\tau]$ is commutative. Now, Theorem 3.3.9 and Lemma 3.4.2(i) yield that $\mathcal{A}[B_0]$ is a C^*-algebra, such that all three elements e, $(e + h^2)^{-1}$ and $h(e + h^2)^{-1}$ belong to it. Further, through the Gelfand representation,

$A[B_0]$ is isometrically $*$-isomorphic to the function C^*-algebra $\mathcal{C}(\mathfrak{M}_0)$, $\mathfrak{M}_0$ being the Gelfand space of $A[B_0]$. Thus, we conclude that $\|h(e + h^2)^{-1}\|_{B_0} \leq 1$, which implies that $h(e + h^2)^{-1} \in B_0$. Since, for every positive real number β, $\beta^{1/2}h \in H(\mathcal{A})$, we therefore have that

$$\beta^{1/2}h(e + \beta h^2)^{-1} = \beta^{1/2}h\big(e + (\beta^{1/2}h)^2\big)^{-1} \in B_0. \tag{4.2.8}$$

Moreover,

$$h - h(e + \beta h^2)^{-1} = h\big(e - (e + \beta h^2)^{-1}\big) = \beta h^3(e + \beta h^2)^{-1}. \tag{4.2.9}$$

Now, the multiplication of $A[\tau]$ is separately continuous; hence, for each 0-neighbourhood V in $\mathcal{A}$, there is a 0-neighbourhood U in $\mathcal{A}$, such that $h^2 U \subset V$. Furthermore, since B_0 is a bounded set, there is a positive number α, such that $\alpha^{1/2}B_0 \subset U$. Taking also into account (4.2.8) and (4.2.9) with $\beta = \alpha$, we conclude that for sufficiently small α,

$$h - h(e + \alpha h^2)^{-1} = \alpha^{1/2}h\big(e + (\alpha^{1/2}h)^2\big)^{-1}\alpha^{1/2}h^2 \in \alpha^{1/2}h^2 B_0 \in h^2 U \subset V,$$

$$\text{therefore } h = \lim_{\alpha \to 0^+} h(e + \alpha h^2)^{-1}. \tag{4.2.10}$$

Observe now that

$$h(e + \alpha h^2)^{-1} = \frac{1}{\alpha^{1/2}}(\alpha^{1/2}h)\big(e + (\alpha^{1/2}h)^2\big)^{-1} \in H(\mathcal{A}[B_0]), \; \alpha > 0,$$

which, by virtue of (4.2.10) and (3.1.1), shows that $A[B_0]$ is dense in $A[\tau]$, taking $\alpha = 1/n$, $n \in \mathbb{N}$. This completes the proof. $\qquad\qquad\square$

From Theorem 4.2.11 and the fact that $A[B_0] \subset A_0$, we have the following

Corollary 4.2.12 (Allan) *Let $A[\tau]$ be a GB*-algebra. Then, A_0 is dense in $A[\tau]$.*

For the proof of Allan for the previous Corollary 4.2.12, see [5, (4.7) Theorem].

Furthermore, from Theorem 4.2.11, Examples 3.3.16(1) and Remark 3.5.7(3) (as an alternative to the latter, see Proposition 3.5.8), we obtain the following (Apostol [8]; see also [60, Theorem 10.23])

Corollary 4.2.13 *Let $A[\tau]$ be a pro-C*-algebra. Then, the C*-algebra $A_b[\| \cdot \|_b]$ (bounded part of $A[\tau]$) is dense in $A[\tau]$.*

Notes Proposition 4.2.3(ii) has been proved by P.G. Dixon in [51] (see also [60, p. 191, Theorem 15.5]). Statement (iii) of the same proposition is an immediate consequence of (ii), since every σ-C^*-algebra is a Fréchet pro-C^*-algebra and as a pro-C^*-algebra has a bounded approximate identity (Inoue; [75, Theorem 2.6], see also [60, p. 137, Theorem 11.5]). Theorem 4.2.11 is due to S.J. Bhatt (cf. [17, (2) Theorem]).

4.3 Equivalent Topologies

The Gelfand–Naimark type theorem for a commutative GB^*-algebra (see Theorem 3.4.9) is clearly a purely algebraic result, in contrast to the classical Gelfand–Naimark theorem for a commutative C^*-algebra (with identity), where the corresponding Gelfand map is an isometric $*$-isomorphism. This is rather due to the fact that the C^*-property is directly related to the given normed topology of a given C^*-algebra, while the GB^*-condition on a locally convex $*$-algebra $\mathcal{A}[\tau]$ is related to a whole collection (namely, the $\mathfrak{B}^*$-collection) of certain bounded subsets of $\mathcal{A}[\tau]$, where the greatest member B_0 of this collection is mostly used. In this way, one obtains the C^*-algebra $\mathcal{A}[B_0]$, through which the study of the GB^*-structure is realized, as it also happens with the Gelfand–Naimark type theorem for a commutative GB^*-algebra.

Concerning C^*-algebras, we know that they *have a unique C^*-norm*. So a natural problem is the following:

Question Do we have a similar situation with a GB^*-topology on a GB^*-algebra? The answer will be investigated in this section. We shall see that if a commutative $*$-algebra with identity is a GB^*-algebra under two distinct topologies τ, τ', then these two topologies have to be connected under a certain equivalence relation (see, for example, Corollary 4.3.10).

Definition 4.3.1 (Allan) Let $\mathcal{A}$ be an algebra, which is a locally convex algebra under two topologies, say τ, τ'. Consider the corresponding collections $(\mathfrak{B}_0)_1$, $(\mathfrak{B}_0)_2$ for $\mathcal{A}[\tau]$ and $\mathcal{A}[\tau']$ respectively, as in Definition 2.2.2. We shall say that τ and τ' are *equivalent* if the collections $(\mathfrak{B}_0)_1$, $(\mathfrak{B}_0)_2$ fulfil the following properties:

(i) *for every $B_1 \in (\mathfrak{B}_0)_1$ there is some $B_2 \in (\mathfrak{B}_0)_2$, such that $B_1 \subseteq B_2$;*
(ii) *the same is true by interchanging $(\mathfrak{B}_0)_1$, $(\mathfrak{B}_0)_2$.*

In this regard, we have Remark 4.3.2 and Proposition 4.3.3.

Remark 4.3.2

(1) The relation introduced in the preceding definition is clearly an "equivalence relation", that we shall denote by "$\sim$". For two equivalent topologies τ and τ' on $\mathcal{A}$ as before, we shall write $\tau \sim \tau'$.
(2) Let us start in Definition 4.3.1 with *a commutative $*$-algebra $\mathcal{A}$* equipped with two topologies τ, τ', under which it becomes *a pseudo-complete locally convex $*$-algebra*. Let $B_0(\tau)$, $B_0(\tau')$ be *the greatest members* in $\mathfrak{B}^*_{\mathcal{A}[\tau]}$ and $\mathfrak{B}^*_{\mathcal{A}[\tau']}$, respectively. Then, it is easily seen that $\tau \sim \tau'$, if and only if, $B_0(\tau) = B_0(\tau')$.

Proposition 4.3.3 *Let $\mathcal{A}$ be a commutative $*$-algebra with an identity. Suppose that $\mathcal{A}$ is also a locally convex algebra under two topologies τ, τ'. Then, the following hold:*

(i) *suppose that $\mathcal{A}[\tau]$ is a GB^*-algebra and B_0 the greatest member in $\mathfrak{B}^*_{\tau}(\equiv \mathfrak{B}^*_{\mathcal{A}[\tau]})$. Then $\tau \sim \tau'$, if and only if, B_0 is also the greatest member in $\mathfrak{B}^*_{\tau'}(\equiv \mathfrak{B}^*_{\mathcal{A}[\tau']})$;*

(ii) *if the involution ∗ is τ'-continuous, $\mathcal{A}[\tau]$ a GB^*-algebra and $\tau \sim \tau'$, then $\mathcal{A}[\tau']$ is also a GB^*-algebra and both topologies determine the same C^*-algebra, which is the commutative C^*-algebra $\mathcal{A}_0$.*

Proof We only sketch the proof of pseudo-completeness of $\mathcal{A}[\tau']$ in (ii); everything else follows easily from the very definitions (see also Lemma 3.3.7(ii)). So, let B' be an arbitrary element in $\mathfrak{B}^*_{\tau'}$. We shall prove that the normed ∗-algebra $\mathcal{A}[B']$ is complete. Since $\tau \sim \tau'$, we have that for the chosen $B' \in \mathfrak{B}^*_{\tau'}$, there is $B \in \mathfrak{B}^*_{\tau}$, such that $B' \subseteq B$, therefore $\mathcal{A}[B'] \subseteq \mathcal{A}[B]$ and $\|\cdot\|_B \leq \|\cdot\|_{B'}$. Hence, if $(x_n)_{n \in \mathbb{N}}$ is a Cauchy sequence in $\mathcal{A}[B']$, it will also be so in $\mathcal{A}[B]$, which is a Banach algebra. So there is $x \in \mathcal{A}[B]$, such that $\|x_n - x\|_B \to 0$. Following now similar arguments to those in the proof of Proposition 2.2.6, we obtain the completeness of $\mathcal{A}[B']$. □

We shall finish this section by describing those topologies on $\mathcal{A}$ that are equivalent to the given topology τ (see Corollary 4.3.10). More precisely, it will be proved that any two topologies τ, τ', that make a commutative ∗-algebra with identity a GB^*-algebra, have to be equivalent in the sense of Definition 4.3.1.

▶ If $\mathcal{A}$ is an algebra, *denote by $\mathcal{A}^*$ the algebraic dual of $\mathcal{A}$. Let $\mathcal{P}(\mathcal{A})$ be the set of all positive linear functionals of $\mathcal{A}$ and $\mathcal{A}^{pf}$ be the linear subspace of $\mathcal{A}^*$ generated by $\mathcal{P}(\mathcal{A})$. If $\mathcal{A}[\tau]$ is a locally convex algebra, let $\mathcal{A}'$ be the* (topological) *dual space of $\mathcal{A}[\tau]$; i.e., $\mathcal{A}' := \{f \in \mathcal{A}^* : f \text{ is continuous}\}$.* The following is an immediate consequence of Corollary 4.2.7.

Remark 4.3.4 Let $\mathcal{A}[\tau]$ be a commutative GB^*-algebra. Then, $\mathcal{A}' \subset \mathcal{A}^{pf}$.

Now given an algebra $\mathcal{A}$, take the pair $(\mathcal{A}, \mathcal{A}^{pf})$ with the bilinear form

$$(\mathcal{A}, \mathcal{A}^{pf}) \ni (x, f) \mapsto f(x) \in \mathbb{C}.$$

Then, according to [74, p. 183, §2, Definition 1], we have a *dual system with respect to the preceding bilinear form*, in the sense that $\mathcal{A}$ separates points of $\mathcal{A}^{pf}$ and $\mathcal{A}^{pf}$ separates points of $\mathcal{A}$; for the latter, in the case when $\mathcal{A}[\tau]$ is a commutative GB^*-algebra, see Corollary 4.2.6. Thus, we may consider the weak topology $\sigma(\mathcal{A}, \mathcal{A}^{pf})$ on $\mathcal{A}$, which is a (Hausdorff) locally convex space topology defined by the family of seminorms $\{q_f\}_{f \in \mathcal{A}^{pf}}$, given as follows (see [74, p. 185])

$$q_f(x) := |f(x)|, \; \forall \, x \in \mathcal{A}. \tag{4.3.11}$$

It is then clear that *each $f \in \mathcal{A}^{pf}$ is continuous with respect to the weak topology $\sigma(\mathcal{A}, \mathcal{A}^{pf})$ on $\mathcal{A}$*. In this aspect, we set the following

Definition 4.3.5 (Allan) If $\tau^{pf} \equiv \sigma(\mathcal{A}, \mathcal{A}^{pf}) \equiv \sigma(\mathcal{A}, \mathcal{P}(\mathcal{A}))$, then the weak topology τ^{pf} on $\mathcal{A}$ is called *positive-functional topology* on $\mathcal{A}$.

From the definition of τ^{pf}, Remark 4.3.4 and the comments just before Definition 4.3.5 we are led to the result that follows.

Proposition 4.3.6 *Let* $\mathcal{A}[\tau]$ *be a commutative GB*-algebra. The following hold:*

(i) *the dual of the locally convex space* $\mathcal{A}[\tau^{pf}]$ *is* $\mathcal{A}^{pf}$;
(ii) *the weak topology* $\sigma(\mathcal{A}, \mathcal{A}')$ *on* $\mathcal{A}$ *is coarser than the positive-functional topology* τ^{pf} *on* $\mathcal{A}$; *i.e.,* $\sigma(\mathcal{A}, \mathcal{A}') \prec \tau^{pf}$.

Theorem 4.3.9 below shows that $\tau \sim \tau^{pf}$. The proof is based on a series of Lemmas that are first proved.

Lemma 4.3.7 *Let* $\mathcal{A}[\tau]$ *and* $\mathcal{A}[\tau^{pf}]$ *be as before. Then,* $\mathcal{A}[\tau^{pf}]$ *is a locally convex* **-algebra.*

Proof In the comments after Remark 4.3.4, we noticed that the algebra $\mathcal{A}[\tau^{pf}]$ is a locally convex space. It remains to show that it also has a separately continuous multiplication and continuous involution. Let $h \in \mathcal{A}^+$ and $f \in \mathcal{P}(\mathcal{A})$. Define the functional

$$f_h(x) := f(hx), \ \forall\, x \in \mathcal{A}.$$

Then, f_h is a positive linear functional on $\mathcal{A}$ (see Theorem 4.1.4(i) too). Now it easily follows (see (4.3.11)) that the map $\mathcal{A}[\tau^{pf}] \to \mathcal{A}[\tau^{pf}] : x \mapsto hx$ is continuous and since any arbitrary $y \in \mathcal{A}$ is a linear combination of positive elements (see (3.1.1) and Theorem 4.1.4(ii)), we conclude that the map $\mathcal{A}[\tau^{pf}] \to \mathcal{A}[\tau^{pf}] : x \mapsto yx$, $y \in \mathcal{A}$, is continuous. This means that the multiplication of $\mathcal{A}[\tau^{pf}]$ is separately continuous. The continuity of involution follows from the definition of τ^{pf} and the fact that every $f \in \mathcal{P}(\mathcal{A})$ is *-preserving (Proposition 4.2.2(i)'). Thus, the proof is complete.   □

Lemma 4.3.8 *Let* $\mathcal{A}[\tau]$, $\mathcal{A}[\tau^{pf}]$ *be as above and*

$$B_1 := \{x \in \mathcal{A} : f(x^*x) \le 1, \ f \in \mathcal{P}(\mathcal{A}) \ \text{with} \ f(e) \le 1\}.$$

Then, B_1 *has the following properties:*

(1) *it is* τ- *and* τ^{pf}-*bounded;*
(2) *it is absolutely convex;*
(3) $B_1^2 \subseteq B_1$; $B_1^* = B_1$; *and* $e \in B_1$.

Proof

(1) Let $f \in \mathcal{P}(\mathcal{A})$, such that $f(e) \le 1$. Then, by the Cauchy–Schwarz inequality (see Proposition 4.2.2(ii)'), we have that

$$|f(x)|^2 \le f(e)f(x^*x) \le 1, \ \forall\, x \in B_1.$$

Hence, B_1 is τ^{pf}-bounded and since $\mathcal{A}' \subseteq \mathcal{A}^{pf}$ (Remark 4.3.4), it follows that B_1 is $\sigma(\mathcal{A}, \mathcal{A}')$-bounded, consequently τ-bounded by [74, p. 209, Theorem 3]. So (1) is shown.

(2) Let $\lambda \in \mathbb{C}$ with $|\lambda| \leq 1$. Then, clearly $\lambda B_1 \subseteq B_1$, so that the proof will be complete if we still show that B_1 is convex. So let $\lambda \in [0, 1]$ and $x, y \in B_1$. Put $z := \lambda x + (1 - \lambda)y$ and take $f \in \mathcal{P}(\mathcal{A})$ with $f(e) \leq 1$. Applying the Cauchy–Schwarz inequality (Proposition 4.2.2(ii)), we obtain that

$$f(z^*z) \leq \lambda^2 f(x^*x) + (1 - \lambda)^2 f(y^*y) + 2\lambda(1 - \lambda)f(x^*x)^{1/2}f(y^*y)^{1/2}$$
$$= \left(\lambda f(x^*x)^{1/2} + (1 - \lambda)f(y^*y)^{1/2}\right)^2 \leq 1.$$

The last inequality occurs, since $x, y \in B_1$. Hence, $z \in B_1$ and B_1 is convex.

(3) It is easily seen that $e \in B_1$ and that $B_1^* = B_1$. So it remains to be proved that $B_1^2 \subseteq B_1$. Consider arbitrary elements $x, y \in B_1$ and put $z := xy$. For $f \in \mathcal{P}(\mathcal{A})$ with $f(e) \leq 1$, define the functional

$$f_y(w) := f(y^*wy), \ \forall\, w \in \mathcal{A}.$$

It is evident that f_y is a positive linear functional on $\mathcal{A}$ and either $f_y = 0$, where in this case one has

$$f(z^*z) = f_y(x^*x) = 0 < 1,$$

or $f_y \neq 0$, that is $f_y(e) \neq 0$, otherwise from the Cauchy–Schwarz inequality we are led to a contradiction. Now, since $y \in B_1$, we clearly have $f_y(e) = f(y^*y) \leq 1$ and since also $x \in B_1$, we conclude that

$$f(z^*z) = f_y(x^*x) \leq 1, \quad \text{which implies } z \in B_1.$$

Thus, $B_1^2 \subseteq B_1$.

$\square$

Theorem 4.3.9 *Let $\mathcal{A}[\tau]$, $\mathcal{A}[\tau^{pf}]$ be as above. Then, $\tau \sim \tau^{pf}$ and the maximal member B_0 in $\mathfrak{B}^*_{\mathcal{A}[\tau]}$ takes the form*

$$B_0 = \left\{x \in \mathcal{A} : f(x^*x) \leq 1, \ f \in \mathcal{P}(\mathcal{A}) \text{ with } f(e) \leq 1\right\}.$$

Proof We first show that B_0 equals to B_1 of Lemma 4.3.5. It is obvious from Lemma 4.3.8 that the τ-closure of B_1 belongs to $\mathfrak{B}^*_{\mathcal{A}[\tau]}$. Hence, $B_1 \subseteq B_0$.

Now, note that $\mathcal{A}_0 = \mathcal{A}[\|\cdot\|_{B_0}]$ is a C^*-algebra and its closed unit ball is B_0. Thus, taking an element $f \in \mathcal{P}(\mathcal{A})$ with $f(e) \leq 1$, then (by Proposition 4.2.3(i))

f restricted to $\mathcal{A}_0$ is a continuous positive linear functional, such that $\|f\| = f(e) \leq 1$. Moreover, $x \in B_0$ implies $x^*x \in B_0$, so that

$$f(x^*x) \leq f(e)\|x^*x\|_{B_0} \leq 1.$$

Thus, $x \in B_1$ and finally $B_1 = B_0$.

By Lemma 4.3.8 $B_0 \in \mathfrak{B}^*_{\mathcal{A}[\tau^{pf}]}$. We show now that B_0 is the greatest member of $\mathfrak{B}^*_{\mathcal{A}[\tau^{pf}]}$. So let B' be an arbitrary element in $\mathfrak{B}^*_{\mathcal{A}[\tau^{pf}]}$. Then, B' is τ^{pf}-bounded, therefore $\sigma(\mathcal{A}, \mathcal{A}')$-bounded by Proposition 4.3.6(ii), which by [128, p. 67, Theorem 1] yields that B' is τ-bounded. It follows that the τ-closure of B' belongs to $\mathfrak{B}^*_{\mathcal{A}[\tau]}$, consequently $B' \subseteq B_0$. Repeating the same argument for the τ^{pf}-closure of B_0 in place of B', we conclude that B_0 is τ^{pf}-closed. Thus, B_0 is the greatest member in $\mathfrak{B}^*_{\mathcal{A}[\tau^{pf}]}$. Proposition 4.3.3(i) yields now that $\tau \sim \tau^{pf}$; this completes the proof. □

Corollary 4.3.10 *Let $\mathcal{A}$ be a commutative $*$-algebra with identity, which is a GB^*-algebra under two topologies τ and τ'. Then, $\tau \sim \tau'$.*

Proof Theorem 4.3.9 implies that $\tau \sim \tau^{pf}$ and $\tau' \sim \tau^{pf}$, therefore $\tau \sim \tau'$. □

For an extension of Corollary 4.3.10 to the noncommutative case, see Corollary 6.3.7 (cf. also the discussion after it). Further, it will be shown that there is a finest locally convex $*$-algebra topology on a given commutative GB^*-algebra $\mathcal{A}[\tau]$ that is equivalent to the initial topology τ. So, taking into account Corollary 4.3.10, we conclude that (see Theorem 4.3.13) *there exists a finest topology, under which $\mathcal{A}$ is a GB^*-algebra*. To prove the announced result, we need definition of the so-called Mackey topology that we first exhibit.

Let $E[\tau]$ be a locally convex space and E' its dual. Denote by $\mathcal{K}$ the family of all subsets K of E' that are absolutely convex and $\sigma(E', E)$-compact. Then, we have (see [74, p. 206, Definition 1]) the following

Definition 4.3.11 The topology of uniform convergence on the members of the family $\mathcal{K}$ is denoted by $\tau(E, E')$ and is called *the Mackey topology*. Namely, $\tau(E, E')$ is the $\mathcal{K}$-topology on E corresponding to the dual system (E, E'), with respect to the bilinear form $\langle x, f \rangle := f(x), x \in E, f \in E'$.

The Mackey topology is a locally convex topology induced by the seminorms q_K, $K \in \mathcal{K}$, given by (cf. [74, p. 195])

$$q_K(x) := \sup\{|\langle x, f \rangle| : f \in K\}, \ \forall \, x \in A. \tag{4.3.12}$$

One important property of the Mackey topology $\tau(E, E')$ reads as follows: *If τ' is a locally convex topology on E, the dual of $E[\tau']$ equals to E', if and only if,*

$$\sigma(E, E') \prec \tau' \prec \tau(E, E'); \tag{4.3.13}$$

for this and further results on the Mackey topology, see [74, p. 206, Proposition 4 and p. 203, §5].

Lemma 4.3.12 *Let $\mathcal{A}[\tau]$ be a locally convex algebra with dual $\mathcal{A}'$. Let $\tau_{\mathcal{K}} := \tau(\mathcal{A}, \mathcal{A}')$. Then, $\mathcal{A}[\tau_{\mathcal{K}}]$ is a locally convex algebra and $\tau_{\mathcal{K}} \sim \tau$.*

Proof We only show that the multiplication in $\mathcal{A}[\tau_{\mathcal{K}}]$ is separately continuous. Everything else holds since $\mathcal{A}[\tau_{\mathcal{K}}]$ is a topological vector space (cf. [74, p. 195, §4]). Thus, let us take a fixed point $y \in \mathcal{A}$, and prove that the map $\mathcal{A}[\tau_{\mathcal{K}}] \to \mathcal{A}[\tau_{\mathcal{K}}] : x \mapsto xy$ is continuous. Consider the map

$$T_y : \mathcal{A}' \to \mathcal{A}' : f \mapsto T_y(f) := f_y, \text{ with } f_y(x) := f(xy), \ x \in \mathcal{A}.$$

Clearly T_y is a well defined $\sigma(\mathcal{A}', \mathcal{A}) - \sigma(\mathcal{A}', \mathcal{A})$ continuous linear operator on $\mathcal{A}'$. Hence, $T_y(K) \in \mathcal{K}$, for every $K \in \mathcal{K}$. On the other hand, if K° is the polar of $K \in \mathcal{K}$, i.e.,

$$K^\circ := \left\{ x \in \mathcal{A} : |f(x)| \le 1, \forall f \in K \right\},$$

then by the very definitions it is easily seen that

$$\forall x \in \mathcal{A}, \ xy \in K^\circ \Leftrightarrow x \in T_y(K)^\circ.$$

For a fixed $y \in \mathcal{A}$, this proves the continuity of the map $\mathcal{A}[\tau_{\mathcal{K}}] \to \mathcal{A}[\tau_{\mathcal{K}}] : x \mapsto xy$, since finite intersections of the sets $\lambda K^\circ, \lambda > 0, \ K \in \mathcal{K}$, form a fundamental system of neighbourhoods of 0 for the topology $\tau_{\mathcal{K}}$ (cf. [74, p. 195, §4]). Thus, the multiplication in $\mathcal{A}[\tau_{\mathcal{K}}]$ is separately continuous.

Taking now an absolutely convex set in $\mathcal{A}$, this is τ-bounded and τ-closed, if and only if, it is $\tau_{\mathcal{K}}$-bounded and $\tau_{\mathcal{K}}$-closed (see [35, Chap. IV, §2, Proposition 4, Corollaire 1 and Théorème 3]). Consequently, the corresponding collections $\mathfrak{B}_{\mathcal{A}[\tau]}$ and $\mathfrak{B}_{\mathcal{A}[\tau_{\mathcal{K}}]}$ are the same, so that $\tau \sim \tau_{\mathcal{K}}$ from Definition 4.3.1. This completes the proof. $\square$

For the notation applied to the next theorem, see discussion before Definition 4.3.5.

Theorem 4.3.13 *Let $\mathcal{A}[\tau]$ be a commutative GB*-algebra. Then, the topology $\tau^\diamond := \tau(\mathcal{A}, \mathcal{A}^{pf})$ is the finest locally convex *-algebra topology on $\mathcal{A}$ that is equivalent to τ. In particular, if τ^*_{lc} is any locally convex *-algebra topology on $\mathcal{A}$, such that all the positive linear functionals on $\mathcal{A}[\tau^*_{lc}]$ are continuous, then*

$$\tau^*_{lc} \sim \tau, \text{ if and only if, } \tau^{pf} \prec \tau^*_{lc} \prec \tau^\diamond.$$

Proof By Theorem 4.3.9 $\tau \sim \tau^{pf}$, therefore we may think of $\mathcal{A}[\tau]$ as being $\mathcal{A}[\tau^{pf}]$, whose dual space is $\mathcal{A}^{pf}$. Hence, from Lemma 4.3.12, we have that $\mathcal{A}[\tau^\diamond]$ is a locally convex algebra and $\tau^\diamond \sim \tau^{pf}$. Thus, $\tau^\diamond \sim \tau$. We must also show that $\mathcal{A}[\tau^\diamond]$ has a continuous involution.

The involution $*$ on $\mathcal{A}$ gives rise to a well-defined, continuous, conjugate linear map

$$\mathcal{A}^{pf}[\sigma(\mathcal{A}^{pf}, \mathcal{A})] \to \mathcal{A}^{pf}[\sigma(\mathcal{A}^{pf}, \mathcal{A})] : f \mapsto f^*, \text{ with } \langle x, f^* \rangle := \overline{\langle x^*, f \rangle}, \ x \in \mathcal{A}, \ f \in \mathcal{A}^{pf}.$$

If K is an absolutely convex, $\sigma(\mathcal{A}^{pf}, \mathcal{A})$-compact subset of $\mathcal{A}^{pf}$, its image $K^* := \{f^* : f \in K\}$ under the preceding (dual) map is also an absolutely convex, $\sigma(\mathcal{A}^{pf}, \mathcal{A})$-compact subset of $\mathcal{A}^{pf}$, so that $K^* \in \mathcal{K}$ and

$$q_K(x^*) := \sup\left\{|\langle x^*, f\rangle| : f \in K\right\} = \sup\{|\langle x, f^*\rangle| : f^* \in K^*\} =: q_{K^*}(x), \ \forall \, x \in \mathcal{A}.$$

This shows that the involution on $\mathcal{A}$ is $\tau^\diamond$-continuous.

Let now τ_{lc}^* be an arbitrary locally convex $*$-algebra topology on $\mathcal{A}$ equivalent to the given one, i.e., $\tau_{lc}^* \sim \tau$. Then, $\tau_{lc}^* \sim \tau^{pf}$ from Theorem 4.3.9, $\mathcal{A}[\tau_{lc}^*]$ is a (commutative) GB^*-algebra from Proposition 4.3.3(ii) and its dual $\mathcal{A}[\tau_{lc}^*]'$ is contained in the dual $\mathcal{A}^{pf}$ of $\mathcal{A}[\tau^{pf}]$, according to Remark 4.3.4. Hence, $\mathcal{A}[\tau_{lc}^*]'[\sigma(\mathcal{A}[\tau_{lc}^*], \mathcal{A})]$ is contained in $\mathcal{A}^{pf}[\sigma(\mathcal{A}^{pf}, \mathcal{A})]$, therefore a weak*-compact subset of the dual $\mathcal{A}[\tau_{lc}^*]'$ of $\mathcal{A}[\tau_{lc}^*]$ is also weak*-compact as a subset of the dual $\mathcal{A}^{pf}[\sigma(\mathcal{A}^{pf}, \mathcal{A})]$ of $\mathcal{A}[\tau^{pf}]$. In conclusion (see also (4.3.13)), we obtain

$$\tau_{lc}^* \prec \tau(\mathcal{A}, \mathcal{A}[\tau_{lc}^*]') \prec \tau(\mathcal{A}, \mathcal{A}^{pf}) = \tau^\diamond.$$

This shows that $\tau^\diamond$ is the finest locally convex $*$-algebra topology on $\mathcal{A}$, equivalent to τ.

For the last claim of the theorem, let σ be an arbitrary locally convex $*$-algebra topology on $\mathcal{A}$, such that every positive linear functional on $\mathcal{A}[\sigma]$ is continuous. Then, Definition 4.3.5 implies that $\tau^{pf} \prec \sigma$. On the other hand, if $\sigma \sim \tau$, then as in the previous paragraph we conclude that $\sigma \prec \tau^\diamond$; consequently, $\tau^{pf} \prec \sigma \prec \tau^\diamond$.

Conversely, if σ is an arbitrary locally convex $*$-algebra topology on $\mathcal{A}$, such that every positive linear functional on $\mathcal{A}[\sigma]$ is continuous and $\tau^{pf} \prec \sigma \prec \tau^\diamond$, then we obtain that the dual of $\mathcal{A}[\sigma]$ coincides with $\mathcal{A}^{pf}$, so [35, Chap. IV, §2, Proposition 4 and Théorèm 3] yields that $\mathfrak{B}_{\mathcal{A}[\sigma]} = \mathfrak{B}_{\mathcal{A}[\tau^{pf}]}$ (see also the last part of the proof of Lemma 4.3.12). It follows that $\sigma \sim \tau^{pf}$ and since $\tau^{pf} \sim \tau$ by Theorem 4.3.9, one finally obtains $\sigma \sim \tau$ and this completes the proof. $\qquad\square$

We close this section by proving that given a commutative GB^*-algebra $\mathcal{A}[\tau]$, the space $\mathcal{A}[\tau^\diamond]$ is barrelled and bornological (Theorem 4.3.16). For this we need some extra definitions and results from the theory of locally convex spaces that we first exhibit (see, e.g., [74, Chapter 3, pp. 208 and 210, as well §6 and §7, respectively], but also [35, 123, 131]).

Definition 4.3.14 Let $E[\tau]$ be a locally convex space.

(1) An absolutely convex subset of E that is absorbing and τ-closed is said to be a *barrel*. If every barrel is a 0-neighbourhood, then $E[\tau]$ is said to be a *barrelled space*.

(2) An absolutely convex subset of E that absorbs every τ-bounded subset of E is
called bornivorous. If every bornivorous subset of E is a 0-neibhourhood, then
$E[\tau]$ is called *a bornological space*.

Note that *every metrizable and complete locally convex space* (i.e., a Fréchet
space, hence a Banach space too) *is barrelled*, while *a metrizable locally convex
space is bornological* (cf., for instance, [74, p. 214, Corollary and p. 222, Proposition 3, resp.]. In particular, *a locally convex space $E[\tau]$ is bornological, if and
only if, τ coincides with the Mackey topology $\tau(E, E')$* (ibid., p. 221, Proposition
2(i)]). Yet, *every quasi-complete* (i.e., all closed and bounded subsets are complete)
bornological locally convex space is barrelled [131, p. 63, Corollary].

Let now $E[\tau]$ be a locally convex space and $\mathfrak{N}$ *the collection of all barrels in E.
The collection $\mathfrak{N}$ can be considered as a base of* 0-neighbourhoods *for a topology,*
say τ', on E. Then, $E[\tau']$ *is a barrelled locally convex space and τ' is called the
associated barrel topology*, with respect to the given topology τ of E. *Then, $\tau \prec \tau'$*
(apply e.g., [74, p. 87, Proposition 4]) *and $E[\tau]$ is barrelled, if and only if, $\tau = \tau'$.*

In a similar way, *the collection $\mathfrak{K}$ of all bornivorous subsets of E, gives a base
of* 0-neighbourhoods *for a locally convex topology*, say τ'', on E that is called *the
associated bornological topology*, with respect to the topology τ of E. Again, one
has that $\tau \prec \tau''$ and that $E[\tau]$ *is bornological, if and only if, $\tau = \tau''$*. Using the
introduced topologies τ', τ'', we have the following

Lemma 4.3.15 *Let $\mathcal{A}[\tau]$ be a locally convex algebra. Then, $\mathcal{A}[\tau']$, $\mathcal{A}[\tau'']$ are both
locally convex algebras. If moreover $\mathcal{A}[\tau]$ is a locally convex $*$-algebra, the same is
true for both $\mathcal{A}[\tau']$ and $\mathcal{A}[\tau'']$.*

Proof First, we must prove that multiplication in $\mathcal{A}$ is separately continuous with
respect to the topologies τ' and τ''; secondly, that if $\mathcal{A}[\tau]$ is endowed with a
continuous involution, then this is also τ'- respectively, τ''-continuous.

Let U be a barrel in $\mathcal{A}[\tau]$ and $y \in \mathcal{A}$. Set $U' := \{x \in \mathcal{A} : xy \in U\}$. Using the
fact that U is a τ-barrel, it is easily seen that U' is a τ-barrel too. This proves that
the map

$$\mathcal{A}[\tau'] \to \mathcal{A}[\tau'] : x \mapsto xy$$

is continuous, i.e., multiplication in $\mathcal{A}[\tau']$ is separately continuous.

Suppose now that $\mathcal{A}[\tau]$ has a continuous involution $*$. Then, taking a barrel U in
$\mathcal{A}[\tau]$, the set $U^* := \{x^* : x \in U\}$ is a τ-barrel and this shows that $* : \mathcal{A}[\tau'] \to
\mathcal{A}[\tau']$ is continuous.

In the same way we work for the case of $\mathcal{A}[\tau'']$. Namely, if V is a bornivorous
subset in $\mathcal{A}[\tau]$ and $V' := \{x \in \mathcal{A} : xy \in V\}$, y a fixed element in $\mathcal{A}$, then, V'
is absolutely convex and if S is a τ-bounded subset in $\mathcal{A}$, the set Sy is also τ-
bounded, since the map $\mathcal{A}[\tau] \to \mathcal{A}[\tau] : x \mapsto xy$ is continuous. It follows that
Sy will be absorbed by every 0-neighbourhood, so that there exists $\alpha > 0$, with
$\alpha Sy \subseteq V$, which yields $\alpha S \subseteq V'$. Hence, V' absorbs τ-bounded sets, consequently

V' is bornivorous. It is now evident that the map $\mathcal{A}[\tau''] \to \mathcal{A}[\tau''] : x \mapsto xy$ is continuous. So $\mathcal{A}[\tau'']$ is a locally convex algebra.

On the other hand, if $\mathcal{A}[\tau]$ has a continuous involution and V is a τ-bornivorous set in $\mathcal{A}$, the same is true for the set V^*, i.e., involution in $\mathcal{A}[\tau'']$ is continuous.   □

Theorem 4.3.16 *Let $\mathcal{A}[\tau]$ be a commutative GB*-algebra and let $\tau^\diamond :=$ $\tau(\mathcal{A}, \mathcal{A}^{pf})$. Then, $\mathcal{A}[\tau^\diamond]$ as a locally convex space is barrelled and bornological.*

Proof Recall the associated barrel and bornological topologies τ', respectively τ'', with respect to the given topology τ on $\mathcal{A}$, introduced just before Lemma 4.3.15. It will be shown that $\tau' = \tau^\diamond = \tau''$.

Observe that $\tau^{pf} \sim \tau$ (Theorem 4.3.9) and that $\tau \sim \tau^\diamond$ (Theorem 4.3.13), consequently $\tau^\diamond \sim \tau$. From Proposition 4.3.3 and Theorem 4.3.13, we conclude that $\mathcal{A}[\tau^\diamond]$ is a commutative GB^*-algebra with the corresponding C^*-subalgebra $\mathcal{A}_0(= \mathcal{A}[B_0])$ exactly the same as that of $\mathcal{A}[\tau]$ (recall that $B_0(\tau) = B_0(\tau^\diamond)$).

Thus, we consider the greatest member B_0 in $\mathfrak{B}^*_{\mathcal{A}[\tau^\diamond]}$ and let U be an arbitrary $\tau^\diamond$-barrel in $\mathcal{A}$ (i.e., U is a 0-neighbourhood in $\mathcal{A}[\tau']$). Then, U is $\tau^\diamond$-closed, therefore $U \cap \mathcal{A}_0$ is $\| \cdot \|_{B_0}$-closed in the C^*-algebra $\mathcal{A}_0$. Since U is absorbing and absolutely convex, the same is true for $U \cap \mathcal{A}_0$, so that $U \cap \mathcal{A}_0$ is a barrel in $\mathcal{A}_0$. But, $\mathcal{A}_0$ as a Banach space is barrelled (cf. comments after Definition 4.3.14), consequently $U \cap \mathcal{A}_0$ is a 0-neighbourhood in $\mathcal{A}_0$ and thus absorbs the closed unit ball B_0 of $\mathcal{A}_0$. It follows that U absorbs B_0, therefore B_0 is τ'-bounded. Moreover, B_0 is $\tau^\diamond$-closed and since $\tau^\diamond \prec \tau'$ (see discussion before Lemma 4.3.15 about τ') it is also τ'-closed. So $B_0 \in \mathfrak{B}^*_{\mathcal{A}[\tau']}$. If now B is an arbitrary member in $\mathfrak{B}^*_{\mathcal{A}[\tau']}$, then B is $\tau^\diamond$-bounded and its $\tau^\diamond$-closure belongs to $\mathfrak{B}^*_{\mathcal{A}[\tau^\diamond]}$. Hence, $B \subseteq B_0$ and so B_0 is the greatest member in $\mathfrak{B}^*_{\mathcal{A}[\tau']}$. Thus, from Proposition 4.3.3(i) and Theorem 4.3.13 we obtain $\tau' \sim \tau^\diamond \sim \tau$. But (ibid.) $\tau^\diamond$ is the finest locally convex *-algebra topology on $\mathcal{A}$ that is equivalent to the given topology τ and also $\tau^\diamond \prec \tau'$ as we noticed before. Consequently, by Lemma 4.3.15, we finally obtain $\tau^\diamond = \tau'$ and this proves that $\mathcal{A}[\tau^\diamond]$ is barrelled.

Suppose again that B_0 is the greatest member in $\mathfrak{B}^*_{\mathcal{A}}[\tau^\diamond]$. Then, using the terms 'bornivorous' and 'bornological' instead of 'barrel' and 'barrelled' we conclude, as before, that $B_0 \in \mathfrak{B}^*_{\mathcal{A}[\tau'']}$. For proving that $\mathcal{A}[\tau'']$ is bornological we argue exactly as in the case of barrelledness above, replacing τ' with τ''. So the proof is complete.   □

For a similar result in the noncommutative case, see Theorem 6.2.6 in Sect. 6.2.

4.4   A *-Algebra of Functions with no GB*-Topology

From the information given in Chaps. 3 and 6, we may briefly say that:

(1) a GB^*-algebra $\mathcal{A}[\tau]$, whose topology τ is defined by a complete norm, is a C^*-algebra (see Corollary 3.3.11(1));

(2) a commutative GB^*-algebra $\mathcal{A}[\tau]$ is $*$-isomorphic to a $*$-algebra $\widehat{\mathcal{A}}$ of $\mathbb{C}^*$-valued continuous functions on a Hausdorff compact space $\mathfrak{M}_0$, that is nothing other than the maximal ideal space of the C^*-algebra $\mathcal{A}[B_0] = \mathcal{A}_0$; moreover, the $*$-algebra $\widehat{\mathcal{A}}$ contains the function C^*-algebra $\mathcal{C}(\mathfrak{M}_0)$ (cf. Theorem 3.4.9);

(3) an arbitrary GB^*-algebra is $*$-isomorphic to a $*$-algebra of unbounded operators in a Hilbert space (see Theorem 6.3.5). In Example 4.4.1 below, it is shown that not every $*$-algebra of functions is realized as in (2) above.

▶ In what follows we use the term *"almost everywhere"* (abbreviated to a.e.) in the topological sense, meaning that an equality holds *almost everywhere, except on a set of first category.*

The following example is due to P.G. Dixon [50, p. 160, Example 1], see also [28, XI.10, Theorem 12, Corollary 1], or [XI.7, Exercise 5], in the 3rd Edition of the same book.

Example 4.4.1 (Dixon) Let $\mathcal{A}$ be the $*$-algebra of all $\mathbb{C}$-valued Borel functions on the unit interval $[0, 1]$, modulo equality a.e., in the preceding topological sense. Denote by **1** the identity element of $\mathcal{A}$. We shall say that a function $f \in \mathcal{A}$ *is essentially bounded* if there is a positive number k such that $|f(t)| \leq k$, for all $t \in [0, 1]$, except on a subset of first category. For $f \in \mathcal{A}$ being essentially bounded, let

$$f^{-1}(k, \infty) \equiv \{t \in [0, 1] : |f(t)| > k\} \quad \text{and}$$

$$U_f^{ess} \equiv \{k \text{ a positive number} : f^{-1}(k, \infty) \text{ is a set of first category}\}$$

be the set of essential upper bounds of f. Then, as usually, one defines

$$ess \sup f := \inf U_f^{ess}, \text{ if } U_f^{ess} \neq \emptyset \text{ and } ess \sup f := \infty, \text{ otherwise.}$$

Consider now the $*$-subalgebra of $\mathcal{A}$

$$\mathcal{A}_b^{ess} \equiv \{f \in \mathcal{A} : f \text{ is essentially bounded}\}$$

and put

$$\|f\|_b^{ess} := ess \sup f, \ \forall \ f \in \mathcal{A}_b^{ess}.$$

Then, $\mathcal{A}_b^{ess}$ is a C^*-algebra endowed with the C^*-norm $\| \cdot \|_b^{ess}$ and it is a $*$-subalgebra of $\mathcal{A}$, such that for each f in $\mathcal{A}$, the elements $(\mathbf{1} + f^*f)^{-1}$ and $f(\mathbf{1} + f^*f)^{-1}$ belong to $\mathcal{A}_b^{ess}$.

We shall show that $\mathcal{A}$ has no GB^*-topology. Suppose that $\mathcal{A}$ admits a GB^*-topology, say τ, with B_0 the greatest member in $\mathfrak{B}^*_{\mathcal{A}[\tau]}$. Then, since $\mathcal{A}[\tau]$ is a commutative GB^*-algebra, it follows that (Lemma 3.3.7(ii))

$$\mathcal{A}[B_0] = \mathcal{A}_0 = \mathcal{A}_b^{ess},$$

where $\mathcal{A}_0$ is the set of all bounded elements of $\mathcal{A}[\tau]$. Using the Gelfand $*$-isomorphism (extension of the Gelfand $*$-isomorphism of $\mathcal{A}_b^{ess}$ to $\mathcal{A}$; cf. Theorem 3.4.9), one can see that the projections (i.e., the self-adjoint idempotents) of $\mathcal{A}$ belong to B_0. These projections are exactly the characteristic functions of the Borel subsets of $[0, 1]$. Take now a non-zero projection p corresponding to a Borel set $P \subseteq [0, 1]$. Then, one can find a strictly decreasing sequence of nonempty Borel sets $P_1, P_2, P_3, \ldots$ in $[0, 1]$, with characteristic functions $p_1, p_2, p_3, \ldots$, such that $\bigcap_{n=1}^{\infty} P_n = P$; we remind the reader that we work with equivalence classes, with respect to the equivalence relation "a.e. except on a set of first category".

Now, according to the correspondence $P_n \leftrightarrow p_n$, $n \in \mathbb{N}$ and properties of the P_n's, the infinite sum in the definition $f(t) := \sum_{n=1}^{\infty} p_n(t)$, $t \in [0, 1]$, is in fact finite. Hence, f belongs to $\mathcal{A}[\tau]$. By the (separate) continuity of multiplication in $\mathcal{A}[\tau]$, we have that for each 0-neighbourhood U in $\mathcal{A}[\tau]$, there is another 0-neighbourhood V in $\mathcal{A}[\tau]$, such that $f(V) \subseteq U$. On the other hand, since B_0 is τ-bounded, there is a positive number N with $B_0 \subseteq NV$, therefore $N^{-1} f B_0 \subseteq U$. Now $p_N \equiv N^{-1} f g$, where $0 \leq g \leq p_N$, as a projection belongs to B_0, therefore so does g. As a result, we obtain that

$$p_N \in N^{-1} f B_0 \subseteq U.$$

So for each non-zero projection p and each 0-neighbourhood U in $\mathcal{A}[\tau]$, we can find a projection p_N in U, such that $0 < p_N \leq p$. Moreover, for each $m \geq N$, we find a projection $p_m \equiv N^{-1} f(g p_m)$ with $p_m \leq p_N$ and $p_N p_m = p_m$. On the other hand, $g, p_m \in B_0$, therefore $g p_m \in B_0^2 \subseteq B_0$. Thus,

$$p_m \equiv N^{-1} f(g p_m) \subseteq N^{-1} f B_0 \subseteq U. \tag{4.4.14}$$

In a similar way, one can show that *every decreasing sequence with null intersection converges to 0, with respect to τ.*

Now, let Q_i, $i = 1, 2, 3, \ldots$, be an enumeration of the open intervals in $[0, 1]$, with rational end-points. Let the q_i's be the associated to Q_i's projections. Suppose that, for each i, we define a projection $s_i \leq q_i$, such that the corresponding Borel sets S_i satisfy the inclusion

$$S_i \subseteq Q_i \setminus \bigcup_{j=1}^{i-1} S_j \quad \text{and} \quad S_i = \emptyset, \quad \text{only if}, \quad Q_i \setminus \bigcup_{j=1}^{i-1} S_j = \emptyset,$$

(always modulo sets of first category). Using the fact that every Borel set is congruent to an open set, modulo sets of first category (see [100, § 11, III]), we obtain that $\bigcup_{j=1}^{\infty} S_j = [0, 1]$. The Borel sets S_i are disjoint by their choice, therefore

$$\sum_{j=1}^{n} s_j \to 1, \quad \text{as } n \to \infty; \quad \text{hence,} \quad \sum_{j=1}^{\infty} s_j = 1, \qquad (4.4.15)$$

with respect to τ. We show (4.4.15). The projections s_i correspond to the Borel sets S_i, so that $1 - \sum_{j=1}^{n} s_j$ corresponds to $[0, 1] \setminus \bigcup_{j=1}^{\infty} S_j = \varnothing$. It follows that $\left((1 - \sum_{j=1}^{n} s_j)_n\right)$ is a decreasing sequence of projections with null intersection, therefore it converges to 0, with respect to τ. Thus, (4.4.15) has been proven.

Now, suppose that U is a closed 0-neighbourhood in $\mathcal{A}[\tau]$ that does not contain the identity $\mathbf{1}$. Since τ is a locally convex topology, we may find a sequence of 0-neighbourhoods in $\mathcal{A}[\tau]$, such that

$$U = U_0 \supseteq U_1 \supseteq U_2 \supseteq \dots \quad \text{with } U_j + U_j \subseteq U_{j-1}, \ \forall \ j.$$

According to (4.4.14), s_j may be chosen, such that $s_j \in U_j$, for all j's. But then, $\sum_{j=1}^{\infty} s_j = \mathbf{1} \in U$, which is a contradiction. Thus, indeed not every *-algebra of functions admits a GB^*-topology.

Notes The results of Sects. 4.1–4.3 are due to G.R. Allan and for these the reader is referred to [5, Sections 3, 4 and 5]). The content of Sect. 4.4 consists of results that are owed to P.G. Dixon and can be found in [50].

Chapter 5
Extended C*-Algebras and Extended W*-Algebras

In this chapter, we discuss GB^*-algebras of closable operators in a Hilbert space. Thus, the reader has the chance to encounter the basics of unbounded *-representations (Sect. 5.1), with various uniform topologies defined on the *-preserving vector space $\mathcal{L}^\dagger(\mathcal{D}, \mathcal{H})$ of all linear operators T with domain a dense subspace $\mathcal{D}$ of a Hilbert space $\mathcal{H}$ and values in $\mathcal{H}$, in such a way that $\mathcal{D}$ is contained in the domain $\mathcal{D}(T^*)$ of the adjoint operator T^* of T. The restriction of T^* on $\mathcal{D}$ defines an involution on $\mathcal{L}^\dagger(\mathcal{D}, \mathcal{H})$, detoted by † (Sect. 5.2). Sections 5.3 and 5.4 deal with EC^*-algebras and EW^*-algebras that correspond to unbounded generalizations of the classical C^*- and W^*-algebras, respectively. The latter algebras occur naturally in the analysis of an unbounded generalization of left Hilbert algebras (see Sect. 5.4 and [84]). An interesting contribution of EC^*-algebras, to the context of GB^*-algebras, the reader can enjoy in Theorem 6.3.5. For more information about unbounded generalizations of von Neumann algebras, the reader is referred to [7] and [85].

5.1 $\mathcal{O}^*$-Algebras and Unbounded *-Representations

$\mathcal{O}^*$-algebras play an important role in this chapter. So, first we present the definition of an $\mathcal{O}^*$-algebra and some basic facts on their theory. For more detail the reader is referred to [7, 85, 136, 138]; see also the Notes at the end of Sect. 5.1.

Let $\mathcal{D}$ be a dense subspace of a Hilbert space $\mathcal{H}$. For a given Hilbert space $\mathcal{H}$, the symbol $\langle\,,\,\rangle$ will stand for the *inner product* of $\mathcal{H}$, while $\|\cdot\|$ will denote the norm on $\mathcal{H}$ resulting from this inner product.

Let $\mathcal{L}^\dagger(\mathcal{D})$ be the set of all linear operators T from $\mathcal{D}$ to $\mathcal{D}$, such that the domain $\mathcal{D}(T^*)$ of the adjoint T^* of T contains $\mathcal{D}$ and $T^*\mathcal{D} \subset \mathcal{D}$. Then, $\mathcal{L}^\dagger(\mathcal{D})$ is a *-algebra under the usual algebraic operations and the involution given by the map $T \mapsto T^\dagger := T^*\!\restriction_\mathcal{D}$.

© The Author(s), under exclusive license to Springer Nature Switzerland AG 2022
M. Fragoulopoulou et al., *Generalized B*-Algebras and Applications*, Lecture Notes in Mathematics 2298, https://doi.org/10.1007/978-3-030-96433-7_5

In this regard, if T, S are two linear operators in $\mathcal{H}$, we say that S is an *extension* of T and we write $T \subset S$, if $\mathcal{D}(T) \subset \mathcal{D}(S)$ and $T\xi = S\xi$, for every $\xi \in \mathcal{D}(T)$.

Furthermore, a linear operator T in $\mathcal{H}$ is said to be *closed* if its graph $\mathcal{G}(T)$ is closed in $\mathcal{H} \oplus \mathcal{H}$, while T is called *closable* if it has a closed extension.

Every closable operator has a minimal closed extension, called *closure* of T, denoted by $\overline{T}$.

Definition 5.1.1 Let T be a closed operator in a Hilbert space $\mathcal{H}$. A subspace $\mathcal{D}$ of the domain $\mathcal{D}(T)$ of T is called *a core* for T if $\overline{T \upharpoonright \mathcal{D}} = T$.

A densely defined linear operator T in $\mathcal{H}$ is said to be *symmetric* if $T \subset T^*$ (i.e., T^* is an extension of T). If $T^* = T$, then T is called *self-adjoint*.

A symmetric operator T in $\mathcal{H}$ is called *essentially self-adjoint* if $\overline{T} = T^*$, i.e., $\mathcal{D}(T)$ is a core for T^*.

A $*$-subalgebra $\mathcal{M}$ of $\mathcal{L}^{\dagger}(\mathcal{D})$ is called an $\mathcal{O}^*$-*algebra on $\mathcal{D}$ in $\mathcal{H}$*. The locally convex topology on $\mathcal{D}$ induced by the seminorms

$$\{\|\cdot\|_T : T \in \mathcal{M}\}, \quad \text{where } \|\xi\|_T := \|\xi\| + \|T\xi\|, \ \xi \in \mathcal{D}, \tag{5.1.1}$$

is called the *graph topology* on $\mathcal{D}$ and is denoted by $t_{\mathcal{M}}$. If the locally convex space $\mathcal{D}[t_{\mathcal{M}}]$ is complete, then $\mathcal{M}$ is called a *closed $\mathcal{O}^*$- algebra*. Denote by $\widetilde{\mathcal{D}}(\mathcal{M})$ the completion of $\mathcal{D}[t_{\mathcal{M}}]$. Then,

$$\widetilde{\mathcal{D}}(\mathcal{M}) = \bigcap_{T \in \mathcal{M}} \mathcal{D}(\overline{T}),$$

where $\overline{T}$ is the *closure of the operator* T. Put now,

$$\widetilde{T} := \overline{T} \upharpoonright_{\widetilde{\mathcal{D}}(\mathcal{M})}, \ T \in \mathcal{M} \text{ and } \widetilde{\mathcal{M}} := \{\widetilde{T} : T \in \mathcal{M}\}.$$

Then, $\widetilde{\mathcal{M}}$ is the smallest closed extension of $\mathcal{M}$, which is called the *closure* of $\mathcal{M}$. It is easily shown that

$$\mathcal{M} \text{ is closed } \Leftrightarrow \mathcal{M} = \widetilde{\mathcal{M}} \Leftrightarrow \mathcal{D} = \widetilde{\mathcal{D}}(\mathcal{M}) = \bigcap_{T \in \mathcal{M}} \mathcal{D}(\overline{T}).$$

Let $\mathcal{M}$ be an $\mathcal{O}^*$-algebra on $\mathcal{D}$ in $\mathcal{H}$. Then, $\mathcal{M}$ is said to be *self-adjoint* if $\mathcal{D}^*(\mathcal{M}) := \bigcap_{T \in \mathcal{M}} \mathcal{D}(T^*) = \mathcal{D}$. It is clear that if $\mathcal{M}$ is self-adjoint, then it is closed. A closed $\mathcal{O}^*$-algebra $\mathcal{M}$ is said to be *integrable* if $T^* = \overline{T^{\dagger}}$, for every $T \in \mathcal{M}$. For integrable $\mathcal{O}^*$-algebras we have the following, for the proof of which we refer to Theorem 7.1 in [125] and in [78].

Proposition 5.1.2 *Let $\mathcal{M}$ be a closed $\mathcal{O}^*$-algebra on $\mathcal{D}$ in $\mathcal{H}$. The following are equivalent:*

(i) *$\mathcal{M}$ is integrable.*

(ii) $\overline{T}$ *is a self-adjoint operator, for each* T *in the set*

$$H(\mathcal{M}) := \{T \in \mathcal{M} : T^\dagger = T\}.$$

Thus, if one of the above equivalent statements (i) and (ii) holds, then $\mathcal{M}$ is self-adjoint and $\overline{\mathcal{M}} := \{\overline{T} : T \in \mathcal{M}\}$ is a *-algebra of closed operators in $\mathcal{H}$ equipped with the *strong sum* $\overline{T} \,\dot{+}\, \overline{S} := \overline{\overline{T} + \overline{S}} = \overline{T + S}$, the *strong scalar multiplication*

$$\alpha \cdot \overline{T} := \begin{cases} \alpha\overline{T}, & \text{if } \alpha \neq 0 \\ 0, & \text{if } \alpha = 0 \end{cases} = \overline{\alpha T}, \ \alpha \in \mathbb{C},$$

the *strong product* $\overline{T} \cdot \overline{S} := \overline{\overline{T}\,\overline{S}} = \overline{TS}$ and the involution $T \mapsto T^* := \overline{T^\dagger}$.

Now, let $\mathcal{A}$ be a *-algebra and $\mathcal{D}$ be a dense subspace of a Hilbert space $\mathcal{H}$. A *-*representation* of $\mathcal{A}$ on $\mathcal{D}$ is a *-homomorhism π from $\mathcal{A}$ into $\mathcal{L}^\dagger(\mathcal{D})$, such that $\pi(e) = I$, when $\mathcal{A}$ has an identity e and I is the identity operator on $\mathcal{D}$. The space $\mathcal{D}$ is called *the domain of* π and it will be denoted by $\mathcal{D}(\pi)$. The respective Hilbert space, in which the domain of π is dense, is accordingly denoted by $\mathcal{H}_\pi$ [136, p. 38]. A *-representation π of $\mathcal{A}$ is said to be *faithful* if it is injective. If two *-representations π_1, π_2 of $\mathcal{A}$ are given, we write $\pi_1(a) \subset \pi_2(a), a \in \mathcal{A}$, if $\mathcal{D}(\pi_1) \subset \mathcal{D}(\pi_2)$ and $\pi_1(a)\xi = \pi_2(a)\xi$, for every $\xi \in \mathcal{D}(\pi_1)$. If $\pi_1(a) \subset \pi_2(a)$, for all $a \in \mathcal{A}$, we say that π_2 *is an extension* of π_1 and we denote it by $\pi_1 \subset \pi_2$.

Let π be a *-representation of a *-algebra $\mathcal{A}$. Denote by t_π the graph topology $t_{\pi(\mathcal{A})}$ on $\mathcal{D}(\pi)$ with respect to the $\mathcal{O}^*$-algebra $\pi(\mathcal{A})$. If $\mathcal{D}(\pi)[t_\pi]$ is complete, i.e., the $\mathcal{O}^*$-algebra $\pi(\mathcal{A})$ is closed, then π is said to be *closed*. Denote by $\widetilde{\mathcal{D}}(\pi)$ the completion of $\mathcal{D}(\pi)[t_\pi]$ and put

$$\widetilde{\pi}(x) \equiv \overline{\pi(x)}\,|_{\widetilde{\mathcal{D}}(\pi)}, \ x \in \mathcal{A}.$$

Then $\widetilde{\pi}$ is a closed *-representation of $\mathcal{A}$, which is the smallest closed extension of π and it is called the *closure* of π.

A closed *-representation π of a *-algebra $\mathcal{A}$ is said to be *self-adjoint* (resp. *integrable*), if the $\mathcal{O}^*$-algebra $\pi(\mathcal{A})$ is self-adjoint (resp. integrable).

Example 5.1.3 Let $\mathcal{D}$ be a dense subspace of a Hilbert space $\mathcal{H}$.

(1) Let T be a self-adjoint element of $\mathcal{L}^\dagger(\mathcal{D})$, that is $T^\dagger = T$ and $\mathcal{P}(T)$ be the polynomial algebra generated by T. Then, the closure $\widetilde{\mathcal{P}}(T)$ of the $\mathcal{O}^*$-algebra $\mathcal{P}(T)$ is integrable, if and only if, $\widetilde{\mathcal{P}}(T)$ is self-adjoint, if and only if, T^n is essentially self-adjoint, for every $n \in \mathbb{N}$.

(2) Let S and T be self-adjoint elements in $\mathcal{L}^\dagger(\mathcal{D})$, such that $ST = TS$. Let $\mathcal{P}(S, T)$ be the commutative $\mathcal{O}^*$-algebra on $\mathcal{D}$ generated by S and T. Then, the closure $\widetilde{\mathcal{P}}(S, T)$ of the $\mathcal{O}^*$-algebra $\mathcal{P}(S, T)$ is integrable, if and only if, there exists a normal operator, which is an extension of the operator $S + iT$.

Example 5.1.4 Let $\mathcal{H} := L^2(\mathbb{R})$ and $\mathcal{D} := C_0^\infty(\mathbb{R})$, the space of all compactly supported smooth functions on $\mathbb{R}$. Define $\mathcal{M}$, as follows:

$$\mathcal{M} := \left\{ \sum_{k=0}^m \sum_{l=0}^n \alpha_{kl} t^k \left(\frac{d}{dt}\right)^l : \alpha_{kl} \in \mathbb{C}, \ n, m \in \mathbb{N} \cup \{0\} \right\}.$$

Then, $\widetilde{\mathcal{D}}(\mathcal{M})$ coincides with the Schwartz space $S(\mathbb{R})$ of all rapidly decreasing smooth functions on $\mathbb{R}$ and $\widetilde{\mathcal{M}}$ (as in the discussion after (5.1.1)) is a closed $\mathcal{O}^*$-algebra on $\mathcal{D}$ in $\mathcal{H}$. Let $\mathcal{A}$ be the *CCR*-algebra for one degree of freedom (see, for instance, [53, p. 69, Definition 4.10]), that is the *-algebra generated by the identity e and two self-adjoint elements p, q satisfying the Heisenberg commutation relation $pq - qp = -ie$. The *-representation π of $\mathcal{A}$ on $S(\mathbb{R})$ in the Hilbert space $L^2(\mathbb{R})$, defined by

$$\pi(e) = I, \ \big(\pi(p)f\big)(t) = tf(t), \ \big(\pi(q)f\big)(t) = -i\frac{df}{dt}, \ f \in S(\mathbb{R}), \ t \in \mathbb{R},$$

is called *Schrödinger representation*. Then, we have $\widetilde{\mathcal{M}} = \pi(\mathcal{A})$, so $\widetilde{\mathcal{M}}$ is not integrable.

Notes Definition 5.1.1 is found in [7, pp. 9, 11]. For the Example 5.1.3(1) see [89, Theorem 2.1], for (2) of the same example see [89, Theorem 3.2], while for the Example 5.1.4 cf. [83, Example 1.1.5]. The original examples of $\mathcal{O}^*$-algebras were described by Borchers (1962) [33] and Uhlmann (1962) [147], who studied examples of $\mathcal{O}^*$-algebras, called Borchers algebras, arising from the Wightman axioms of quantum field theory. Powers (1971) [125] and Lassner (1972) [103] began a systematic study of algebras of unbounded operators, whose $\mathcal{O}^*$-algebras are an important subclass.

5.2   Uniform Topologies on $\mathcal{O}^*$-Algebras

Let $\mathcal{D}$ be a dense subspace of a Hilbert space $\mathcal{H}$ and $\mathcal{M}$ a closed $\mathcal{O}^*$-algebra on $\mathcal{D}$ in $\mathcal{H}$. We shall introduce on $\mathcal{M}$ the weak, strong and strong* topology (see [85, p. 23, 1.6(A)]).

Consider the set $\mathcal{L}^\dagger(\mathcal{D}, \mathcal{H})$ of all linear operators T from $\mathcal{D}$ to $\mathcal{H}$, such that $\mathcal{D}$ is contained in the domain $\mathcal{D}(T^*)$ of the adjoint operator T^* of T. Then, $\mathcal{L}^\dagger(\mathcal{D}, \mathcal{H})$ becomes a *-preserving vector space endowed with the usual addition and scalar-multiplication and with the involution $T^\dagger := T^* \restriction_\mathcal{D}, \ T \in \mathcal{L}^\dagger(\mathcal{D}, \mathcal{H})$. Consider the following families of seminorms:

$$p_{\xi,\eta}(T) := |\langle T\xi, \eta\rangle|, \; T \in \mathcal{L}^\dagger(\mathcal{D}, \mathcal{H}) \text{ and } \xi, \eta \in \mathcal{D},$$

$$p^\xi(T) := \|T\xi\|, \; T \in \mathcal{L}^\dagger(\mathcal{D}, \mathcal{H}) \text{ and } \xi \in \mathcal{D}, \tag{5.2.1}$$

$$p_*^\xi(T) := p^\xi(T) + p^\xi(T^\dagger), \; T \in \mathcal{L}^\dagger(\mathcal{D}, \mathcal{H}) \text{ and } \xi \in \mathcal{D}.$$

The topologies induced by the families of seminorms

$$\left\{p_{\xi,\eta} : \xi, \eta \in \mathcal{D}\right\}, \; \left\{p^\xi : \xi \in \mathcal{D}\right\} \text{ and } \left\{p_*^\xi : \xi \in \mathcal{D}\right\}$$

are called the *weak, strong and strong* topology* on $\mathcal{L}^\dagger(\mathcal{D}, \mathcal{H})$, respectively and they will be denoted, respectively by τ_w, τ_s and τ_s^*. The locally convex spaces corresponding to the preceding topologies will be denoted by $\mathcal{L}^\dagger(\mathcal{D}, \mathcal{H})[\tau_\mathrm{w}]$, $\mathcal{L}^\dagger(\mathcal{D}, \mathcal{H})[\tau_s]$ and $\mathcal{L}^\dagger(\mathcal{D}, \mathcal{H})[\tau_s^*]$, respectively. The latter locally convex space is complete. The restrictions of the topologies τ_w, τ_s and τ_s^* on $\mathcal{M}$ are called *weak, strong and strong* topology on $\mathcal{M}$*. It is easy to see that $\mathcal{M}[\tau_\mathrm{w}]$ is a locally convex $*$-algebra, but $\mathcal{M}[\tau_s]$ and $\mathcal{M}[\tau_s^*]$ are not necessarily locally convex algebras. This happens, for instance, with $\mathcal{L}^\dagger(\mathcal{D})[\tau_s]$ and $\mathcal{L}^\dagger(\mathcal{D})[\tau_s^*]$. For this reason, we define the following quasi-strong (resp. quasi-strong*) topology on $\mathcal{M}$, under which $\mathcal{M}$ becomes a locally convex (resp. a locally convex $*$-)algebra.

Let $\xi \in \mathcal{D}$ and $T \in \mathcal{L}^\dagger(\mathcal{D})$. We put

$$p_T^\xi(S) := \|TS\xi\|, \; \forall S \in \mathcal{M},$$

$$p_{*,T}^\xi(S) := \|TS\xi\| + \|TS^\dagger\xi\|, \; \forall S \in \mathcal{M}. \tag{5.2.2}$$

Each of the preceding functions is a seminorm on $\mathcal{M}$. Let $\mathcal{A}$ be an $\mathcal{O}^*$-algebra on $\mathcal{D}$ containing the $\mathcal{O}^*$-algebra $\mathcal{M}_I$ on $\mathcal{D}$ generated by $\mathcal{M}$ and the identity operator I. We call the locally convex topology on $\mathcal{M}$ induced by the family of seminorms $\left\{p_T^\xi, \; T \in \mathcal{A}, \; \xi \in \mathcal{D}\right\}$ *quasi-strong topology on $\mathcal{M}$ for $\mathcal{A}$* and we denote it by $\tau_{qs}^\mathcal{A}$. The locally convex topology on $\mathcal{M}$ defined by the family of seminorms $\left\{p_{*,T}^\xi, \; T \in \mathcal{A}, \; \xi \in \mathcal{D}\right\}$ is called *quasi-strong* topology on $\mathcal{M}$ for $\mathcal{A}$* and we denote it by $\tau_{qs}^{*,\mathcal{A}}$. Such topologies depend, in general, on $\mathcal{A}$. In particular, the topology $\tau_{qs}^{\mathcal{L}^\dagger(\mathcal{D})}$ (resp. $\tau_{qs}^{*,\mathcal{L}^\dagger(\mathcal{D})}$) is the strongest and $\tau_{qs}^{\mathcal{M}_I}$ (resp. $\tau_{qs}^{*,\mathcal{M}_I}$) the weakest among such quasi-strong (resp. quasi-strong*) topologies.

Now note that for any $T \in \mathcal{A}$ and $\xi \in \mathcal{D}$, we have

$$p_T^\xi(S_1 S_2) = \|TS_1 S_2\xi\| = p_T^{S_2\xi}(S_1) = p_{TS_1}^\xi(S_2), \; \forall S_1, S_2 \in \mathcal{M}.$$

Since $S_2\xi \in \mathcal{D}$ and $TS_1 \in \mathcal{A}$ (as $\mathcal{M} \subset \mathcal{A}$), it follows that $\mathcal{M}[\tau_{qs}^\mathcal{A}]$ is a *locally convex algebra* and $\mathcal{M}[\tau_{qs}^{*,\mathcal{A}}]$ is a *locally convex $*$-algebra*.

▶ *The quasi-strong topology* $\tau_{qs}^{\mathcal{L}^{\dagger}(\mathcal{D})}$ *and* *the quasi-strong* topology $\tau_{qs}^{*,\mathcal{L}^{\dagger}(\mathcal{D})}$, *for simplicity's sake, from now on will be denoted by* τ_{qs}, τ_{qs}^*, *respectively.*

The following diagram shows the relationship of all the foregoing locally convex topologies defined on $\mathcal{M}$ in this section:

$$
\begin{array}{ccc}
\tau_w & \prec \tau_s \prec & \tau_s^* \\
& \curlywedge \qquad \curlywedge & \\
\tau_{qs}^{\mathcal{M}_I} & \prec & \tau_{qs}^{*,\mathcal{M}_I} \\
& \curlywedge \qquad \curlywedge & \\
\tau_{qs} & \prec & \tau_{qs}^*.
\end{array}
\tag{5.2.3}
$$

We next define various uniform topologies on a closed $\mathcal{O}^*$-algebra $\mathcal{M}$ on $\mathcal{D}$. Let $\mathfrak{M}$ denote *the set of all* $t_{\mathcal{L}^{\dagger}(\mathcal{D})}$-*bounded subsets of* $\mathcal{D}$. Then, for any $M \in \mathfrak{M}$ we define:

$$
p_M(T) := \sup_{\xi,\eta \in M} |\langle T\xi, \eta \rangle|, \ \forall \, T \in \mathcal{L}^{\dagger}(\mathcal{D}),
$$

$$
p^M(T) := \sup_{\xi \in M} \|T\xi\|, \ \forall \, T \in \mathcal{L}^{\dagger}(\mathcal{D}),
\tag{5.2.4}
$$

$$
p_*^M(T) := p^M(T) + p^M(T^{\dagger}), \ \forall \, T \in \mathcal{L}^{\dagger}(\mathcal{D}).
$$

The families of seminorms $\{p_M : M \in \mathfrak{M}\}$, $\{p^M : M \in \mathfrak{M}\}$ and $\{p_*^M : M \in \mathfrak{M}\}$ define locally convex topologies on $\mathcal{L}^{\dagger}(\mathcal{D})$, denoted respectively by τ_u, τ^u, τ_*^u and called *uniform topologies*. Since $\mathcal{M}$ is closed, it follows from [136, Corollary 2.3.11] that *a subset M of $\mathcal{D}$ is* $t_{\mathcal{L}^{\dagger}(\mathcal{D})}$-*bounded* (i.e., $M \in \mathfrak{M}$), *if and only if, M is $t_{\mathcal{M}}$-bounded*. Therefore, the induced locally convex topologies on $\mathcal{M}$ by the uniform topologies τ_u, τ^u and τ_*^u of $\mathcal{L}^{\dagger}(\mathcal{D})$, will be also denoted by the preceding symbols.

Using the very definitions, it follows easily that $\mathcal{M}[\tau_u]$ *is a locally convex ∗-algebra*, but $\mathcal{M}[\tau^u]$ and $\mathcal{M}[\tau_*^u]$ *are not necessarily locally convex algebras.* Conditions under which $\mathcal{M}[\tau^u]$ is a locally convex algebra and $\mathcal{M}[\tau_*^u]$ is a locally convex ∗-algebra can be found in Proposition 5.2.1(2)(3) and Corollary 5.2.2, below.

For this, as in the case of the quasi-strong and quasi-strong* topologies, we introduce some extra uniform topologies on $\mathcal{M}$. Let $T \in \mathcal{L}^{\dagger}(\mathcal{D})$ and $M \in \mathfrak{M}$

be arbitrary. Define

$$p_T^M(S) := p^M(TS) = \sup_{\xi \in M} \|TS\xi\|, \ \forall\, S \in \mathcal{M},$$

$$p_{*,T}^M(S) := p_T^M(S) + p_T^M(S^\dagger), \ \forall\, S \in \mathcal{M}. \tag{5.2.5}$$

Let $\mathcal{A}$ be an $\mathcal{O}^*$-algebra on $\mathcal{D}$ containing $\mathcal{M}_I$. The topologies induced by the families of seminorms $\{p_T^M : T \in \mathcal{A}, M \in \mathfrak{M}\}$ and $\{p_{*,T}^M : T \in \mathcal{A}, M \in \mathfrak{M}\}$, respectively, are called *quasi-uniform* and *quasi-uniform** topologies on $\mathcal{M}$ for $\mathcal{A}$ and will be denoted by $\tau_{qu}^{\mathcal{A}}$ and $\tau_{qu}^{*,\mathcal{A}}$, respectively. Then it is shown that $\mathcal{M}[\tau_{qu}^{\mathcal{A}}]$ is a locally convex algebra and $\mathcal{M}[\tau_{qu}^{*,\mathcal{A}}]$ a locally convex $*$-algebra. Such topologies depend on $\mathcal{A}$. In particular, $\tau_{qu}^{\mathcal{L}^\dagger(\mathcal{D})}$ (resp. $\tau_{qu}^{*,\mathcal{L}^\dagger(\mathcal{D})}$) is the strongest topology and $\tau_{qu}^{\mathcal{M}_I}$ (resp. $\tau_{qu}^{*,\mathcal{M}_I}$) the weakest topology among such quasi-uniform topologies. Here we call $\tau_{qu}^{\mathcal{L}^\dagger(\mathcal{D})}$ and $\tau_{qu}^{*,\mathcal{L}^\dagger(\mathcal{D})}$ the *quasi-uniform* and the *quasi-uniform** topology and denote them by τ_{qu} and τ_{qu}^*, respectively.

It is clear that when $\mathcal{M}$ is a $*$-algebra of bounded operators, the uniform and quasi-uniform topologies coincide with the usual uniform topology on $\mathcal{M}$. In the unbounded case, this does not happen, in general, as is shown below.

Proposition 5.2.1

(1)

$$\tau_u \prec \tau^u \prec \tau_*^u$$

$$\wedge \qquad \wedge$$

$$\tau_{qu}^{\mathcal{M}_I} \prec \tau_{qu}^{*,\mathcal{M}_I}$$

$$\wedge \qquad \wedge$$

$$\tau_{qu} \prec \tau_{qu}^*.$$

(2) $\mathcal{M}[\tau^u]$ *is a locally convex algebra, if and only if,* $\tau^u = \tau_{qu}^{\mathcal{M}_I}$.
(3) $\mathcal{M}[\tau_*^u]$ *is a locally convex $*$-algebra, if and only if,* $\tau_*^u = \tau_{qu}^{*,\mathcal{M}_I}$.
(4) *If* $t_{\mathcal{M}} = t_{\mathcal{L}^\dagger(\mathcal{D})}$, *in particular, if* $\mathcal{D}[t_{\mathcal{M}}]$ *is a Fréchet space, then* $\tau_{qu} = \tau_{qu}^{\mathcal{M}_I}$ *and* $\tau_{qu}^* = \tau_{qu}^{*,\mathcal{M}_I}$.
(5) $\mathcal{L}^\dagger(\mathcal{D})[\tau_{qu}^*]$ *is complete.*

Proof

(1) is easily shown by the very definitions.
(2) Suppose that $\mathcal{M}[\tau^u]$ is a locally convex algebra. Then, for any $S \in \mathcal{M}$, the map $T \in \mathcal{M}[\tau^u] \mapsto ST \in \mathcal{M}[\tau^u]$ is continuous, which implies that $\tau_{qu}^{\mathcal{M}_I} \prec \tau^u$. The converse follows, since $\mathcal{M}[\tau_{qu}^{\mathcal{M}_I}]$ is a locally convex algebra.

(3) is shown in a similar way as (2).

(4) Take arbitrary $T \in \mathcal{L}^{\dagger}(\mathcal{D})$ and $M \in \mathfrak{M}$. Since $t_{\mathcal{M}} = t_{\mathcal{L}^{\dagger}(\mathcal{D})}$ there exists an operator $K \in \mathcal{M}$ such that $\|T\xi\| \le \|K\xi\|$, for all $\xi \in \mathcal{D}$. Hence,

$$p_T^M(S) \le p_K^M(S) \text{ and } p_{*,T}^M(S) \le p_{*,K}^M(S), \ \forall \, S \in \mathcal{M},$$

so that, taking also into account the corresponding definitions, we conclude that $\tau_{qu} = \tau_{qu}^{\mathcal{M}_I}$ and $\tau_{qu}^* = \tau_{qu}^{*,\mathcal{M}_I}$.

Suppose now that $\mathcal{D}[t_{\mathcal{M}}]$ is a Fréchet space. Then, by the closed graph theorem, every operator $T \in \mathcal{L}^{\dagger}(\mathcal{D})$ is continuous from the Fréchet space $\mathcal{D}[t_{\mathcal{M}}]$ into the Hilbert space $\mathcal{H}$. Thus, we obtain $t_{\mathcal{M}} = t_{\mathcal{L}^{\dagger}(\mathcal{D})}$, hence the conclusion.

(5) Take an arbitrary Cauchy net $(T_{\nu})_{\nu \in \Lambda}$ in $\mathcal{L}^{\dagger}(\mathcal{D})[\tau_{qu}^*]$. Hence, for every $S \in \mathcal{L}^{\dagger}(\mathcal{D})$ and every $\xi \in \mathcal{D}$, the sequence $(ST_{\nu}\xi)_{\nu \in \Lambda}$ converges strongly in $\mathcal{H}$, since $\tau_{qu}^* \succ \tau_s^*$. It follows that $\lim_{\nu} ST_{\nu}\xi$ exists in $\mathcal{H}$. Put $ST\xi :=$ $\lim_{\nu} ST_{\nu}\xi$, $\xi \in \mathcal{D}$. Then, for $T \in \mathcal{L}^{\dagger}(\mathcal{D}, \mathcal{H})$, we have that

$$|\langle S^*\eta, T\xi \rangle| = \lim_{\nu} |\langle \eta, ST_{\nu}\xi \rangle| \le \|\eta\| \, \| \lim_{\nu} ST_{\nu}\xi \|,$$

for each $S \in \mathcal{L}^{\dagger}(\mathcal{D})$ and $\eta \in \mathcal{D}(S^*)$. Therefore, since $\mathcal{L}^{\dagger}(\mathcal{D})$ is closed by the closedness of $\mathcal{M}$, we have $T\xi \in \bigcap_{S \in \mathcal{L}^{\dagger}(\mathcal{D})} \mathcal{D}(\bar{S}) = \mathcal{D}$, from which we conclude that $T \in \mathcal{L}^{\dagger}(\mathcal{D})$. Furthermore, it can be shown that $\lim_{\nu} p_{*,S}^M(T_{\nu} - T) = 0$, for all $S \in \mathcal{L}^{\dagger}(\mathcal{D})$ and $M \in \mathfrak{M}$. Thus, $\mathcal{L}^{\dagger}(\mathcal{D})[\tau_{qu}^*]$ is complete. For a more detailed proof the reader is referred to [136, Proposition 3.3.15].   □

The next corollary provides another condition under which $\mathcal{M}[\tau^u]$ is a locally convex algebra and $\mathcal{M}[\tau_*^u]$ is a locally convex *-algebra.

Corollary 5.2.2 *Suppose that there exists a subset $\mathcal{C}$ of $\mathcal{L}^{\dagger}(\mathcal{D})$, such that $ST = TS$, for all $T \in M$ and $S \in \mathcal{C}$ and $t_{\mathcal{M}} = t_{\mathcal{C}}$ (take, for example, $\mathcal{M}$ to be commutative). Then, $\mathcal{M}[\tau^u]$ is a locally convex algebra and $\mathcal{M}[\tau_*^u]$ is a locally convex *-algebra.*

Proof Take an arbitrary operator $T \in M$ and an arbitrary set M from $\mathfrak{M}$. Since $t_{\mathcal{M}} = t_{\mathcal{C}}$, there exists a finite subset $\{K_1, K_2, \ldots, K_n\}$ of $\mathcal{C}$, such that

$$\|T\xi\| \le \sum_{k=1}^{n} \|K_k\xi\|, \ \forall \, \xi \in \mathcal{D}.$$

This implies that

$$p_T^M(S) \le \sum_{k=1}^{n} p^{K_k M}(S) \text{ and } p_{*,T}^M(S) \le \sum_{k=1}^{n} p_*^{K_k M}(S), \ \forall \, S \in \mathcal{M}.$$

Since $K_k M \in \mathfrak{M}$, for $k = 1, 2, \ldots, n$, we conclude that $\tau^u = \tau_{qu}$ and $\tau_*^u = \tau_{qu}^*$. Applying now Proposition 5.2.1(1)(2)(3), we have the conclusion. $\qquad\square$

Further, we study conditions under which all uniform and all quasi uniform topologies are identical.

Proposition 5.2.3 *Let $\mathcal{M}$ be a closed $\mathcal{O}^*$-algebra on $\mathcal{D}$ in $\mathcal{H}$. The following statements are equivalent:*

(i) *$\mathcal{M}[\tau_u]$ has jointly continuous multiplication;*
(ii) *$\tau_u = \tau^u$;*
(iii) *$\tau_u = \tau_{qu}^*$.*

Proof (i) $\Rightarrow$ (iii) Take arbitrary $M \in \mathfrak{M}$ and $T \in \mathcal{L}^\dagger(\mathcal{D})$. Then by (i) there is an element $M_1 \in \mathfrak{M}$ (see also (5.2.4), (5.2.5)), such that

$$p_T^M(S)^2 = p_M(S^\dagger T^\dagger T S) \le p_{M_1}(TS)^2$$

$$= \sup_{\xi, \eta \in M_1} |\langle S\xi, T^\dagger \eta\rangle|^2$$

$$\le \sup_{\xi, \eta \in M_1 \cup T^\dagger M_1} |\langle S\xi, \eta\rangle|^2$$

$$= p_{M_1 \cup T^\dagger M_1}(S)^2, \ \forall S \in \mathcal{M}.$$

Since $M_1 \cup T^\dagger M_1 \in \mathfrak{M}$, we obtain that $\tau_u = \tau_{qu}$. But the involution $T \to T^\dagger$ is τ_u-continuous, therefore using (5.2.5) we conclude that $\tau_u = \tau_{qu} = \tau_{qu}^*$.

(iii) $\Rightarrow$ (ii) Follows from (iii) and Proposition 5.2.1(1).
(ii) $\Rightarrow$ (i) Let $M \in \mathfrak{M}$. Then,

$$p_M(TS) = \sup_{\xi, \eta \in M} |\langle S\xi, T^\dagger \eta\rangle| \le \sup_{\xi, \eta \in M} \|T^\dagger \eta\| \|S\xi\|$$

$$= p^M(T^\dagger) p^M(S), \ \forall T, S \in \mathcal{M}.$$

The assertion now follows from (ii) and the τ_u-continuity of the involution $\dagger$. This completes the proof. $\qquad\square$

The example that follows, provides a closed $\mathcal{O}^*$-algebra, on which all uniform and all quasi uniform topologies coincide.

Example 5.2.4 Let $(\mathcal{M}_n)_{n \in \mathbb{N}}$ be a sequence of infinite dimensional $*$-algebras of bounded operators acting on the Hilbert spaces $\mathcal{H}_n$, $n \in \mathbb{N}$. Suppose that each $\mathcal{M}_n$ contains the identity operator. Denote by $\mathcal{H}$ the Hilbert space (orthogonal) direct sum of the Hilbert spaces $\mathcal{H}_n$, $n \in \mathbb{N}$ and let

$$\mathcal{D} := \Big\{\xi = (\xi_n) \in \Pi_{n \in \mathbb{N}} \mathcal{H}_n : \xi_n = 0, \ \forall n \in \mathbb{N},$$

$$\text{except for a finite number of them}\Big\}.$$

Then, the product $\mathcal{M} := \Pi_{n \in \mathbb{N}} \mathcal{M}_n$ is a closed $\mathcal{O}^*$-algebra on $\mathcal{D}$ in $\mathcal{H}$, under the following algebraic operations:

$$T + S := (T_n + S_n), \quad \lambda T := (\lambda T_n),$$

$$TS := (T_n S_n), \quad T^\dagger := (T_n^*),$$

where $T = (T_n)$, $S = (S_n)$ in $\mathcal{M}$ and $\lambda \in \mathbb{C}$. It is shown that for any $M \in \mathfrak{M}$ there exists a positive integer N such that

$$M \subseteq \{\xi = (\xi_n) \in \mathcal{D} : \xi_n = 0, \ \forall n > N\}.$$

This implies that $\tau_u = \tau^u$; therefore Proposition 5.2.3 gives that all uniform and all quasi uniform topologies are equal on $\mathcal{M}$.

Furthermore, all previous topologies are also equal to the *locally uniform topology* τ_{lu} defined by the family $\{\|\cdot\|_n : n \in \mathbb{N}\}$ of seminorms, where $\|T\|_n := \|T_n\|$ (the operator norm of T_n), for all $T = (T_n) \in \mathcal{M}$. For more details the reader is referred to [87, Proposition 3.5].

Example 5.2.5 Let $\mathcal{D}$ be a dense subspace of a Hilbert space $\mathcal{H}$ and $\mathcal{P}(T)$ *the* $\mathcal{O}^*$-*algebra of all polynomials generated by an unbounded operator T in* $\mathcal{L}^\dagger(\mathcal{D})$. *Let* Γ_∞ *be the set of all positive sequences (r_n) with $1 \le r_0 \le r_1 \le \ldots$. We introduce the topology τ_∞ defined by all the seminorms*

$$\|p(T)\|_{(r_n)} := \sum_{n=0}^{\infty} r_n |\alpha_n|, \ (r_n) \in \Gamma_\infty \ \text{and} \ p(T) = \sum_{n=0}^{k} \alpha_n T^n \in \mathcal{P}(T).$$

Then, τ_∞ *is the finest locally convex topology on the* $\mathcal{O}^*$-*algebra* $\mathcal{P}(T)$. Suppose that $\mathcal{P}(T)$ is closed. Then, we have that

$$\tau_s = \tau_s^* = \tau_u = \tau^u = \tau_*^u = \tau_{qu}^* = \tau_\infty.$$

We next prove that $\tau_u \neq \tau^u$, for every maximal $\mathcal{O}^*$-algebra $\mathcal{L}^\dagger(\mathcal{D})$.

Example 5.2.6 Let $\mathcal{D}$ be a dense proper subspace of a Hilbert space $\mathcal{H}$. Suppose that $\mathcal{L}^\dagger(\mathcal{D})$ is closed. Then, we have the following:

(1) *The multiplication of $\mathcal{L}^\dagger(\mathcal{D})$ is not jointly continuous, under any of the uniform and quasi uniform topologies.*
(2) $\tau_u \neq \tau^u$, $\tau^u \neq \tau_{qu}$ *and* $\tau_*^u \neq \tau_{qu}^*$.
(3) $\mathcal{L}^\dagger(\mathcal{D})[\tau_u]$ *and* $\mathcal{L}^\dagger(\mathcal{D})[\tau_*^u]$ *are not complete.*

Indeed, suppose that multiplication in $\mathcal{L}^\dagger(\mathcal{D})[\tau_{qu}]$ is jointly continuous. Take an arbitrary $\eta \in \mathcal{D}$ with $\|\eta\| = 1$. Then, for $M = \{\eta\} \in \mathfrak{M}$, there exists $T \in \mathcal{L}^\dagger(\mathcal{D})$

and $M_1 \in \mathfrak{M}$ such that

$$p^M(S(\xi \otimes \overline{\eta})) = \|S\xi\|$$

$$\leq p_T^{M_1}(S) p_T^{M_1}(\xi \otimes \overline{\eta})$$

$$= p_T^{M_1}(S) \sup_{\zeta \in M_1} \|\langle \zeta, \eta \rangle \, T\xi\|$$

$$\leq p_T^{M_1}(S)(\sup_{\zeta \in M_1} \|\zeta\|)\|T\xi\|, \ \forall \, S \in \mathcal{L}^\dagger(\mathcal{D}) \text{ and } \xi \in \mathcal{D},$$

where $\xi \otimes \overline{\eta}$ is the operator $\zeta \mapsto \langle \zeta, \eta \rangle \xi$, $\zeta \in \mathcal{H}$. Thus,

$$\mathcal{D}(\overline{T}) = \bigcap_{S \in \mathcal{L}^\dagger(\mathcal{D})} \mathcal{D}(\overline{S}) = \mathcal{D} = \mathcal{D}(T),$$

therefore T is closed. Hence, $\mathcal{D}$ becomes a Hilbert space, denoted by $\mathcal{H}_T$, under the inner product

$$\langle \xi, \eta \rangle_T = \langle T\xi, T\eta \rangle, \ \forall \, \xi, \eta \in \mathcal{D}.$$

Now, for any $x \in \mathcal{H}$, we put

$$f(\xi) := \langle T\xi, x \rangle, \ \forall \, \xi \in \mathcal{H}_T.$$

Then, f is a continuous linear functional on the Hilbert space $\mathcal{H}_T$, so that by the Riesz representation theorem there exists an element $\eta \in \mathcal{H}_T$, such that

$$f(\xi) := \langle T\xi, x \rangle = \langle T\xi, T\eta \rangle = \langle \xi, T^\dagger T\eta \rangle, \ \forall \, \xi \in \mathcal{D} = \mathcal{H}_T.$$

It follows that $x \in \mathcal{D}(T^*)$, hence $\mathcal{D}(T^*) = \mathcal{H}$. By the closed graph theorem T^* is then a bounded operator, therefore $T = T^{**}$ is a bounded operator on $\mathcal{H}$. Hence, $\mathcal{D} = \mathcal{H}$, which is a contradiction according to our hypothesis. So, the multiplication of $\mathcal{L}^\dagger(\mathcal{D})[\tau_{qu}]$ is not jointly continuous.

In the same way, we can show that the multiplication is not jointly continuous if we consider any of the other uniform and quasi uniform topologies on $\mathcal{L}^\dagger(\mathcal{D})$. Thus, (1) is proved.

Now, by Proposition 5.2.3 and (1) we obtain that $\tau_u \neq \tau^u$. We show that $\tau_*^u \neq \tau_{qu}^*$. Suppose we have equality. Take arbitrary $T \in \mathcal{L}^\dagger(\mathcal{D})$ and $\eta \in \mathcal{D}$ with $\|\eta\| = 1$. Then, for $M = \{\eta\} \in \mathfrak{M}$ there exists $M_1 \in \mathfrak{M}$, such that

$$p_T^M(\xi \otimes \overline{\eta}) = \|T\xi\| \leq p^{M_1}(\xi \otimes \overline{\eta}) + p^{M_1}((\xi \otimes \overline{\eta})^\dagger)$$

$$\leq 2(\sup_{\zeta \in M} \|\zeta\|)\|\xi\|, \ \forall \, \xi \in \mathcal{D}.$$

Hence, T is bounded. Consequently, $\mathcal{D} = \bigcap_{T \in \mathcal{L}^{\dagger}(\mathcal{D})} \mathcal{D}(\overline{T}) = \mathcal{H}$, which is a contradiction. Thus, $\tau_*^u \neq \tau_{qu}^*$.

We show now (3). Let $x \in \mathcal{H} \backslash \mathcal{D}$. Then, for any $M \in \mathfrak{M}$ and $\xi \in \mathcal{D}$ we have

$$p_M\big((\xi \otimes \overline{\xi_n}) - (\xi \otimes \overline{x})\big) \leq \Big(\sup_{\zeta \in M_1} \|\zeta\| \Big)^2 \|\xi\| \|\xi_n - x\| \text{ and}$$

$$p_*^M\big((\xi \otimes \overline{\xi_n}) - (\xi \otimes \overline{x})\big) \leq 2\Big(\sup_{\zeta \in M_1} \|\zeta\| \Big) \|\xi\| \|\xi_n - x\|,$$

where (ξ_n) is a sequence in $\mathcal{D}$ converging to x. Hence, $(\xi \otimes \overline{\xi_n})$ is a Cauchy sequence in both $\mathcal{L}^{\dagger}(\mathcal{D})[\tau_u]$ and $\mathcal{L}^{\dagger}(\mathcal{D})[\tau_*^u]$, but has no limit in any of them. This completes the proof of (3).

Notes The Example 5.2.6 is Proposition 3.3.19 in [136]. The uniform topologies τ_u, τ^u, τ_*^u and the quasi uniform topologies τ_{qu}, τ_{qu}^* were first defined by G. Lassner in [103]. The preceding notations, used in this section, are different from those of Lassner. There are another two uniform topologies, called ρ-*topology* and λ-*topology*, introduced by J.P. Jurzuk (see, [92–94]), but we do not treat these topologies here. The example 5.2.5 is due to K. Schmüdgen [134]; see also [87, Proposition 3.8]. For almost all the results in this section, we refer the reader to [136] and [87].

5.3 Extended C*-Algebras

In this section, we investigate conditions under which the locally convex algebra $\mathcal{M}[\tau_u]$ of a closed $\mathcal{O}^*$-algebra $\mathcal{M}$, becomes a (locally convex) GB^*-algebra of Dixon (Definition 3.3.5). We first introduce the so-called extended C^*-algebras. *Throughout this section, let $\mathcal{M}$ be a closed $\mathcal{O}^*$-algebra on $\mathcal{D}$ in $\mathcal{H}$, containing the identity operator I, where $\mathcal{D}$ is a dense subspace of a Hilbert space $\mathcal{H}$. Let $\mathcal{M}_b$ be the bounded part of a closed $\mathcal{O}^*$-algebra $\mathcal{M}$, that is, $\mathcal{M}_b := \{T \in \mathcal{M} : \overline{T} \in \mathcal{B}(\mathcal{H})\}$.*

Definition 5.3.1 A closed $\mathcal{O}^*$-algebra $\mathcal{M}$ is called *symmetric* if $(I + T^{\dagger}T)^{-1}$ belongs to $\mathcal{M}_b$, for all $T \in \mathcal{M}$. If $\mathcal{M}$ is symmetric and $\overline{\mathcal{M}_b} := \{\overline{T} : T \in \mathcal{M}_b\}$ is a C^*-algebra (resp. a von Neumann algebra), then $\mathcal{M}$ is said to be an *EC*-algebra* (abbreviation of extended C^*-algebra), respectively an *EW*-algebra* (abbreviation of 'extended W^*-algebra') *on $\mathcal{D}$ over $\overline{\mathcal{M}_b}$*.

By Theorem 2.3 in [78] we have the following

Proposition 5.3.2 *A closed symmetric $\mathcal{O}^*$-algebra $\mathcal{M}$ on $\mathcal{D}$ in $\mathcal{H}$, in particular, an EC*-algebra $\mathcal{M}$ is integrable. By Proposition 5.1.2, $\overline{\mathcal{M}} := \{\overline{T} : T \in \mathcal{M}\}$ is a *-algebra of closed operators in $\mathcal{H}$ equipped with the strong sum, the strong scalar multiplication, the strong product and the involution $T \mapsto \overline{T^{\dagger}}$.*

Consider now a closed $\mathcal{O}^*$-algebra on $\mathcal{D}$ in $\mathcal{H}$ and let τ be one of the topologies

$$\tau_w, \quad \tau_{qs}^*, \quad \tau_{qs}^{*,\mathcal{M}}, \quad \tau_u, \quad \tau_{qu}^*, \quad \tau_{qu}^{*,\mathcal{M}}$$

on $\mathcal{M}$. Then, by the discussion after (5.2.1), as well as by (5.2.3) and Proposition 5.2.1(3), $\mathcal{M}[\tau]$ is a locally convex $*$-algebra and $\tau_w \prec \tau$.

Recall that $\mathfrak{B}^*_{\mathcal{M}[\tau]}$ is the collection of all closed, bounded and absolutely convex subsets B of $\mathcal{M}[\tau]$, such that (see Definition 3.3.1)

$$I \in B, \quad B^2 \subset B \text{ and } B^* = B.$$

In this regard, we have the following

Lemma 5.3.3 *The set* $U(\mathcal{M}) := \{T \in \mathcal{M}_b : \|\overline{T}\| \le 1\}$ *is a greatest member in* $\mathfrak{B}^*_{\mathcal{M}[\tau]}$.

Proof It is easily shown that $U(\mathcal{M}) \in \mathfrak{B}^*_{\mathcal{M}[\tau]}$. We still show that $U(\mathcal{M})$ is a greatest element in $\mathfrak{B}^*_{\mathcal{M}[\tau]}$. Take arbitrary $B \in \mathfrak{B}^*_{\mathcal{M}[\tau]}$. Suppose that there exists $S \in B$ such that $\|\overline{S}\| > 1$. Then, there is an element $\xi \in \mathcal{D}$ with $\|\xi\| = 1$ and $\|S\xi\| > 1$, therefore

$$\lim_{n \to \infty} \left| \langle (S^\dagger S)^{2^n} \xi, \xi \rangle \right| \ge \lim_{n \to \infty} \|S\xi\|^{2^{n+1}} = \infty.$$

On the other hand, since $\tau_w \prec \tau$ and B is τ-bounded, it will be τ_w-bounded too. Moreover, $(S^\dagger S)^{2^n} \in B$, for every $n \in \mathbb{N}$. Hence, $\overline{\lim}_{n \to \infty} \left| \langle (S^\dagger S)^{2^n} \xi, \xi \rangle \right| < \infty$, which is a contradiction. Therefore, $B \subseteq U(\mathcal{M})$ and this completes the proof. □

Lemma 5.3.3 implies the following

Theorem 5.3.4 *Let* $\mathcal{M}$ *be a closed* $\mathcal{O}^*$-*algebra. Then,* $\mathcal{M}$ *is an* EC^*-*algebra, if and only if,* $\mathcal{M}[\tau]$ *is a* (locally convex) GB^*-*algebra of Dixon, where* τ *is one of the topologies* $\tau_w, \tau_{qs}^*, \tau_{qs}^{*,\mathcal{M}}, \tau_u, \tau_{qu}^*, \tau_{qu}^{*,\mathcal{M}}$ *on* $\mathcal{M}$.

Proof It is immediate from Definition 5.3.1, and Lemma 5.3.3. □

For another characterization of a certain closed $\mathcal{O}^*$-algebra, as an EC^*-algebra, in terms of the set of 'commutatively quasi-positive' elements of a 'locally convex quasi C^*-algebra', see [13, Proposition 4.5].

An $\mathcal{O}^*$-algebra is said to be C^*-*like* if it is a C^*-like locally convex $*$-algebra (see Definition 3.5.1).

Proposition 5.3.5 *Every* C^*-*like* $\mathcal{O}^*$-*algebra is an* EC^*-*algebra.*

Proof Let $\mathcal{M}$ be a C^*-like $\mathcal{O}^*$-algebra on $\mathcal{D}$ in $\mathcal{H}$ with a C^*-like family $\Gamma = \{p_v\}_{v \in \Lambda}$ of seminorms. By Theorem 3.5.3 $\mathcal{M}$ is a GB^*-algebra, whose closed unit ball $U(p_\Lambda)$ is the greatest element in $\mathfrak{B}^*_\Lambda$. For any $\xi \in \mathcal{D}$ we can define a positive linear functional ω_ξ on the C^*-algebra $\mathcal{D}(p_\Lambda)$ by $\omega_\xi(T) = \langle T\xi, \xi \rangle$, $T \in \mathcal{D}(p_\Lambda)$. Then, $\|T\xi\| \le p_\Lambda(T)\|\xi\|$, for each $T \in \mathcal{D}(p_\Lambda)$ and $\xi \in \mathcal{D}$. It follows that

$\mathcal{D}(p_\Lambda) \subset \overline{\mathcal{M}_b}$ and p_Λ equals the operator norm. Furthermore, we can prove as in the proof of Theorem 3.5.3 that $\mathcal{D}(p_\Lambda)$ is dense in $\overline{\mathcal{M}_b}$ with respect to the operator norm. Therefore, it turns out that $\mathcal{D}(p_\Lambda) = \overline{\mathcal{M}_b}$, consequently $\mathcal{M}$ is an EC^*-algebra. This completes the proof.   □

By Theorems 3.5.3 and 6.3.5 (in Sect. 6.3) we have the following

Theorem 5.3.6 *Every C^*-like locally convex $*$-algebra is isomorphic to a C^*-like $\mathcal{O}^*$-algebra.*

Notes P.G. Dixon defined EC^*-algebras and EW^*-algebras of closed operators in a Hilbert space, in [48, 49]. But, it is difficult to introduce a locally convex topology on EC^*-algebras, in general. From this viewpoint, in this chapter, we dealt with EC^*-algebras and EW^*-algebras of closable operators on a common dense subspace in a Hilbert space, defined by A. Inoue in [78]. For the results of this section the reader is referred to [78, 87]. Proposition 5.3.5 and Theorem 5.3.6 are found in [88]. For a study of irreducible representations of EC^*-algebras and EW^*-algebras, the reader may consult [21].

5.4 Left Extended W*-Algebras of Unbounded Hilbert Algebras

It is well known that semifinite von Neumann algebras are related to Hilbert algebras; for the latter term, see [46, Chapter 5]. That is, if $\mathfrak{A}_0$ is a *Hilbert algebra*, then the left von Neumann algebra $\mathcal{U}_0(\mathfrak{A}_0)$ is defined and it is a *semifinite von Neumann algebra*. Conversely, if $\mathcal{M}_0$ is a semifinite von Neumann algebra, then it is isomorphic to the left von Neumann algebra $\mathcal{U}_0(\mathfrak{A}_0)$ of a Hilbert algebra $\mathfrak{A}_0$ (see [46, Part I, Chapter 6, Theorem 2]). In this section, we extend the preceding results to EW^*-algebras (see Definition 5.3.1). Our starting point will be an unbounded extension of Hilbert algebras.

Let $\mathfrak{A}$ be a $*$-algebra, which is also a pre-Hilbert space with inner product $\langle \, , \, \rangle$. Let $\mathcal{H}$ be the completion of $\mathfrak{A}$ with respect to the norm induced by the inner product. Suppose that $\mathfrak{A}$ satisfies the following properties:

$$\text{(i)} \quad \langle \xi, \eta \rangle = \langle \eta^*, \xi^* \rangle, \ \forall \, \xi, \eta \in \mathfrak{A};$$

$$\text{(ii)} \quad \langle \xi\eta, \zeta \rangle = \langle \eta, \xi^*\zeta \rangle, \ \forall \, \xi, \eta, \zeta \in \mathfrak{A}.$$

Now we define the operators $\pi(\xi)$ and $\pi'(\xi)$, $\xi \in \mathfrak{A}$, by

$$\pi(\xi)(\eta) := \xi\eta \ \text{ and } \ \pi'(\xi)(\eta) := \eta\xi, \ \forall \, \eta \in \mathfrak{A}.$$

Then, π is a $*$-representation of $\mathfrak{A}$ in $\mathcal{H}$ with domain $\mathcal{D}(\pi) = \mathfrak{A}$ and $\pi(\xi^*)$ $\subset \pi(\xi)^*$, for every $\xi \in \mathfrak{A}$, while π' is a linear map of $\mathfrak{A}$ into $\mathcal{L}^{\dagger}(\mathfrak{A})$, such that

$$\pi'(\xi^*) \subset \pi'(\xi)^* \text{ and } \pi'(\xi\xi_1) = \pi'(\xi_1)\pi'(\xi), \ \forall\, \xi, \xi_1 \in \mathfrak{A}.$$

Such a π' is called a $*$-*antirepresentation*.

Definition 5.4.1 If $\mathfrak{A}$ satisfies the properties (i), (ii) and

(iii) $\mathfrak{A}_0^2$ is dense in $\mathcal{H}$, where $\mathfrak{A}_0 := \{\xi \in \mathfrak{A} : \overline{\pi(\xi)} \in \mathcal{B}(\mathcal{H})\}$,

then $\mathfrak{A}$ is called an *unbounded Hilbert algebra over* $\mathfrak{A}_0$ in $\mathcal{H}$. In particular, if $\mathfrak{A}_0 \neq \mathfrak{A}$, then $\mathfrak{A}$ is said to be *pure*. If $\mathfrak{A} = \mathfrak{A}_0$, then $\mathfrak{A}$ *is named a* Hilbert algebra *in* $\mathcal{H}$. The closure of the $*$-representation π (resp. π') is called *the left* (resp. right) *regular representation of* $\mathfrak{A}$ and it is denoted by $\pi_{\mathfrak{A}}$ (resp. $\pi'_{\mathfrak{A}}$).

For the left and right regular representation of an unbounded Hilbert algebra, we have the following

Theorem 5.4.2 *Let* $\mathfrak{A}$ *be an unbounded Hilbert algebra over* $\mathfrak{A}_0$ *in* $\mathcal{H}$. *Then, the left regular representations* $\pi_{\mathfrak{A}}$ *and* $\pi'_{\mathfrak{A}}$ *of* $\mathfrak{A}$ *are integrable. Hence,* $\overline{\pi_{\mathfrak{A}}(\mathfrak{A})}$ *is* $*$-*algebra of closed operators in* $\mathcal{H}$ *under the algebraic operations of strong sum, strong scalar multiplication, strong product and the involution given by* † *(see, Proposition 5.3.2).*

Proof It is clear that $\mathfrak{A}_0$ is a Hilbert algebra in $\mathcal{H}$. Let π_0 (resp. π'_0) be the left (resp. right) regular representation of the Hilbert algebra $\mathfrak{A}_0$. More precisely, for any $x \in \mathcal{H}$ we define $\pi_0(x)$ and $\pi'_0(x)$ by

$$\pi_0(x)\xi := \overline{\pi'_0(\xi)}x \text{ and } \pi'_0(x)\xi := \overline{\pi_0(\xi)}x, \ \forall\, \xi \in \mathfrak{A}_0.$$

Then, $\pi_0(x)$ and $\pi'_0(x)$ are linear operators in $\mathcal{H}$ with domain $\mathfrak{A}_0$. The involution on $\mathfrak{A}$ is extended to an involution on $\mathcal{H}$, which is also denoted by $*$. Then, for $\xi \in \mathfrak{A}$, by the definition of $\pi_{\mathfrak{A}}$ and π_0 we have

$$\pi_{\mathfrak{A}}(\xi)\eta = \pi'_0(\eta)\xi = \pi_0(\xi)\eta, \ \forall\, \eta \in \mathfrak{A}_0,$$

so that $\pi_0(\xi) \subset \pi_{\mathfrak{A}}(\xi)$. Furthermore, from [119, Theorem 3], it follows that

$$\pi_0(x)^* = \overline{\pi_0(x^*)} \text{ and } \pi'_0(x)^* = \overline{\pi'_0(x^*)}, \ x \in \mathcal{H}, \tag{5.4.1}$$

which implies that

$$\pi_{\mathfrak{A}}(\xi)^* \subset \pi_0(\xi)^* = \overline{\pi_0(\xi^*)} \subset \overline{\pi_{\mathfrak{A}}(\xi^*)} \subset \pi_{\mathfrak{A}}(\xi)^*, \ \xi \in \mathfrak{A}.$$

Hence, we have

$$\pi_{\mathfrak{A}}(\xi)^* = \overline{\pi_{\mathfrak{A}}(\xi^*)} = \overline{\pi_{\mathfrak{A}}(\xi)}^\dagger, \ \xi \in \mathfrak{A}, \tag{5.4.2}$$

which means that $\pi_{\mathfrak{A}}$ is integrable. Similarly, we can show that $\pi'_{\mathfrak{A}}$ is integrable. This completes the proof.   □

In the sequel, we shall construct GB^*-algebras through unbounded Hilbert algebras. The first step for this is the construction of an L^ω-space by using the noncommutative integration theory of I.E. Segal [140]. The reader is referred to [117, 119, 140] for noncommutative integration of semifinite von Neumann algebras.

Let $\mathcal{U}_0(\mathfrak{A}_0)$ (resp. $\mathcal{V}_0(\mathfrak{A}_0)$) be the left (resp. right) von Neumann algebra of a given Hilbert algebra $\mathfrak{A}_0$. It is well known that both are semifinite. Let φ_0 be the natural trace on $\mathcal{U}_0(\mathfrak{A}_0)_+$. If

$$(\mathfrak{A}_0)_b := \left\{ x \in \mathcal{H} : \overline{\pi_0(x)} \in \mathcal{B}(\mathcal{H}) \right\},$$

then $(\mathfrak{A}_0)_b$ is a Hilbert algebra containing $\mathfrak{A}_0$. When $\mathfrak{A}_0 = (\mathfrak{A}_0)_b$, then $\mathfrak{A}_0$ is called a *maximal* (or *achieved*) *Hilbert algebra* in $\mathcal{H}$. Let $\mathfrak{M}[\mathcal{U}_0(\mathfrak{M}_0)]$ be the set of all *measurable operators* in $\mathcal{H}$ (in the sense of I.G. Segal) with respect to the von Neumann algebra $\mathcal{U}_0(\mathfrak{A}_0)$ and $\mathfrak{M}[\mathcal{U}_0(\mathfrak{A}_0)]_+$ the set of all *positive self-adjoint operators* in $\mathfrak{M}[\mathcal{U}_0(\mathfrak{A}_0)]$.

For any $T \in \mathfrak{M}[\mathcal{U}_0(\mathfrak{A}_0)]_+$ we put

$$\mu_0(T) := \sup \left\{ \varphi_0(\overline{\pi_0(\xi)}) : 0 \le \overline{\pi_0(\xi)} \le T, \ \xi \in (\mathfrak{A}_0)_b^2 \right\} \text{ and}$$

$$L^p(\varphi_0) := \left\{ T \in \mathfrak{M}[\mathcal{U}_0(\mathfrak{A}_0)] : \|T\|_p := \mu_0(|T|^p)^{1/p} < \infty \right\}, \ 1 \le p < \infty.$$

Then, $\|T\|_p$ is called L^p-*norm of* T *in* $L^p(\varphi_0)$ *and* μ_0 *is called integral on* $L^p(\varphi_0)$. If $p = \infty$, we identify $L^\infty(\varphi_0)$ with $\mathcal{U}_0(\mathfrak{A}_0)$ and denote by $\|T\|$ or $\|T\|_\infty$, the operator norm of $T \in \mathcal{U}_0(\mathfrak{A}_0)$. It is well known that $L^p(\varphi_0)[\| \cdot \|_p]$ is a Banach space for $1 \le p \le \infty$.

Now, for $2 \le p \le \infty$, we define

$$L_2^p(\mathfrak{A}_0) := \left\{ x \in \mathcal{H} : \overline{\pi_0(x)} \in L^p(\varphi_0) \right\} \text{ and } \|x\|_{p \wedge 2} := \max \left\{ \|x\|_p, \|x\|_2 \right\},$$

$$x \in L_2^p(\mathfrak{A}_0), \text{ where } \|x\|_p :- \|\overline{\pi_0(x)}\|_p.$$

In this regard, we have the next result

Lemma 5.4.3 *The following hold:*

(1) $L_2^p(\mathfrak{A}_0)[\| \cdot \|_{p \wedge 2}]$ *is a Banach space and*

$$\mathcal{H} = L_2^2(\mathfrak{A}_0) \supseteq L_2^p(\mathfrak{A}_0) \supseteq L_2^q(\mathfrak{A}_0) \supseteq L_2^\omega(\mathfrak{A}_0) \supseteq L_2^\infty(\mathfrak{A}_0) = (\mathfrak{A}_0)_b,$$

$$2 \le p < q, \; where \; L_2^\omega(\mathfrak{A}_0) := \bigcap_{2 \le p < \infty} L_2^p(\mathfrak{A}_0).$$

(2) *The following statements are equivalent:*

(i) $\mathcal{H} \neq (\mathfrak{A}_0)_b.$
(ii) $L_2^\omega(\mathfrak{A}_0) \neq (\mathfrak{A}_0)_b.$

Proof

(1) It is clear that $L_2^p(\mathfrak{A}_0)[\| \cdot \|_{p \wedge 2}]$ is a Banach space. Let $2 \le p < q$ and $x \in L_2^q(\mathfrak{A}_0)$. Let $\overline{\pi_0(x)} = U|\overline{\pi_0(x)}|$ be the polar decomposition of the operator $\overline{\pi_0(x)}$ and $|\overline{\pi_0(x)}| = \int_0^\infty \lambda dE(\lambda)$ its spectral resolution. Then, we have

$$\|x\|_p^p = \|\overline{\pi_0(x)}\|_p^p = -\int_0^\infty \lambda^p d\varphi_0\big(E(\lambda)^\perp\big)$$

$$= -\int_0^1 \lambda^p d\varphi_0\big(E(\lambda)^\perp\big) - \int_1^\infty \lambda^p d\varphi_0\big(E(\lambda)^\perp\big) \qquad (5.4.3)$$

$$\le -\int_0^1 \lambda^2 d\varphi_0\big(E(\lambda)^\perp\big) - \int_1^q \lambda^q d\varphi_0\big(E(\lambda)^\perp\big)$$

$$\le \|x\|_2^2 + \|x\|_q^q < \infty,$$

where $E(\lambda)^\perp := I - E(\lambda)$, which implies (1).

(2) We show (i) $\Rightarrow$ (ii). Suppose that $x \in \mathcal{H} \backslash (\mathfrak{A}_0)_b$. Let $|\overline{\pi_0(x)}| = \int_0^\infty \lambda dE(\lambda)$ be the spectral resolution of the operator $|\overline{\pi_0(x)}|$. Since $\overline{\pi_0(x)} \notin \mathcal{B}(\mathcal{H})$, $E(n+1) - E(n) \neq 0$, for infinite many n, therefore we may suppose that $E(n+1) - E(n) \neq 0$, for every $n \in \mathbb{N}$. Thus, since

$$\|E(n+1) - E(n)\|_2^2 = \varphi_0\big(E(n+1) - E(n)\big)$$

$$= -\int_n^{n+1} d\varphi_0\big(E(\lambda)^\perp\big)$$

$$\le -\int_n^{n+1} \lambda^2 d\varphi_0\big(E(\lambda)^\perp\big)$$

$$\le \|\overline{\pi_0(x)}\|_2^2 = \|x\|_2^2,$$

it follows that $E(n + 1) - E(n) \in L^\infty(\varphi_0) \cap L^2(\varphi_0)$. Hence, there exists an element e_n of $(\mathfrak{A}_0)_b$, such that $e_n^2 = e_n = e_n^*$ (such an e_n is called a *projection* in $(\mathfrak{A}_0)_b$) and $E(n + 1) - E(n) = \pi_0(e_n)$. Clearly, (e_n) is a sequence of non-zero mutually orthogonal projections in $(\mathfrak{A}_0)_b$. Furthermore, we have that $\sum_{n=1}^\infty \|e_n\|_2^2 < \infty$. Indeed, since

$$\sum_{k=n}^m \|e_k\|_2^2 = \sum_{k=n}^m \varphi_0\big(\pi_0(e_k)\big)$$

$$= \sum_{k=n}^m \varphi_0\big(E(k + 1) - E(k)\big)$$

$$= \varphi_0\big(E(m + 1) - E(n)\big)$$

and the sequence $\big(E(m + 1) - E(n)\big)$ converges σ-weakly to 0 as $m, n \to \infty$; the σ-weak continuity of φ_0 implies that $\lim_{m,n \to \infty} \sum_{k=n}^m \|e_k\|_2^2 = 0$. Hence,

$$\sum_{n=1}^\infty \|e_n\|_2^2 < \infty.$$

Now, take a positive integer n_0, such that $\sum_{k=n_0}^\infty \|e_k\|_2^2 < 1$ and put

$$\alpha_n := \log\Big(\sum_{k=n_0+n}^\infty \|e_k\|_2^2\Big), \quad n = 0, 1, 2, \ldots \quad \text{and} \quad a = \sum_{k=n_0}^\infty \alpha_k\, e_{n_0+k}\,.$$

Then, for every $n \in \mathbb{N}$, we obtain

$$\sum_{k=0}^\infty |\alpha_n|^n \|e_{n_0+k}\|_2^2 < \int_0^1 |\log t|^n dt = n!, \quad \text{hence}$$

$$\lim_{m,n \to \infty} \|\sum_{k=n}^m \alpha_k\, e_{n_0+k}\|_2^2 \leq \lim_{m,n \to \infty} \sum_{k=n}^m |\alpha_k|^2 \|e_{n_0+k}\|_2^2 = 0,$$

which implies that $a \in \bigcap_{n \in \mathbb{N}} L_2^n(\mathfrak{A}_0) = L_2^\omega(\mathfrak{A}_0)$. On the other hand, since

$$\lim_{n \to \infty} \alpha_n = \infty \quad \text{and} \quad \|e_{k_0+n}\|_2 \neq 0, \ n \in \mathbb{N},$$

we obtain that $\overline{\pi_0(a)} \notin \mathcal{B}(\mathcal{H})$, therefore $a \notin (\mathfrak{A}_0)_b$. Thus, we have proved that $a \in L_2^\omega(\mathfrak{A}_0) \backslash (\mathfrak{A}_0)_b$.

(ii) $\Rightarrow$ (i) Follows from (1).

$\square$

Theorem 5.4.4 $L_2^\omega(\mathfrak{A}_0)$ *is an unbounded Hilbert algebra over* $(\mathfrak{A}_0)_b$, *which is maximal among all unbounded Hilbert algebras containing* $\mathfrak{A}_0$. *Moreover,* $L_2^\omega(\mathfrak{A}_0)$ *is pure, if and only if,* $\mathcal{H}$ *is not a Hilbert algebra, that is* $\mathcal{H} \neq (\mathfrak{A}_0)_b$.

Proof It is clear that $L_2^\omega(\mathfrak{A}_0)$ is a pre-Hilbert space in $\mathcal{H}$. Take arbitrary elements $x, y \in L_2^\omega(\mathfrak{A}_0)$. Then, by the generalized Hölder inequality, we have

$$\|\overline{\pi_0(x)} \cdot \overline{\pi_0(y)}\|_p \leq \|\overline{\pi_0(x)}\|_{2p} \|\overline{\pi_0(y)}\|_{2p}, \ \forall \, p \geq 2.$$

Furthermore, $\{\overline{\pi_0(x)} : x \in \mathcal{H}\}$ and $L^2(\varphi_0)$ are Hilbert spaces isometric to $\mathcal{H}$. Hence, there exists an element $z \in \mathcal{H}$, such that $\overline{\pi_0(x)} \cdot \overline{\pi_0(y)} = \overline{\pi_0(z)}$. Applying (5.4.1) we obtain

$$\begin{aligned}
\langle \pi_0(x^*)\xi, y \rangle &= \varphi_0\big((\pi_0(y)^* \cdot \overline{\pi_0(x^*)})\overline{\pi_0(\xi)}\big) \\
&= \varphi_0\big((\overline{\pi_0(x)} \cdot \overline{\pi_0(y)})^* \overline{\pi_0(\xi)}\big) \\
&= \varphi_0\big(\overline{\pi_0(z)}^* \overline{\pi_0(\xi)}\big) = \langle \xi, z \rangle, \ \forall \, \xi \in \mathfrak{A}_0,
\end{aligned}$$

which implies that $y \in \mathcal{D}\big(\pi_0(x^*)^*\big) = \mathcal{D}\big(\overline{\pi_0(x)}\big)$ and $\overline{\pi_0(x)}y = z$. Thus, $xy := \overline{\pi_0(x)}y$ is defined and belongs to $L_2^\omega(\mathfrak{A}_0)$. Furthermore, since

$$\|\overline{\pi_0(x)}\|_p = \|\overline{\pi_0(x)}^*\|_p = \|\overline{\pi_0(x^*)}\|_p, \ \forall \, x \in L_2^\omega(\mathfrak{A}_0) \text{ and } 2 \leq p < \infty,$$

it follows that $L_2^\omega(\mathfrak{A}_0)$ is $*$-invariant. Thus, $L_2^\omega(\mathfrak{A}_0)$ is a $*$-algebra. Moreover, by (5.4.1), we have

$$\begin{aligned}
\langle x, y \rangle &= \langle y^*, x^* \rangle \text{ and} \\
\langle xy, z \rangle &= \langle \overline{\pi_0(x)}y, z \rangle \\
&= \langle y, \overline{\pi_0(x)}^* z \rangle \\
&= \langle y, \overline{\pi_0(x^*)}z \rangle \\
&= \langle y, x^* z \rangle, \ \forall \, x, y, z \in L_2^\omega(\mathfrak{A}_0).
\end{aligned}$$

Thus, $L_2^\omega(\mathfrak{A}_0)$ is an unbounded Hilbert algebra over $(\mathfrak{A}_0)_b$ in $\mathcal{H}$.

Let now $\mathfrak{A}$ be an unbounded Hilbert algebra in $\mathcal{H}$ containing $\mathfrak{A}_0$. Then, using a similar argument as before, it is shown that

$$xy = \overline{\pi_0(x)}y, \ \forall \, x, y \in \mathfrak{A}, \text{ which implies } x \in L_2^\omega(\mathfrak{A}_0), \ \forall \, x \in \mathfrak{A}.$$

Hence, $\mathfrak{A} \subseteq L_2^\omega(\mathfrak{A}_0)$. Suppose now that $(\mathfrak{A}_0)_b \neq \mathcal{H}$. Then, from Lemma 5.4.3(2)(i), $L_2^\omega(\mathfrak{A}_0)$ is pure and this completes the proof. $\qquad\square$

Let $\mathfrak{A}$ be an unbounded Hilbert algebra over $\mathfrak{A}_0$ in a Hilbert space $\mathcal{H}$ and φ_0 (resp. ψ_0) the natural trace on $\mathcal{U}_0(\mathfrak{A}_0)_+$ (resp. $\mathcal{V}_0(\mathfrak{A}_0)_+$). If J denotes the involution on $\mathcal{H}$, for every $x \in \mathcal{H}$, we have

$$J\overline{\pi_0(x)}J = \overline{\pi_0'(x^*)} \quad \text{and} \quad J\overline{\pi_0'(x)}J = \overline{\pi_0(x^*)}, \tag{5.4.4}$$

which implies that

$$JL_2^\omega(\mathfrak{A}_0)J = L_2^\omega(\psi_0)$$

$$L_2^\omega(\mathfrak{A}_0) = \left\{ x \in \mathcal{H} : \overline{\pi_0'(x)} \in L_2^\omega(\psi_0) \right\}$$

and moreover

$$\mathcal{U}_0(\mathfrak{A}_0)L_2^\omega(\mathfrak{A}_0) \subseteq L_2^\omega(\mathfrak{A}_0), \quad \mathcal{V}_0(\mathfrak{A}_0)L_2^\omega(\mathfrak{A}_0) \subseteq L_2^\omega(\mathfrak{A}_0). \tag{5.4.5}$$

Now, using (5.4.2), we obtain

$$\overline{\pi_{L_2^\omega(\mathfrak{A}_0)}(\xi)} = \overline{\pi_\mathfrak{A}(\xi)} = \overline{\pi_0(\xi)} \; \eta \; \mathcal{U}_0(\mathfrak{A}_0), \quad \overline{\pi'_{L_2^\omega(\mathfrak{A}_0)}(\xi)} = \overline{\pi'_\mathfrak{A}(\xi)} = \overline{\pi_0'(\xi)} \; \eta \; \mathcal{V}_0(\mathfrak{A}_0),$$

where $\overline{\pi_0(\xi)} \; \eta \; \mathcal{U}_0(\mathfrak{A}_0)$ means that the operator $\overline{\pi_0(\xi)}$ is affiliated with the left von Neumann algebra $\mathcal{U}_0(\mathfrak{A}_0)$. Hence,

$$\pi_{L_2^\omega(\mathfrak{A}_0)}(\mathfrak{A})'_w = \pi_\mathfrak{A}(\mathfrak{A})'_w = \mathcal{V}_0(\mathfrak{A}_0) \quad \text{and} \quad \pi'_{L_2^\omega(\mathfrak{A}_0)}(\mathfrak{A})'_w = \pi'_\mathfrak{A}(\mathfrak{A})'_w = \mathcal{U}_0(\mathfrak{A}_0),$$

which in view of (5.4.4) implies that the closure $\mathcal{U}(\mathfrak{A})$ (resp. $\mathcal{V}(\mathfrak{A})$) of the $\mathcal{O}^*$-algebra on $L_2^\omega(\mathfrak{A}_0)$ generated by $\pi_{L_2^\omega(\mathfrak{A}_0)}(\mathfrak{A})$ and $\mathcal{U}_0(\mathfrak{A}_0) \upharpoonright_{L_2^\omega(\mathfrak{A}_0)}$ (resp. $\pi_{L_2^\omega(\mathfrak{A}_0)}(\mathfrak{A})$ and $\mathcal{V}_0(\mathfrak{A}_0) \upharpoonright_{L_2^\omega(\mathfrak{A}_0)}$) is an EW^*-algebra over $\mathcal{U}_0(\mathfrak{A}_0)$ (resp. $\mathcal{V}_0(\mathfrak{A}_0)$), and it is called *left* (resp. *right*) EW^*-algebra of $\mathfrak{A}$. Further, let $\mathcal{D}(\mathcal{U}(\mathfrak{A}))$ (resp. $\mathcal{D}(\mathcal{V}(\mathfrak{A}))$) be the domain of the $\mathcal{O}^*$-algebra $\mathcal{U}(\mathfrak{A})$ (resp. $\mathcal{V}(\mathfrak{A})$), that is

$$\mathcal{D}(\mathcal{U}(\mathfrak{A})) = \bigcap_{T \in \mathcal{U}(\mathfrak{A})} \mathcal{D}(\overline{T}) \quad \text{and} \quad \mathcal{D}(\mathcal{V}(\mathfrak{A})) = \bigcap_{T \in \mathcal{V}(\mathfrak{A})} \mathcal{D}(\overline{T}).$$

The left and right EW^*-algebras $\mathcal{U}(\mathfrak{A})$, $\mathcal{V}(\mathfrak{A})$ are related in the following way

$$J\mathcal{D}(\mathcal{V}(\mathfrak{A})) = \mathcal{D}(\mathcal{U}(\mathfrak{A})) \quad \text{and} \quad \mathcal{V}(\mathfrak{A}) = J\mathcal{U}(\mathfrak{A})J.$$

In particular, for the left and right EW^*-algebras of the maximal unbounded Hilbert algebra $L_2^\omega(\mathfrak{A}_0)$ we have

$$\mathcal{D}(\mathcal{U}(L_2^\omega(\mathfrak{A}_0))) = \mathcal{D}(\pi_{L_2^\omega(\mathfrak{A}_0)}),$$

$$\mathcal{U}(L_2^\omega(\mathfrak{A}_0)) = \mathcal{U}_0(\mathfrak{A}_0) \upharpoonright_{\mathcal{D}(\pi_{L_2^\omega(\mathfrak{A}_0)})} + \pi_{L_2^\omega(\mathfrak{A}_0)}(L_2^\omega(\mathfrak{A}_0)), \tag{5.4.6}$$

$$\mathcal{D}\big(\mathcal{V}\big(L_2^\omega(\mathfrak{A}_0)\big)\big) = \mathcal{D}(\pi'_{L_2^\omega(\mathfrak{A}_0)}),$$

$$\mathcal{V}\big(L_2^\omega(\mathfrak{A}_0)\big) = \mathcal{V}_0(\mathfrak{A}_0)\,\lceil_{\mathcal{D}(\pi'_{L_2^\omega(\mathfrak{A}_0)})} + \pi'_{L_2^\omega(\mathfrak{A}_0)}\big(L_2^\omega(\mathfrak{A}_0)\big). \tag{5.4.7}$$

Indeed, by (5.4.5) it is shown that the $\mathcal{O}^*$-algebra on $L_2^\omega(\mathfrak{A}_0)$ generated by $\mathcal{U}_0(\mathfrak{A}_0)\,\lceil_{L_2^\omega(\mathfrak{A}_0)}$ and $\pi_{L_2^\omega(\mathfrak{A}_0)}\big(L_2^\omega(\mathfrak{A}_0)\big)\,\lceil_{L_2^\omega(\mathfrak{A}_0)}$ equals to

$$\mathcal{U}_0(\mathfrak{A}_0)\,\lceil_{L_2^\omega(\mathfrak{A}_0)} + \pi_{L_2^\omega(\mathfrak{A}_0)}\big(L_2^\omega(\mathfrak{A}_0)\big)\,\lceil_{L_2^\omega(\mathfrak{A}_0)}$$

and this implies (5.4.6). The relation (5.4.7) is similarly shown.

Thus, we obtain the following

Theorem 5.4.5 *Let $\mathfrak{A}$ be an unbounded Hilbert algebra over $\mathfrak{A}_0$ in $\mathcal{H}$. Then, the left and right EW*-algebras $\mathcal{U}(\mathfrak{A})$ and $\mathcal{V}(\mathfrak{A})$ of $\mathfrak{A}$ are respectively defined, such that $\mathcal{V}(\mathfrak{A}) = J\mathcal{U}(\mathfrak{A})J$. In particular,*

$$\mathcal{U}\big(L_2^\omega(\mathfrak{A}_0)\big) = \mathcal{U}_0(\mathfrak{A}_0)\,\lceil_{\mathcal{D}(\pi_{L_2^\omega(\mathfrak{A}_0)})} + \pi_{L_2^\omega(\mathfrak{A}_0)}\big(L_2^\omega(\mathfrak{A}_0)\big)\ and$$

$$\mathcal{V}\big(L_2^\omega(\mathfrak{A}_0)\big) = \mathcal{V}_0(\mathfrak{A}_0)\,\lceil_{\mathcal{D}(\pi'_{L_2^\omega(\mathfrak{A}_0)})} + \pi'_{L_2^\omega(\mathfrak{A}_0)}\big(L_2^\omega(\mathfrak{A}_0)\big).$$

A consequence of Theorems 5.4.4 and 5.4.5 is the following

Corollary 5.4.6 *Let $\mathfrak{A}_0$ be a maximal Hilbert algebra in a Hilbert space $\mathcal{H}$. Suppose that $\mathfrak{A}_0 \neq \mathcal{H}$. Then, $L_2^\omega(\mathfrak{A}_0)$ is a pure unbounded Hilbert algebra over $\mathfrak{A}_0$ in $\mathcal{H}$ and the respective left and right EW*-algebras of $L_2^\omega(\mathfrak{A}_0)$ are defined, in such a way that*

$$\mathcal{U}\big(L_2^\omega(\mathfrak{A}_0)\big) = \mathcal{U}_0(\mathfrak{A}_0)\,\lceil_{\mathcal{D}(\pi_{L_2^\omega(\mathfrak{A}_0)})} + \pi_{L_2^\omega(\mathfrak{A}_0)}\big(L_2^\omega(\mathfrak{A}_0)\big)\ and$$

$$\mathcal{V}\big(L_2^\omega(\mathfrak{A}_0)\big) = \mathcal{V}_0(\mathfrak{A}_0)\,\lceil_{\mathcal{D}(\pi'_{L_2^\omega(\mathfrak{A}_0)})} + \pi'_{L_2^\omega(\mathfrak{A}_0)}\big(L_2^\omega(\mathfrak{A}_0)\big) = J\mathcal{U}\big(L_2^\omega(\mathfrak{A}_0)\big)J.$$

Definition 5.4.7 An EW*-algebra $\mathcal{M}$ is called *standard* if there exists a pure unbounded Hilbert algebra $\mathfrak{A}$, such that $\mathcal{M} = \mathcal{U}(\mathfrak{A})$.

Every left EW*-algebra of an unbounded Hilbert algebra is standard.

In the sequel, we consider when an EW*-algebra is (algebraically) isomorphic to a standard EW*-algebra.

Let $\mathcal{M}_0$ be a semifinite von Neumann algebra on a Hilbert space $\mathcal{H}$ and τ_0 a faithful semifinite trace on $(\mathcal{M}_0)_+$. Let $\mathfrak{M}[\mathcal{M}_0]$ denote *the set of all measurable operators with respect to $\mathcal{M}_0$*. For every $T \in \mathfrak{M}[\mathcal{M}_0]_+$, we put

$$\mu(T) := \sup\big\{\tau_0(S) : S \in (\mathcal{M}_0)_+,\ \text{such that}\ \tau_0(S) < \infty\big\}\ \text{and}$$

$$L^p(\tau_0) := \big\{T \in \mathfrak{M}[\mathcal{M}_0] : \|T\|_p := \mu\big((|T|^p)^{1/p}\big) < \infty\big\},\ 1 \le p < \infty,$$

$$L^\infty(\tau_0) := \mathcal{M}_0.$$

Then, $L_2^\infty(\tau_0) := L^\infty(\tau_0) \bigcap L^2(\tau_0)$ *is a maximal Hilbert algebra in the Hilbert space* $L^2(\tau_0)$ *under the inner product*

$$\langle S, T \rangle := \mu(T^* \cdot S), \ S, \ T \in L_2^\infty(\tau_0).$$

Moreover, $L_2^\omega(\tau_0) := \bigcap_{2 \le p < \infty} L^p(\tau_0)$ *is a maximal unbounded Hilbert algebra over* $L_2^\infty(\tau_0)$. *Furthermore,* $\mathcal{M}_0 + L_2^\omega(\tau_0)$ *is a *-algebra of closed operators on* $\mathcal{H}$ *generated by* $\mathcal{M}_0$ *and* $L_2^\omega(\tau_0)$, *equipped with the algebraic operations given by the strong sum, strong scalar multiplication, strong product and the involution* $T^* := \overline{T^\dagger}$, *for* $T \in \mathcal{M}_0 + L_2^\omega(\tau_0)$.

Suppose now that $\mathcal{M}$ is a pure EW^*-algebra over $\mathcal{M}_0$ contained in $\mathcal{M}_0 + L_2^\omega(\tau_0)$. Then, it is shown that $\overline{\mathcal{M} \bigcap L_2^\omega(\tau_0)}$ is a pure unbounded Hilbert algebra over $L_2^\infty(\tau_0)$ and $\mathcal{M}$ is isomorphic to $\mathcal{U}(\overline{\mathcal{M} \bigcap L_2^\omega(\tau_0)})$. Thus, we obtain the following

Theorem 5.4.8 *Let* $\mathcal{M}_0$ *be a semifinite von Neumann algebra on a Hilbert space* $\mathcal{H}$ *and* τ_0 *a faithful normal semifinite trace on* $(\mathcal{M}_0)_+$. *Then, every pure EW^*-algebra over* $\mathcal{M}_0$ *contained in* $\mathcal{M}_0 + L_2^\infty(\tau_0)$ *is isomorphic to a standard EW^*-algebra.*

Finally, we investigate unbounded Hilbert algebras from the point of view of the theory of locally convex *-algebras. So, let $\mathfrak{A}_0$ be a Hilbert algebra in a Hilbert space $\mathcal{H}$, such that $(\mathfrak{A}_0)_b \ne \mathcal{H}$.

Definition 5.4.9 The locally convex topology on $L_2^\omega(\mathfrak{A}_0)$ defined by the family of L^p-norms $\{\| \cdot \|_p\}_{2 \le p < \infty}$ is called the L_2^ω-*topology* and is denoted by τ_2^ω.

Note that, the next Theorem 5.4.10 in the presence of identity, shows that $L_2^\omega(\mathfrak{A}_0)[\tau_2^\omega]$ becomes a Fréchet GB^*-algebra.

Theorem 5.4.10 $L_2^\omega(\mathfrak{A}_0)[\tau_2^\omega]$ *is a Fréchet *-algebra, whose the subset* $\left(L_2^\omega(\mathfrak{A}_0)[\tau_2^\omega]\right)_0$, *of all* (Allan) *bounded elements, equals to* $(\mathfrak{A}_0)_b$ *and* $\overline{(\mathfrak{A}_0)_b}^{\tau_2^\omega} = L_2^\omega(\mathfrak{A}_0)$. *In particular, if* $\mathfrak{A}_0$ *has an identity element* e, *then* $L_2^\omega(\mathfrak{A}_0)[\tau_2^\omega]$ *is a Fréchet GB^*-algebra over* $U\left(\mathcal{U}_0(\mathfrak{A}_0)\right)e$, *where* $U\left(\mathcal{U}_0(\mathfrak{A}_0)\right)$ *is the closed unit ball of* $\mathcal{U}_0(\mathfrak{A}_0)$.

Proof Let x, y be elements of $L_2^\omega(\mathfrak{A}_0)$ and $2 \le p < \infty$. Then, since

$$\|xy\|_p \ \le \ \|x\|_{2p} \|y\|_{2p} \ \text{and} \ \|x^*\|_p = \|x\|_p$$

and similarly to (5.4.3)

$$\|x\|_p^p \ \le \ \|x\|_n^n + \|x\|_{n+1}^{n+1}, \ \text{with} \ n \in \mathbb{N} \ \text{and} \ n \le p < n+1,$$

it follows that τ_2^ω coincides with the topology defined by the sequence of the L^n-norms $\{\|\cdot\|_n\}_{n\geq 2}$, so that $L_2^\omega(\mathfrak{A}_0)[\tau_2^\omega]$ is a Fréchet *-algebra. We show that $(\mathfrak{A}_0)_b = \left(L_2^\omega(\mathfrak{A}_0)[\tau_2^\omega]\right)_0$. Take an arbitrary $\xi \in (\mathfrak{A}_0)_b$. Let

$$\overline{|\pi_0(\xi)|} = \int_0^{\|\xi\|_\infty} \lambda \, d E_\xi(\lambda)$$

be the spectral resolution of the operator $\overline{|\pi_0(\xi)|}$, where $\|\xi\|_\infty := \|\overline{\pi_0(\xi)}\|$ (see also discussion before Lemma 5.4.3). For $2 \leq p < \infty$ and $0 < \lambda < 1/\|\xi\|_\infty$ we obtain

$$\|\lambda^n \xi^n\|_p = \lambda^n \left\|\overline{\pi_0(\xi)}^n\right\|_p \leq \lambda^n \|\xi\|_\infty^{n-1} \|\xi\|_p$$
$$= (\lambda \|\xi\|_\infty)^n (\|\xi\|_p/\|\xi\|_\infty) \leq \|\xi\|_p/\|\xi\|_\infty,$$

which yields $\xi \in \left(L_2^\omega(\mathfrak{A}_0)[\tau_2^\omega]\right)_0$. Now, for any $\xi \in (\mathfrak{A}_0)_b$ we have that

$$p_\xi(x) := |\langle\overline{\pi_0(x)}\xi, \xi\rangle| \leq \|\xi\| \|\xi\|_\infty \|x\|, \ \forall \, x \in L_2^\omega(\mathfrak{A}_0).$$

Therefore, the topology τ_ω on $L_2^\omega(\mathfrak{A}_0)$ defined by the family of seminorms $\{p_\xi : \xi \in (\mathfrak{A}_0)_b\}$, is weaker than the topology τ_2^ω. Let now $B \in \mathfrak{B}^*_{L_2^\omega(\mathfrak{A}_0)}$. Since, $\tau_\omega \leq \tau_2^\omega$, B is τ_ω-bounded. Take an arbitrary $x \in B$, such that $\|\overline{\pi_0(x)}\| > 1$. Then, there exists an element ξ in $(\mathfrak{A}_0)_b$ with $\|\xi\| = 1$ and $\|\pi_0(x)\xi\| > 1$. But, $(x^*x)^{2^n} \in B$ and

$$\lim_{n\to\infty} p_\xi\left((x^*x)^{2^n}\right) = \lim_{n\to\infty} \left\langle\left(\pi_0(x)^*\overline{\pi_0(x)}\right)^{2^n}\xi, \xi\right\rangle \geq \lim_{n\to\infty} \|\overline{\pi_0(x)}\xi\|^{2^{n+1}} = \infty,$$

hence it follows that B is not τ_ω-bounded, a contradiction. Therefore, $\|\overline{\pi_0(x)}\| \leq 1$ and

$$B \subseteq \left\{x \in (\mathfrak{A}_0)_b : \|x\|_\infty \leq 1\right\}. \tag{5.4.8}$$

We show that $\left(L_2^\omega(\mathfrak{A}_0)[\tau_2^\omega]\right)_0 \subseteq (\mathfrak{A}_0)_b$. Take $x \in \left(L_2^\omega(\mathfrak{A}_0)[\tau_2^\omega]\right)_0$, arbitrary. Then, there exists $\lambda > 0$, such that $\{(\lambda x)^n : n \in \mathbb{N}\}$ is a bounded subset of $L_2^\omega(\mathfrak{A}_0)[\tau_2^\omega]$. Let $\overline{\pi_0(x)} = U|\overline{\pi_0(x)}|$ be the polar decomposition of $\overline{\pi_0(x)}$. It is easily shown that $\{(\lambda U^*x)^n : n \in \mathbb{N}\}$ is a bounded subset of $L_2^\omega(\mathfrak{A}_0)[\tau_2^\omega]$. Suppose that $\overline{\pi_0(x)}$ is unbounded. Then, there exists an element $\xi \in (\mathfrak{A}_0)_b$, such that $\|\xi\| = 1$ and $\|\pi_0(y)\xi\| > 1$, where $y := \lambda U^*x$. Hence,

$$\lim_{n\to\infty} \left\langle\left(\overline{\pi_0(y)}\right)^{2^n}\xi, \xi\right\rangle \geq \lim_{n\to\infty} \|\overline{\pi_0(y)}\|^{2^{n+1}} = \infty,$$

which contradicts the fact that $\{y^n : n \in \mathbb{N}\}$ is a bounded subset of $L_2^\omega(\mathfrak{A}_0)[\tau_2^\omega]$. Consequently, $\left(L_2^\omega(\mathfrak{A}_0)[\tau_2^\omega]\right)_0 \subseteq (\mathfrak{A}_0)_b$. Thus, we have $\left(L_2^\omega(\mathfrak{A}_0)[\tau_2^\omega]\right)_0 = (\mathfrak{A}_0)_b$.

We next show that $\overline{(\mathfrak{A}_0)_b}^{\tau_2^\omega} = L_2^\omega(\mathfrak{A}_0)$. Let x be any element of $L_2^\omega(\mathfrak{A}_0)$. Let $\overline{\pi_0(x)} = U|\overline{\pi_0(x)}|$ be the polar decomposition of the operator $\overline{\pi_0(x)}$ and $|\overline{\pi_0(x)}| = \int_0^\infty \lambda\, dE_x(\lambda)$ be the spectral resolution of $|\overline{\pi_0(x)}|$. Since,

$$ S_n := \int_0^n \lambda\, dE_x(\lambda) \in \mathcal{U}_0(\mathfrak{A}_0) \bigcap L^2(\varphi_0), \ n \in \mathbb{N}, $$

there exists a sequence $(\xi_n)_{n\in\mathbb{N}}$ in $(\mathfrak{A}_0)_b$, such that $S_n = \overline{\pi_0(\xi_n)}$, $n \in \mathbb{N}$. Then, for $p \geq 2$

$$ \|U\xi_n - x\|_p = \|\,U\overline{\pi_0(\xi_n)} - U|\overline{\pi_0(x)}|\,\|_p $$

$$ \leq \|\,\overline{\pi_0(\xi_n)} - |\overline{\pi_0(x)}|\,\|_p = -\left(\int_n^\infty \lambda^p d\varphi_0(E_x(\lambda)^\perp)\right)^{1/p} \underset{n\to\infty}{\to} 0 $$

and $U\xi_n \in (\mathfrak{A}_0)_b$, since $U\overline{\pi_0(\xi_n)} = \overline{\pi_0(U\xi_n)}$, $n \in \mathbb{N}$. Hence, it follows that $(\mathfrak{A}_0)_b$ is dense in $L_2^\omega(\mathfrak{A}_0)[\tau_2^\omega]$.

Suppose now that $\mathfrak{A}_0$ has an identity element e. Then, it is easily shown that $(\mathfrak{A}_0)_b = \mathcal{U}_0(\mathfrak{A}_0)e$. Moreover, the set $B_0 \equiv U(\mathcal{U}_0(\mathfrak{A}_0))e := \{Ae : A \in U(\mathcal{U}_0(\mathfrak{A}_0))\}$ is a bounded absolutely convex subset of $L_2^\omega(\mathfrak{A}_0)[\tau_2^\omega]$, such that $B_0^2 \subseteq B_0$ and $B_0^* = B_0$. Furthermore, since B_0 is τ_ω-closed and $\tau_\omega \prec \tau_2^\omega$, it follows that B_0 is τ_2^ω-closed, therefore $B_0 \in \mathfrak{B}_{L_2^\omega(\mathfrak{A}_0)}^*$. Thus, by Theorem 7.1.2 (in Sect. 7.1), we conclude that $L_2^\omega(\mathfrak{A}_0)[\tau_2^\omega]$ is a GB^*-algebra over $U(\mathcal{U}_0(\mathfrak{A}_0))e$.   □

An immediate consequence of Theorem 5.4.10 is the following

Corollary 5.4.11 *Let $\mathfrak{A}$ be a pure unbounded Hilbert algebra over $\mathfrak{A}_0$. Then, $\mathfrak{A}$ is a $*$-subalgebra of the Fréchet GB^*-algebra $L_2^\omega(\mathfrak{A}_0)[\tau_2^\omega]$ with $(\mathfrak{A}[\tau_2^\omega])_0 = \mathfrak{A}_0$. In particular, if $\mathfrak{A}_0$ is a maximal Hilbert algebra with an identity element e, then $\mathfrak{A}[\tau_2^\omega]$ is a GB^*-algebra over $\mathcal{U}_0(\mathfrak{A}_0)_1 e$.*

Recall that $\mathcal{U}_0(\mathfrak{A}_0)_1$ denotes the unitization of the left von Neumann algebra $\mathcal{U}_0(\mathfrak{A}_0)$.

Notes Theorem 5.4.2 comes from Lemma 2.1 and Proposition 2.3 in [79]. For Lemma 5.4.3 see [81, Lemma 3.1 and Theorem 3.4]. Theorem 5.4.4 is due to Proposition 3.6 in [79] and Theorem 3.4 in [81]. Theorem 5.4.5 is due to Theorem 4.2 in [81], while Theorem 5.4.8 is due to Corollary 4.6 in [81]. Finally Theorem 5.4.10 is due to Theorems 3.2, 3.3 and 4.1 in [80].

Chapter 6
Generalized B*-Algebras: Unbounded *-Representation Theory

In representing a noncommutative C^*-algebra as a norm closed $*$-subalgebra of bounded linear operators on some Hilbert space, positive linear functionals play an important role. In ensuring that the involved $*$-representation is faithful, one requires that there are enough positive linear functionals in the sense that they separate the points of the C^*-algebra. That this is so for C^*-algebras is well known, and relies on the fact that the positive cone of a C^*-algebra is closed. It is not known if the positive cone is closed for GB^*-algebras, in general, but we prove in Sect. 6.2 that the positive cone of a GB^*-algebra $\mathcal{A}[\tau]$ is closed in some stronger topology T making $\mathcal{A}$ a GB^*-algebra. In Sect. 4.2 (see Corollary 4.2.6), it was established (as in the case of C^*-algebras), that there are enough positive linear functionals on a commutative GB^*-algebra, that separate its points. A partial analogue to this in the noncommutative case, we may say that it is given by Theorem 6.3.4. It is this result that we use to prove a noncommutative algebraic Gelfand–Naimark type theorem for GB^*-algebras (see Theorem 6.3.5). More precisely, we show that any GB^*-algebra can be faithfully represented as an EC^*-algebra (see Definition 5.3.1). For a topological analogue of the Gelfand–Naimark type theorem for GB^*-algebras, see Theorem 6.3.11 and comments before Definition 6.3.8.

6.1 A Functional Calculus

We have seen in Chaps. 3 and 4 that for the study of commutative GB^*-algebras, an essential role is played by the Gelfand–Naimark type theorem (cf. Theorem 3.4.9) and the functional calculus given by Theorem 4.1.2. The Gelfand–Naimark type theorem just mentioned, can always be applied 'locally' in the noncommutative case too. Indeed, having a normal element x in an arbitrary GB^*-algebra $\mathcal{A}[\tau]$, we may take a maximal commutative $*$-subalgebra $\mathcal{B}$ of $\mathcal{A}$ containing x. This

© The Author(s), under exclusive license to Springer Nature Switzerland AG 2022
M. Fragoulopoulou et al., *Generalized B*-Algebras and Applications*, Lecture Notes in Mathematics 2298, https://doi.org/10.1007/978-3-030-96433-7_6

is clearly closed and contains the identity e of $\mathcal{A}$, so it is a commutative GB^*-algebra (see Proposition 3.3.19). In this way, an analogous functional calculus for noncommutative GB^*-algebras is stated in Theorem 6.1.3 by just using, locally, Theorem 4.1.2. In this regard, see [48, p. 696, Section 4] together with Remarks 3.3.10 and 3.4.3. Dixon [48, Section 4], presented a noncommutative functional calculus for his 'topological' GB^*-algebras (Definition 3.3.3), using locally the commutative functional calculus of Allan exhibited in Sect. 4.1. For this he used in the place of $S \equiv \sigma(\cdot)$, the spectrum $\sigma^D(\cdot)$ (see (3.4.17)), taking into account Remarks 3.3.10 and 3.4.3.

> ▶ As we have noticed in Chap. 3, throughout this book, *we treat GB*-algebras in the sense of* Allan (Definition 3.3.2), *unless mention is made to the contrary; the same we also do in the whole* Chap. 6 and only where there is need, we apply Dixon's strategy, based on the aforementioned Remarks 3.3.10 and 3.4.3.

Recall that $\mathbb{C}^*$ denotes the one point compactification of the complex plane $\mathbb{C}$ (see beginning of Sect. 2.3).

Let $\mathcal{A}[\tau]$ be a GB^*-algebra, $x \in \mathcal{A}$ and $\sigma(x)$ the spectrum of x. As in Sect. 4.1, we put $S \equiv \sigma(x)$ and denote by $\mathcal{C}_1(S)$ the set of all continuous $\mathbb{C}$-valued functions f on the finite part of $S \equiv \sigma(x)$, such that for a non-negative integer n, depending on f, the function

$$\lambda \; \rightarrow \; \frac{f(\lambda)}{(1 + |\lambda|^2)^n}$$

extends to a bounded continuous function on the whole of $\sigma(x)$.

Lemma 6.1.1 *Let* $\mathcal{A}[\tau]$ *be a* GB^*-*algebra and* x *a normal element of* $\mathcal{A}$. *Let* $\mathcal{C}$ *be a maximal commutative* *-*subalgebra of* $\mathcal{A}$ *containing* x. *Then, we have the following:*

  (i) $\mathcal{C}$ *is a closed* *-*subalgebra of* $\mathcal{A}[\tau]$;
  (ii) $\mathcal{C}[\tau \lceil_{\mathcal{C}}]$ *is a* GB^*-*algebra having* $B_0 \cap \mathcal{C}$ *in* $\mathfrak{B}_{\mathcal{C}}^*$ *as a maximal member.*
  (iii) $\mathcal{A}[B_0] \cap \mathcal{C}$ *is a* $\| \cdot \|_{B_0}$-*closed* *-*subalgebra of* $\mathcal{A}[B_0]$;
  (iv) $\sigma_{\mathcal{A}}(y) = \sigma_{\mathcal{C}}(y)$, *for every* $y \in \mathcal{C}$.

Proof

  (i) is obvious.
  (ii) By (i) and Proposition 3.3.19, we conclude that $\mathcal{C}$ is a GB^*-algebra, with $B_0 \cap \mathcal{C}$ a maximal member in $\mathfrak{B}_{\mathcal{C}}^*$.
  (iii) It is easily seen that the set $B_0 \cap \mathcal{C}$ belongs to $\mathfrak{B}_{\mathcal{A}}^*$ and that $\mathcal{A}[B_0 \cap \mathcal{C}] = \mathcal{C} \cap \mathcal{A}[B_0]$. The *-norm on $\mathcal{A}[B_0 \cap \mathcal{C}]$ is just the restriction of the C^*- norm

$\| \cdot \|_{B_0}$ of $\mathcal{A}[B_0]$. Thus, it follows that $\mathcal{A}[B_0 \cap C]$ is a Banach $*$-algebra from which (iii) follows.

(iv) See Proposition 2.3.2.

$\square$

Lemma 6.1.2 *Let $\mathcal{A}[\tau]$ be a GB^*-algebra and let C_1, C_2 be maximal commutative $*$-subalgebras of $\mathcal{A}$. If $C \equiv C_1 \cap C_2$, then C fulfills all the assertions (i)–(iv) of Lemma 6.1.1.*

Proof It follows easily from Proposition 3.3.19. $\square$

The following theorem gives a functional calculus for a not necessarily commutative GB^*-algebra. For the notation applied, see Theorem 4.1.2.

Theorem 6.1.3 (Dixon) *Let $\mathcal{A}[\tau]$ be a GB^*-algebra and x a normal element in $\mathcal{A}$. Then, there exists a unique $*$-isomorphism $\Phi : C_1(S) \to \mathcal{A}$, such that:*

(i) *if $u_0(\lambda) = \lambda$, then $u_0(x) = e$;*
(ii) *if $u_1(\lambda) = \lambda$, then $u_1(x) = x$;*
(iii) *for every maximal commutative $*$-subalgebra C of $\mathcal{A}$ containing x and every $f \in C_1(S)$, one has that*

$$f(x) \in C \quad and \quad \widehat{f(x)}(\phi) = f\big(\widehat{x}(\phi)\big), \ \forall \phi \in \mathfrak{M}_0,$$

$\mathfrak{M}_0$ *being the carrier space of the commutative C^*-algebra $C[B_0 \cap C]$.*

Proof Fix a maximal commutative $*$-subalgebra B_1 of $\mathcal{A}$ containing x. Then, Theorem 4.1.2 yields the existence of a unique $*$-isomorphism $\Phi_1 : C_1(S) \to B_1$ that fulfills (i) and (ii). Assume now that B_2 is a second maximal commutative $*$-subalgebra of $\mathcal{A}$ containing x that gives rise to a corresponding $*$-isomorphism $\Phi_2 : C_1(S) \to B_2$. Then, by Lemma 6.1.2, $B_3 := B_1 \cap B_2$ is a commutative GB^*-subalgebra of $\mathcal{A}$ containing x. Thus, we obtain a third $*$-isomorphism $\Phi_3 : C_1(S) \to B_3$. Applying the uniqueness claim of Theorem 4.1.2 for B_1, respectively B_2, we conclude that $\Phi_1 = \Phi_3$, respectively $\Phi_2 = \Phi_3$. Consequently, $\Phi = \Phi_1$ satisfies (iii) for any maximal, commutative $*$-subalgebra B of $\mathcal{A}$ containing x. $\square$

Remark 6.1.4 In the sequel, we shall often apply (iii) of the preceding theorem, taking the Gelfand representation of a maximal commutative $*$-subalgebra of a GB^*-algebra $\mathcal{A}[\tau]$, containing a given normal element x of $\mathcal{A}$.

6.2 Positive Elements

Using the technique of Theorem 6.1.3(iii) and applying Theorem 3.4.9 and Corollary 3.4.10, one easily obtains the following

Proposition 6.2.1 *Let $\mathcal{A}[\tau]$ be a GB^*-algebra, and $x \in A$. The following statements are equivalent:*

(i) x is normal and $\sigma(x) \subset \{\lambda \in \mathbb{C}^* : \lambda \geq 0\}$;
(ii) x is normal and $\widehat{x} \geq 0$;
(iii) $x = h^2$, for some $h \in H(A)$;
(iv) $x = y^* y$, for some $y \in A$.

Definition 6.2.2 If $A[\tau]$ is a GB^*-algebra, then we say that an element $x \in A$ is *positive*, denoted by $x \geq 0$, if x satisfies the equivalent conditions of Proposition 6.2.1. The set of positive elements is denoted by A^+. If x, y are in A, we write $x \leq y$ to mean that $y - x \geq 0 \Leftrightarrow y - x \in A^+$.

Note that, in the case when, $A[\tau]$ is commutative, then Definition 6.2.2 coincides with Definition 4.1.3.

Proposition 6.2.3 *Let $A[\tau]$ be a GB^*-algebra and h a self-adjoint element in A. Then, there exist positive elements h^+, h^- in A, such that*

$$h = h^+ - h^- \quad and \quad h^+ h^- = 0.$$

Proof Take a maximal commutative *-subalgebra B of A containing h. Then, B is a commutative GB^*-algebra with identity according to Propositions 2.3.2 and 3.3.19. Thus, applying Theorem 4.1.4(ii), there exist positive elements h^+, h^- in B, such that the conclusion of our proposition is true.   □

Since the set of positive elements of a C^*-algebra forms a convex cone, it would be interesting to know if this is also the case for GB^*-algebras, in general. That this is the case is proven below in Theorem 6.2.5. In order to prove it, we require the following lemma, which is an immediate consequence of Proposition 6.2.1(iv).

Lemma 6.2.4 *Let $A[\tau]$ be a GB^*-algebra. Then, for every $x, y \in A^+$, we have that $xyx \in A^+$.*

Theorem 6.2.5 (Dixon) *Let $A[\tau]$ be a GB^*-algebra. Then, A^+ is a convex cone, with the property $A^+ \cap (-A^+) = \{0\}$.*

Proof If $x \in A^+$ and $\lambda > 0$, then clearly $\lambda x \in A^+$. Next we prove that $x + y \in A^+$, for every $x, y \in A^+$.

We first show that $(A[B_0])^+ = A[B_0] \cap A^+$. Readily, $(A[B_0])^+ \subset A[B_0] \cap A^+$. Let now $x \in A[B_0] \cap A^+$. Then, its Gelfand transform $\widehat{x}$ is a bounded positive function on the maximal ideal space $\mathfrak{M}_0$ of a maximal commutative *-subalgebra C of A containing x. Since the restriction of the Gelfand map to $A[B_0] \cap C$ is the Gelfand representation of a maximal commutative *-subalgebra of the C^*-algebra $A[B_0]$, it follows that $x \in (A[B_0])^+$, thereby establishing the equality $(A[B_0])^+ = A[B_0] \cap A^+$.

If now $h \in H(A)$, applying Remark 6.1.4, it follows from Theorem 3.3.9 that $(e + h^2)^{-1} \in A^+ \cap A[B_0]$. Using this and Lemma 6.2.4, one has that, for every $x, y \in A^+$,

$$z \equiv (e + y^2)^{-1} x (e + y^2)^{-1} \in A^+.$$

Therefore, by appealing to Lemma 6.2.4 once more, we obtain that

$$(e + z^2)^{-1}z(e + z^2)^{-1} \in \mathcal{A}^+. \tag{6.2.1}$$

Applying again Remark 6.1.4, by Theorem 3.3.9 and Lemma 3.4.2, we have that

$$(e + z^2)^{-1}z(e + z^2)^{-1} \in \mathcal{A}[B_0]. \tag{6.2.2}$$

From (6.2.1) and (6.2.2), we obtain now that

$$\begin{aligned} x_1 &\equiv (e + z^2)^{-1}(e + y^2)^{-1}x(e + y^2)^{-1}(e + z^2)^{-1} \\ &= (e + z^2)^{-1}z(e + z^2)^{-1} \in (\mathcal{A}[B_0])^+ = \mathcal{A}^+ \cap \mathcal{A}[B_0]. \end{aligned}$$

Since $y \in \mathcal{A}^+$, again from (6.2.1) and (6.2.2), we conclude that

$$(e + y^2)^{-1}y(e + y^2)^{-1} \in \mathcal{A}^+ \cap \mathcal{A}[B_0] = (\mathcal{A}[B_0])^+.$$

Consequently, as in the reasoning of (6.2.2), we have

$$y_1 \equiv (e + z^2)^{-1}(e + y^2)^{-1}y(e + y^2)^{-1}(e + z^2)^{-1} \in \mathcal{A}[B_0].$$

By Lemma 6.2.4, it follows now that $y_1 \in \mathcal{A}^+ \cap \mathcal{A}[B_0] = (\mathcal{A}[B_0])^+$. Since the positive cone of $\mathcal{A}[B_0]$ is convex [45, p. 15, Proposition 1.6.1], we conclude that $x_1 + y_1 \in (\mathcal{A}[B_0])^+ \subset \mathcal{A}^+$, therefore by applying twice Lemma 6.2.4, we have

$$x + y = (e + y^2)(e + z^2)(x_1 + y_1)(e + z^2)(e + y^2) \in \mathcal{A}^+.$$

Furthermore, it is clear by Proposition 6.2.1 that $\mathcal{A}^+ \cap (-\mathcal{A}^+) = \{0\}$. Thus, $\mathcal{A}^+$ is a convex cone. This completes the proof. $\qquad\square$

We next consider whether the positive cone $\mathcal{A}^+$ of a GB*-algebra $\mathcal{A}[\tau]$ is τ-closed (see Theorem 6.2.11).

Let $\mathcal{A}[\tau]$ be a GB^*-algebra and let U be an absolutely convex 0-neighbourhood in $\mathcal{A}[\tau]$. Fix an element y in $\mathcal{A}$. Then, by the continuity of the map

$$\mathcal{A}[\tau] \;\rightarrow\; \mathcal{A}[\tau] : x \;\mapsto\; xy,$$

we have that for every x in $\mathcal{A}$ there exists a 0-neighbourhood V_x, such that $V_x y \subseteq U$. Furthermore, using the 0-neighbourhood V_x and fixing now an element x in $\mathcal{A}$, we obtain that for every y in $\mathcal{A}$ there is a 0-neighbourhood $W_{x,y}$, such that $x W_{x,y} \subseteq V_x$, concluding thus, that $x W_{x,y} y \subseteq V_x y \subseteq U$. Since B_0 is a bounded subset of $\mathcal{A}$, there exists $\delta_{x,y} \equiv \delta(x, y) > 0$, such that $\delta_{x,y} B_0 \subset W_{x,y}$. Let ΔS denote the absolute convex hull of a subset S of $\mathcal{A}$. Then, taking into account the preceding discussion we have that

$$N(\delta) := \Delta \bigcup_{x,y \in \mathcal{A}} \delta_{x,y} x\, B_0\, y \subset U. \tag{6.2.3}$$

Now, *we define a topology T on $\mathcal{A}$ to be the topology given by the base of neighbourhoods $N(\delta)$, where δ ranges over all strictly positive functions of x, y in $\mathcal{A}$.* It is clear from (6.2.3) that *the topology T is finer than the topology τ.*

▶ Observe that $N(\delta)$ is absorbent (since $e \in B_0$) and absolutely convex from its construction; *therefore T is a locally convex topology.*

Theorem 6.2.6 *If $\mathcal{A}[\tau]$ is a GB^*-algebra, then $\mathcal{A}[T]$ is a barrelled and bornological GB^*-algebra, with $B_0(T) = B_0(\tau)$, i.e., the greatest member B_0 of $\mathfrak{B}^*_{\mathcal{A}}$, is the same with respect to both topologies T and τ on $\mathcal{A}$.*

Proof If $\delta^*_{x,y} = \delta_{y^*,x^*}$, for all $x, y \in \mathcal{A}$, we have that $N(\delta^*) = N(\delta)^*$, which implies that the involution on $\mathcal{A}$ is T-continuous. Given $z \in \mathcal{A}$ and $\delta > 0$, we define $\delta_z(x, y) := \delta_{zx,y}$, for all $x, y \in \mathcal{A}$. Hence, (cf. (6.2.3)),

$$zN(\delta_z) = \Delta \bigcup_{x,y \in \mathcal{A}} \delta_{zx,y} zx\, B_0\, y$$

$$\subset \Delta \bigcup_{w,y \in \mathcal{A}} \delta_{w,y} w\, B_0\, y$$

$$= N(\delta).$$

Therefore, the left multiplication is continuous. Similarly, the right multiplication is continuous. We now show that the conditions (i)–(iii) of Definition 3.3.3 are satisfied.

(i) Since $B_0(\tau) \subset \frac{1}{\delta_{e,e}} N(\delta)$, we get that $B_0(\tau)$ is T-bounded. Furthermore, since the topology τ is weaker than the topology T, it follows that $B_0(\tau)$ is T-closed, and hence $B_0(\tau) \in \mathfrak{B}_{\mathcal{A}[T]}$. This implies that $B_0(\tau) \subset B_0(T)$. If $B \in \mathfrak{B}_{\mathcal{A}[T]}$, then B is T-bounded, and therefore, since the topology τ is weaker than the topology T, we obtain that B is also τ-bounded. Therefore, the τ-closure $\overline{B}^\tau$ of B is in $\mathfrak{B}_{\mathcal{A}[\tau]}$, so that $B \subset \overline{B}^\tau \subset B_0(\tau)$. It follows that $B_0(\tau)$ is the greatest member of $\mathfrak{B}_{\mathcal{A}[T]}$ with respect to set inclusion, and hence $B_0(T) \subset B_0(\tau)$, implying that $B_0(T) = B_0(\tau)$.

(ii) Since $\mathcal{A}[\tau]$ is a GB^*-algebra and $B_0(T) = B_0(\tau)$, we obtain that the element $(e + x^*x)^{-1}$, $x \in \mathcal{A}$, sits in $\mathcal{A}[B_0(\tau)] = \mathcal{A}[B_0(T)]$ and this implies symmetry of $\mathcal{A}[T]$.

(iii) Since $\mathcal{A}[\tau]$ is a GB^*-algebra, $\mathcal{A}[B_0(\tau)]$ is complete, and hence $\mathcal{A}[B_0(T)]$ (being equal to $\mathcal{A}[B_0(\tau)]$) is also complete. It follows that $\mathcal{A}[T]$ is a GB^*-algebra.

The fact that $\mathcal{A}[T]$ is barrelled and bornological follows from the proof of Theorem 4.3.16 by applying similar steps. $\quad\square$

Remark 6.2.7 In the sense of Definition 4.3.1 (see also Remark 4.3.2(2)) the topologies T and τ of Theorem 6.2.6 are "equivalent". Clearly, this does not mean that the identity map $id_{\mathcal{A}} : \mathcal{A}[\tau] \to \mathcal{A}[T]$ is a homeomorphism; it simply means that the C^*-algebras $\mathcal{A}[B_0(\tau)]$ and $\mathcal{A}[B_0(T)]$ (through which $\mathcal{A}[\tau]$ and $\mathcal{A}[T]$ are studied) are identical (see also comments after Corollary 6.3.7).

Let E, F, G be topological vector spaces. A bilinear map $f : E \times F \to G$ is called *hypocontinuous*, if f is separately continuous and for each bounded subset B of E and each 0-neighbourhood W in G, there is a 0-neighbourhood V in F, such that $f(B \times V) \subseteq W$; see [131, p. 89]. In this regard, see also [74, p. 358].

Every locally convex algebra having jointly continuous multiplication, hence every Fréchet locally convex algebra, has hypocontinuous multiplication.

Lemma 6.2.8 (Dixon) *Let $\mathcal{A}[\tau]$ be a barrelled locally convex algebra. Then, given a bounded subset B and a 0-neighbourhood U of $\mathcal{A}[\tau]$, there exists a 0-neighbourhood V in $\mathcal{A}[\tau]$, such that*

$$BV \subseteq U \text{ and } VB \subseteq U.$$

Proof We may choose the 0-neighbourhood U to be closed and absolutely convex. Consider the sets

$$V_1 \equiv \{x \in \mathcal{A} : B \cdot x \subseteq U\}, \quad V_2 \equiv \{x \in \mathcal{A} : x \cdot B \subseteq U\}.$$

Note that, for every $y \in \mathcal{A}$, the sets $B \cdot y$, and $y \cdot B$ are bounded from the separate continuity of multiplication of $\mathcal{A}[\tau]$. Therefore, $B \cdot y \subseteq \lambda U$ for some $\lambda > 0$. This implies $\lambda^{-1} y \in V_1$, which shows that V_1 is absorbing. Moreover, using the definition of V_1 and the fact that U is absolutely convex, it is easily seen that V_1 is absolutely convex too. Finally, using again the separate continuity of multiplication of $\mathcal{A}[\tau]$ and the closedness of U we conclude that V_1 is also closed. Hence, V_1 is a barrel. Analogously for V_2. Taking now $V \equiv V_1 \cap V_2$, we have V to be a barrel and so a 0-neighbourhood in $\mathcal{A}[\tau]$, such that

$$BV \subseteq U \text{ and } VB \subseteq U.$$

Observe that V is nonempty. Indeed, given that B is bounded there is $\lambda > 0$, such that $B \subseteq \lambda U$, hence $\frac{1}{\lambda} e \in V_1 \cap V_2 \equiv V$, with e the identity in $\mathcal{A}$. So the proof is complete. $\quad\square$

An immediate consequence of Theorem 6.2.6 and Lemma 6.2.8 is the following

Corollary 6.2.9 *For every 0-neighbourhood U with respect to* τ, *there is a 0-neighbourhood V with respect to* τ, *such that*

$$B_0 V \subset U \quad \text{and} \quad V B_0 \subset U. \tag{[A]}$$

We shall refer to the previous property as *property* [A] *of the topology T.* Property [A] *is equivalent to the following*:

If $(x_\lambda)_{\lambda \in \Lambda}$, $(y_\lambda)_{\lambda \in \Lambda}$ are nets in $\mathcal{A}$, such that $x_\lambda \in B_0$,

$\forall \lambda \in \Lambda$ and $y_\lambda \to 0$, then $x_\lambda y_\lambda \to 0$ and $y_\lambda x_\lambda \to 0$.

▶ Observe that the property [A] *is a weaker property than that of multiplication being hypocontinuous*; that is *every locally convex algebra having jointly continuous multiplication, hence every Fréchet locally convex algebra, has property* [A]. The multiplication of every barrelled locally convex algebra is hypocontinuous [74, Theorem 2, p. 360].

Lemma 6.2.10 *Let* $\mathcal{A}[\tau]$ *be a GB*-algebra, such that* τ *satisfies property* [A]. *If* $(y_\mu)_{\mu \in M}$ *is a net in* $\mathcal{A}^+$ *converging to zero, then* $y_\mu^{1/2}$ *converges also to zero.*

Proof Let

$$f(\lambda) = \begin{cases} 1 & \text{if } 0 \le \lambda \le 1 \\ \lambda^{-1/2} & \text{if } \lambda > 1 \end{cases} \quad \text{and} \quad g(\lambda) = \begin{cases} \lambda^{1/2} - \lambda & \text{if } 0 \le \lambda \le 1 \\ 0 & \text{if } \lambda > 1. \end{cases}$$

We have that $|f(\lambda)| \le 1$ and $|g(\lambda)| \le 1$, for all $0 \le \lambda < \infty$, and therefore $f(y_\mu), g(y_\mu) \in B_0$ for all $\mu \in M$ (see Theorem 6.1.3 and Remark 6.1.4). Furthermore,

$$\lambda^{1/2} = \lambda f(\lambda) + g(\lambda), \quad \forall 0 \le \lambda < \infty;$$

hence, again applying the last reference, we conclude that

$$y_\mu^{1/2} = y_\mu f(y_\mu) + g(y_\mu), \quad \forall \mu \in M.$$

Since the topology τ has the property [A], and since $y_\mu \to 0$ and $f(y_\mu) \in B_0$, for all $\mu \in M$, it follows that $y_\mu f(y_\mu) \to 0$, i.e., for every 0-neighbourhood U, we have that $y_\mu f(y_\mu) \in U$, for μ large enough. Therefore, $y_\mu^{1/2} \in U + B_0$, for μ large enough. If V is a 0-neighbourhood, then there exists $\varepsilon > 0$ such that $\varepsilon B_0 \subset \frac{1}{2} V$.

Let $U = \frac{1}{2}\varepsilon^{-1}V$. Since $\varepsilon^{-1}y_\mu \to 0$, the above observation implies that

$$\varepsilon^{-1}y_\mu{}^{1/2} \in U + B_0 = \frac{1}{2}\varepsilon^{-1}V + B_0,$$

for μ large enough. Therefore,

$$y_\mu{}^{1/2} \in \frac{1}{2}V + \varepsilon B_0 \subset \frac{1}{2}V + \frac{1}{2}V = V,$$

for μ large enough, and so $y_\mu{}^{1/2} \to 0$. $\square$

Theorem 6.2.11 *If $\mathcal{A}[\tau]$ is a GB*-algebra, such that τ satisfies property* [A], *then $\mathcal{A}^+$ is τ-closed.*

Proof Suppose that $(x_\mu)_{\mu \in M}$ is a net in $\mathcal{A}^+$ and $x_\mu \to x$ with respect to the topology τ and $x \notin \mathcal{A}^+$. Since the involution of $\mathcal{A}[\tau]$ is continuous, we obtain that $x \in H(\mathcal{A})$. Hence, by Proposition 6.2.3, there are elements x^+ and x^- in $\mathcal{A}^+$, such that $x = x^+ - x^-$ and $x^+x^- = 0$. By separate continuity of multiplication,

$$x^-x_\mu x^- \to x^-xx^- = x^-x^+x^- - (x^-)^3 = -(x^-)^3 \in -\mathcal{A}^+.$$

Let $y_\mu = x^-x_\mu x^- + (x^-)^3$, for all $\mu \in M$. By Lemma 6.2.4 and Theorem 6.2.5, we obtain that $y_\mu \in \mathcal{A}^+$ and $y_\mu \geq (x^-)^3 \in \mathcal{A}^+$, for all $\mu \in M$. Furthermore, $y_\mu \to 0$. Let $y = (x^-)^3$. By the proof of Lemma 6.2.4, it follows that

$$(e+y)^{-1}y_\mu(e+y)^{-1} \geq (e+y)^{-1}y(e+y)^{-1} \geq 0,$$

for all $\mu \in M$. By separate continuity of multiplication, it follows that $(e+y)^{-1}y_\mu(e+y)^{-1} \to 0$. Since $x \notin \mathcal{A}^+$, it follows that $x^- \neq 0$, therefore $y > 0$. Hence, $(e+y)^{-1}y(e+y)^{-1} \neq 0$. We show that

$$(e+y)^{-1}y(e+y)^{-1} \in B_0. \tag{6.2.4}$$

The elements y and $(e+y)^{-1}$ are self-adjoint and commuting. Thus, we may consider a maximal commutative $*$-subalgebra $\mathcal{C}$ of $\mathcal{A}[\tau]$ containing them. Then, $\mathcal{C}[\tau\restriction_{\mathcal{C}}]$ is a GB*-algebra with $B_0 \cap \mathcal{C}$ a maximal member in $\mathfrak{B}_{\mathcal{C}}^*$ (Lemma 6.1.1(ii)). Using the commutative Gelfand–Naimark type Theorem 3.4.9 for $\mathcal{C}[\tau\restriction_{\mathcal{C}}]$, we have that

$$\mathcal{C} = \widehat{\mathcal{C}} = \{\widehat{x} : \mathcal{C}_0 \to \mathbb{C}^* : \varphi \mapsto \widehat{x}(\varphi) := \varphi'(x)\},$$

up to an algebraic $*$-isomorphism. We now show that $\varphi'(y) \neq \infty$, where y as above. For this we apply Lemma 3.4.4(ii), to $\mathcal{C}[\tau\restriction_{\mathcal{C}}]$, with y in the place of x, $(e+y^2)^{-1}\big(= (e+y^*y)^{-1}\big)$ in the place of y and that $\varphi \in \mathfrak{M}(\mathcal{C}_0)$, such that $\varphi(e+y^2)^{-1} \neq 0$ (note that $(e+y^2)^{-1} \in \mathcal{C}_0$ from Lemma 3.4.2 and that $\mathcal{C}_0$ is a C*-algebra, since $\mathcal{C}$ is commutative). But then from Lemma 3.4.4(ii) we obtain $\varphi'(y) \neq \infty$. This shows that $\widehat{y} \in \mathcal{C}(\mathfrak{M}(\mathcal{C}_0))$, therefore, $y \in \mathcal{C}_0$. In the same way (taking in the place of x the element $(e + y)^{-1}$ and in the place of y the element $(e + y^2)^{-1}$), we prove that $\varphi'((e + y)^{-1}) \neq \infty$, hence $\widehat{(e + y)^{-1}} \in \mathcal{C}(\mathfrak{M}(\mathcal{C}_0))$, i.e., $(e + y)^{-1} \in \mathcal{C}_0$. Finally $(e + y)^{-1}y(e + y)^{-1} \in \mathcal{C}_0 = \mathcal{C}[\mathcal{C} \cap B_0]$ and since $(e + y)^{-1}y(e + y)^{-1}$ is nonzero self-adjoint and B_0 is balanced we conclude that $(e + y)^{-1}y(e + y)^{-1} \in B_0$ and so (6.2.4) is proved.

Let $z_\mu = (e+y)^{-1}y_\mu(e+y)^{-1}$, for all $\mu \in M$. Let $z \equiv (e+y)^{-1}y(e+y)^{-1} \in B_0$. Then, $z_\mu \to 0$, $z_\mu \geq 0$ and $z_\mu \geq z \geq 0$, with $0 \neq z \in B_0$.

Put $\Lambda := M \times N$, with N a directed set of positive integers. The set M becomes directed by defining $(\lambda_1, n_1) \geq (\lambda_2, n_2)$, if and only if, $\lambda_1 \geq \lambda_2$ and $n_1 \geq n_2$. Then, a new net (w_λ) can be defined as follows

$$w_{(\mu,n)} = z_\mu + \frac{1}{n}e, \quad \text{with } \lambda = (\mu, n) \in \Lambda.$$

By the functional calculus (Remark 6.1.4) and Theorem 3.4.9, every w_λ is invertible and $(w_\lambda^{-1})^{1/2} = (w_\lambda^{1/2})^{-1}$, which we write as $w_\lambda^{-1/2}$. We therefore have that

$$w_\lambda \geq z > 0, \quad w_\lambda^{-1/2} \in \mathcal{A}^+ \quad \text{and} \quad w_\lambda \to 0.$$

By Lemma 6.2.10, it follows that $w_\lambda^{1/2} \to 0$. Using again the last reference and applying Lemma 6.2.4, one easily obtains that $e \geq w_\lambda^{-1/2}zw_\lambda^{-1/2} > 0$, i.e., $w_\lambda^{-1/2}zw_\lambda^{-1/2} \in B_0$. Since $w_\lambda^{1/2} \to 0$ and the fact that the topology τ has property [A], we obtain that $w_\lambda^{-1/2}z = (w_\lambda^{-1/2}zw_\lambda^{-1/2})w_\lambda^{1/2} \to 0$. Now $(e + w_\lambda)^{-1}w_\lambda^{1/2} \in B_0$, for all $\lambda \in \Lambda$, and therefore, by appealing again to the fact that the topology τ has property [A], it follows that

$$(e + w_\lambda)^{-1}z = ((e + w_\lambda)^{-1}w_\lambda^{1/2})(w_\lambda^{-1/2}z) \to 0.$$

Moreover, $(e + w_\lambda)^{-1} \in B_0$ and $z \in B_0$, and so $(e + w_\lambda)^{-1}z \in B_0$. Consequently, since the topology τ has property [A], $w_\lambda(e + w_\lambda)^{-1}z \to 0$, implying that

$$z = (e + w_\lambda)^{-1}z + w_\lambda((e + w_\lambda)^{-1}z) \to 0.$$

Hence, $z = 0$, which contradicts the fact that $z \neq 0$. Therefore $x \in \mathcal{A}^+$ and this completes the proof. $\qquad \square$

From the comment ▶ after Corollary 6.2.9 and Theorem 6.2.11, we have the following

Corollary 6.2.12 *For every Fréchet GB*-algebra $\mathcal{A}[\tau]$, the convex cone $\mathcal{A}^+$ is τ-closed.*

By Theorem 6.2.11 and the fact that topology T has property [A], we have the following result.

Theorem 6.2.13 *If $\mathcal{A}[\tau]$ is a GB*-algebra, then the cone $\mathcal{A}^+$ of positive elements in $\mathcal{A}[\tau]$ is T-closed.*

Notes Almost all results in this chapter have been obtained by P. G. Dixon in [48]. Not all locally convex GB^*-algebras have property [A]. As an example, the *-algebra of all bounded linear operators on a Hilbert space is a GB^*-algebra in the weak-operator topology, which does not satisfy property [A] and multiplication is only separately continuous. Hence, for every GB^*-algebra $\mathcal{A}[\tau]$ we define a stronger topology T making $\mathcal{A}$ a (locally convex) GB^*-algebra of Dixon (Definition 3.3.5), under which the positive cone $\mathcal{A}^+$ is T-closed.

6.3 *-Representations of GB*-Algebras

The theory of *-representations of C^*-algebras is well understood. For instance, every C^*-algebra has a faithful *-representation onto a C^*-algebra of bounded linear operators on some Hilbert space. In this section, we consider a generalization of this result for C^*-algebras to the context of GB*-algebras.

We recall the Gelfand–Naimark–Segal *-representation of a positive linear functional on a *-algebra. Let f be a positive linear functional on a *-algebra $\mathcal{A}$. Then,

$$N_f = \{x \in \mathcal{A} : f(x^*x) = 0\}$$

is a left ideal of $\mathcal{A}$. If $\lambda_f : \mathcal{A} \to \mathcal{A}/N_f$ is the natural quotient map, the quotient space

$$\lambda_f(\mathcal{A}) = \{\lambda_f(x) := x + N_f : x \in \mathcal{A}\}$$

is a pre-Hilbert space with inner product

$$\langle \lambda_f(x), \lambda_f(y) \rangle = f(y^*x), \ \forall \, x, y \in \mathcal{A}.$$

We denote by $\mathcal{H}_f$ the Hilbert space obtained by the completion of the pre-Hilbert space $\lambda_f(\mathcal{A})$. We define a *-representation π_f^0 of $\mathcal{A}$ by

$$\pi_f^0(x)\lambda_f(y) = \lambda_f(xy), \ \forall \, x, y \in \mathcal{A}$$

and denote by π_f the closure of π_f^0. In this regard, we have the following

Proposition 6.3.1 *Let f be a positive linear functional on a *-algebra $\mathcal{A}$. Then, there exists a closed *-representation π_f of $\mathcal{A}$ in the Hilbert space $\mathcal{H}_f$ and a linear*

map λ_f of $\mathcal{A}$ on the domain $\mathcal{D}(\pi_f)$ of π_f, such that $\lambda_f(\mathcal{A})$ is dense in $\mathcal{D}(\pi_f)$ with respect to the graph topology t_{π_f} and

$$\pi_f(x)\lambda_f(y) := \lambda_f(xy), \ \forall \, x, \, y \in \mathcal{A}.$$

The pair (π_f, λ_f) *is uniquely determined by* f *up to unitary equivalence.*

We call the triple $(\pi_f, \lambda_f, \mathcal{H}_f)$ *the GNS-construction for* f. For more details, the reader is referred to [85, Theorem 1.9.1]. For a *GNS* like representation theorem in the context of *-bimodules, see [139], where an algebraic model for *-bimodules is presented and Hilbert space representations of *-bimodules are defined and studied.

We define now the direct sum $\oplus_{i \in I} \pi_i$ of a family of closed *-representations $(\pi_i)_{i \in I}$ of a *-algebra.

Definition 6.3.2 Let $\{\pi_i : i \in I\}$ be a family of closed *-representations of a *-algebra $\mathcal{A}$. The *algebraic direct sum* of the *-representations π_i's, denoted by π, is defined on the pre-Hilbert space

$$\mathcal{D}_\pi = \sum_{i \in I} \mathcal{D}_{\pi_i} := \Big\{ (\xi_i)_{i \in I} \in \Pi_{i \in I} \mathcal{D}_{\pi_i}, \text{ with } \xi_i \neq 0, \text{ for at most finitely}$$

$$\text{many } i \in I \Big\}.$$

The inner product on $\mathcal{D}_\pi$ and $\pi(x)$ on $\mathcal{D}_\pi$, $x \in \mathcal{A}$, are given as follows

$$\big\langle (\xi_i)_{i \in I}), (\eta_i)_{i \in I}) \big\rangle := \sum_{i \in I} \langle \xi_i, \eta_i \rangle_i, \ \forall \, (\xi_i)_{i \in I}, (\eta_i)_{i \in I} \in \mathcal{D}_\pi \text{ and}$$

$$\pi(x)\big((\xi_i)_{i \in I}\big) := \big(\pi_i(x)\xi_i\big)_{i \in I}, \ \forall \, x \in \mathcal{A} \text{ and } (\xi_i)_{i \in I} \in \mathcal{D}_\pi.$$

Considering now the Hilbert spaces $\mathcal{H}_i$'s, the completions of the pre-Hilbert spaces $\mathcal{D}_{\pi_i}$'s, $i \in I$, we take the *Hilbert space direct sum* of $\mathcal{H}_i$'s, given by

$$\mathcal{H}_\pi = \Big\{ \xi = (\xi_i)_{i \in I} \in \Pi_{i \in I} \mathcal{H}_i, \ \sum_{i \in I} \|\xi_i\|^2 < \infty \Big\}.$$

We clearly have that $\mathcal{D}_\pi$ is dense in $\mathcal{H}_\pi$. Now, the closure $\tilde{\pi}$ (see comments before Example 5.1.3), of the *algebraic direct sum* *-representation π, is called *the direct sum* of the closed *-representations $\{\pi_i : i \in I\}$, and it is denoted by $\oplus_{i \in I} \pi_i$.

Note that $\tilde{\pi} = \oplus_{i \in I} \tilde{\pi}_i$, therefore π is closed, if and only if, all π_i's are closed (see [136, p. 213, 8.3.]). Thus, since each π_i, $i \in I$, is closed, that is $\pi_i = \tilde{\pi}_i$, we obtain that $\pi = \tilde{\pi}$, consequently the algebraic direct sum *-representation π is closed too.

Theorem 6.3.5 below is an extension of the classical noncommutative Gelfand–Naimark theorem to GB^*-algebras in the context of $\mathcal{O}^*$-algebras. In order to prove this, we require the following two results.

Lemma 6.3.3 *Let π be a *-representation of a GB^*-algebra $\mathcal{A}[\tau]$ and let $U\big(\pi(\mathcal{A})\big)$* $:= \{\pi(x) \in \pi(\mathcal{A}) \cap \mathcal{B}(\mathcal{H}) : \|\pi(x)\| \leq 1\}$ *be the closed unit ball of $\pi(\mathcal{A})$. Then* $\pi(B_0) \subseteq U\big(\pi(\mathcal{A})\big)$.

Proof If $x \in B_0$ and f is a positive linear functional on $\mathcal{A}$, then $f(x^*x) \leq f(e)$, since $\mathcal{A}[B_0]$ is a C^*-algebra. The map $\mathcal{A} \ni z \mapsto \langle \pi(z)\xi, \xi \rangle \in \mathbb{C}, \xi \in \mathcal{D}(\pi)$, is also a positive linear functional on $\mathcal{A}$, and therefore $\langle \pi(x)^*\pi(x)\xi, \xi \rangle \leq \langle \xi, \xi \rangle$. Hence, $\|\pi(x)\| \leq 1$.   □

If $\mathcal{A}$ is a C^*-algebra, then f is continuous and $\|f\| = f(e)$ (cf. Proposition 4.2.3). The theorem that follows is well known for C^*-algebras and is concluded by a similar proof; for completeness' sake we give its proof.

Theorem 6.3.4 (Dixon) *Let $\mathcal{A}[\tau]$ be a GB^*-algebra. The following hold:*

(i) *If $x \in H(\mathcal{A}) \setminus \mathcal{A}^+$, then there is a positive linear functional f on $\mathcal{A}$, such that $f(x) < 0$;*
(ii) *for every nonzero $x \in \mathcal{A}^+$, there is a positive linear functional f on $\mathcal{A}$, such that $f(x) > 0$.*

Proof

(i) By Theorem 6.2.13, we obtain that $\mathcal{A}^+$ is a T-closed convex set. By the Hahn–Banach theorem, there is a linear functional f on $\mathcal{A}$, such that $f(x) < 0$ and $f(\mathcal{A}^+) \geq 0$.
(ii) This is an immediate consequence of (i) and its proof.

□

As the case is with C^*-algebras, it will be illustrated in the proof of Theorem 6.3.5 that Theorem 6.3.4 is crucial in constructing faithful *-representations on GB^*-algebras.

Besides, it is evident that *Theorem 6.3.4 provides a GB^*-algebra with enough positive linear functionals in order to separate its points*; in this regard, see also Corollary 4.2.6. The next Theorem 6.3.5 offers an algebraic noncommutative Gelfand–Naimark type theorem in the context of GB^*-algebras.

Theorem 6.3.5 (Dixon) *Every GB^*-algebra $\mathcal{A}[\tau]$ has a faithful closed *-representation π as an EC^*-algebra on $\mathcal{D}(\pi)$. Furthermore, the *-representation π can be chosen, such that $\pi(B_0) = U\big(\pi(\mathcal{A})\big)$.*

Proof We construct a faithful *-representation π in the same manner as one would in the C^*-algebra case. Let $\mathcal{P}(\mathcal{A})$ be the set of all positive linear functionals on $\mathcal{A}$. For every $f \in \mathcal{P}(\mathcal{A})$, we define π_f as in Proposition 6.3.1.

Let π be the (algebraic) direct sum *-representation of the family $\{\pi_f : f \in \mathcal{P}(\mathcal{A})\}$. From Proposition 6.3.1, each π_f, $f \in \mathcal{P}(\mathcal{A})$, is closed; so by the comments following Definition 6.3.2, we conclude that π is also closed. We show now that

π is faithful. Let $x \neq 0$. This implies that $x^*x \neq 0$ too. Indeed, suppose that $x^*x = 0$. Then, from the Cauchy–Schwartz inequality we obtain $f(x) = 0$, for all $f \in \mathcal{P}(\mathcal{A})$. On the other hand, $x = h + ik$, with $h, k \in H(\mathcal{A})$ and since every $f \in \mathcal{P}(\mathcal{A})$ preserves involution, we conclude that all $f(h), f(k)$ are real numbers. So, $f(x) = 0$ implies $f(h) = 0 = f(k)$, for all $f \in \mathcal{P}(\mathcal{A})$. Finally, Theorem 6.3.4 gives $h = 0 = k$, i.e., $x = 0$, a contradiction. Thus, having $x^*x \neq 0$, from Theorem 6.3.4(2), there will exist $f \in \mathcal{P}(\mathcal{A})$, with $f(x^*x) \neq 0$. Consequently, $\|\pi_f(x)\lambda_f(e)\|^2 = f(x^*x) \neq 0$. Letting now $\xi = (\xi_g)_{g \in \mathcal{P}(\mathcal{A})}$, where for $g = f$, $\xi_f = \lambda_f(e)$ and for every other $g \in \mathcal{P}(\mathcal{A})$, $\xi_g = 0$, it follows that $\pi(x)\xi \neq 0$, therefore $\pi(x) \neq 0$. Hence, π is injective.

We now show that $\pi(B_0) = U(\pi(\mathcal{A}))$. By Lemma 6.3.3, $\pi(B_0) \subset U(\pi(\mathcal{A}))$, so it remains to prove that $U(\pi(\mathcal{A})) \subset \pi(B_0)$.

We first show that, for any $h \in H(\mathcal{A})$ with $\|\pi(h)\| \leq 1$, we obtain $\|h\|_{B_0} \leq \|\pi(h)\|$. Suppose that $\|\pi(h)\| \leq 1$ and let $f \in \mathcal{P}(\mathcal{A})$. Let $\xi = (\xi_g)_{g \in \mathcal{P}(\mathcal{A})}$ with

$$\xi_g := \begin{cases} \lambda_f(e) & \text{if } g = f \\ 0 & \text{if } g \neq f \end{cases}.$$

Then, $f(x) = \langle \pi(x)\xi, \xi \rangle$, for all $x \in \mathcal{A}$. Since $\|\pi(h)\| \leq 1$, we obtain that $\|\pi(h)\xi\|^2 \leq \|\xi\|^2$; therefore,

$$f(h^2) = \|\pi(h)\xi\|^2 \leq \|\xi\|^2 = f(e), \text{ i.e., } f(e - h^2) \geq 0, \forall f \in \mathcal{P}(\mathcal{A}).$$

By Theorem 6.3.4, it follows that $e - h^2 \geq 0$, i.e., $h^2 \leq e$. Consider the (commutative) GB^*-subalgebra $\mathcal{C}[\tau \upharpoonright_{\mathcal{C}}]$ of $\mathcal{A}[\tau]$ generated by the elements e, h and apply Theorem 3.4.9. Then, $\widehat{h^2} \leq \widehat{e}$. This implies that $\widehat{h}$ is a bounded continuous function on the Gelfand space of $\mathcal{C}_0 = \mathcal{C}[\mathcal{C} \cap B_0]$. It follows that $h \in \mathcal{C}_0 \subset \mathcal{A}[B_0]$, such that $\|h\|_{B_0} \leq 1$. Thus, if $h \in H(\mathcal{A})$ with $\|\pi(h)\| \leq 1$, then $h \in \mathcal{A}[B_0]$ and $\|h\|_{B_0} \leq 1$.

From this, it follows that if $h \in H(\mathcal{A})$ and $\pi(h) \in \mathcal{B}(\mathcal{H}_\pi)$ (for $\mathcal{H}_\pi$, see Definition 6.3.2), then $h \in \mathcal{A}[B_0]$ with $\|h\|_{B_0} \leq \|\pi(h)\|$.

Let $x \in \mathcal{A}$ with $\pi(x) \in U(\pi(\mathcal{A}))$, and x not necessarily in $H(\mathcal{A})$. Then, $\pi(x) \in \mathcal{B}(\mathcal{H}_\pi)$ and there exist $h, k \in H(\mathcal{A})$, such that $x = h + ik$. Therefore, $\pi(h)$ and $\pi(k)$ are in $\mathcal{B}(\mathcal{H}_\pi)$, and so from the above, it follows that $h, k \in \mathcal{A}[B_0]$, consequently $x \in \mathcal{A}[B_0]$. Also, $\|h\|_{B_0} \leq \|\pi(h)\|$ and $\|k\|_{B_0} \leq \|\pi(k)\|$; hence

$$\|x\|_{B_0} \leq \|h\|_{B_0} + \|k\|_{B_0} \leq \|\pi(h)\| + \|\pi(k)\|$$
$$= \frac{1}{2}\|\pi(x^*) + \pi(x)\| + \frac{1}{2}\|\pi(x^*) - \pi(x)\| \leq 2\|\pi(x)\|.$$

Recall that $\|\pi(x)\| \leq \|x\|_{B_0}$, therefore $\|\cdot\|_{B_0}$ and $\|\pi(\cdot)\|$ are equivalent norms on $\mathcal{A}[B_0]$. Since $\mathcal{A}[B_0]$ is a C^*-algebra under $\|\cdot\|_{B_0}$, it is complete. Hence, $\mathcal{A}[B_0]$ is a C^*-algebra too, under the C^*-norm $\|\pi(\cdot)\|$. By standard C^*-algebra theory, $\|\cdot\|_{B_0} = \|\pi(\cdot)\|$. Therefore, if $\pi(x) \in U(\pi(\mathcal{A}))$, we get that $\|x\|_{B_0} \leq 1$, i.e., $x \in B_0$.

Thus, $\pi(x) \in \pi(B_0)$, proving that $U\big(\pi(\mathcal{A})\big) \subset \pi(B_0)$. Since $\pi(B_0) = U\big(\pi(\mathcal{A})\big)$, it follows that

$$\pi(\mathcal{A}[B_0]) = \pi(\mathcal{A})_b := \pi(\mathcal{A}) \cap B(\mathcal{H}_\pi),$$

which implies that $\pi(\mathcal{A})$ is symmetric and $\pi(\mathcal{A})_b$ is a C*-algebra. From this we conclude that $\pi(\mathcal{A})$ is an EC^*-algebra (see Definition 5.3.1). This completes the proof.   □

Remark 6.3.6 Note that the *-representation π of the preceding theorem was constructed by using the positive linear functionals of $\mathcal{A}$ and without involvement of any topological structure of $\mathcal{A}$. Consequently, the construction of π depends only on the algebraic structure of the GB^*-algebra $\mathcal{A}[\tau]$ and the same is also true for the set $\pi^{-1}\big(U\big(\pi(\mathcal{A})\big)\big)$.

Combining Remark 6.3.6 with the equivalence of two locally convex topologies given in Definition 4.3.1 (see also Remark 4.3.2(2)), we obtain an extension of Corollary 4.3.10 in the noncommutative case. More precisely, we have the following

Corollary 6.3.7 (Dixon) *Any two locally convex GB^*-topologies τ_1, τ_2 on a *-algebra $\mathcal{A}$ are equivalent in the sense that the maximal elements of the collections* $\mathfrak{B}^*_{\mathcal{A}[\tau_1]}$, $\mathfrak{B}^*_{\mathcal{A}[\tau_2]}$ *coincide; i.e., $B_0(\tau_1) = B_0(\tau_2)$ and thus $\mathcal{A}[B_0(\tau_1)] = \mathcal{A}[B_0(\tau_2)]$.*

Proof From Theorem 6.3.5 and Remark 6.3.6 we conclude that

$$B_0(\tau_1) = \pi^{-1}\big(U\big(\pi(\mathcal{A})\big)\big) = B_0(\tau_2).$$

Therefore $\tau_1 \sim \tau_2$ on $\mathcal{A}$, according to the references in Chap. 4, mentioned just before.   □

The preceding corollary provides us with very good information. Namely, if a *-algebra $\mathcal{A}$ becomes a GB^*-algebra (of Allan) under two topologies τ_1, τ_2, then the C*-algebras $\mathcal{A}[B_0(\tau_1)]$, $\mathcal{A}[B_0(\tau_2)]$ respectively, coincide. As we have noticed these C*-algebras are the key tools through which the structure of the GB^*-algebras $\mathcal{A}[\tau_1]$, $\mathcal{A}[\tau_2]$ respectively, is investigated. In this sense, we may say that *any two locally convex topologies on a *-algebra $\mathcal{A}$ that make it a GB^*-algebra are equivalent*, having thus an 'analogue', in the context of GB^*-algebras, of the situation that we meet in C*-algebras. In fact, according to the comments we have just mentioned, we can make the equivalence of τ_1, τ_2, even clearer by saying that $\tau_1 \sim \tau_2$, *if and only if, the identity map $\mathcal{A}[B_0(\tau_1)] \to \mathcal{A}[B_0(\tau_2)]$ is bicontinuous, which equivalently means that $\| \cdot \|_{B_0(\tau_1)} = \| \cdot \|_{B_0(\tau_2)}$.*

Coming back to the result of Theorem 6.3.5, we recall that for every C*-algebra $\mathcal{A}$, there is a bicontinuous faithful *-representation $\pi : \mathcal{A} \to \mathcal{B}$, where $\mathcal{B}$ is a C*-algebra on some Hilbert space $\mathcal{H}$. In this regard, Theorem 6.3.5 only provides us with the existence of a faithful *-representation of a GB^*-algebra onto an EC^*-algebra, and not necessarily a bicontinuous faithful *-representation onto an EC^*-algebra with a suitable locally convex topology. The question is therefore if one can obtain such a representation theorem for GB^*-algebras. For this, we require the following notion of an $A\mathcal{O}^*$-algebra, which is a non-normed generalization of a C^*-algebra, first defined by G. Lassner in [103, Definition 4.3]. In this way, Theorem 6.3.11 below. provides us with a necessary and sufficient condition, under which we may say that we obtain a noncommutative Gelfand–Naimark type theorem for GB^*-algebras.

Definition 6.3.8 *An $A\mathcal{O}^*$-algebra is a complete locally convex $*$-algebra, which is topologically $*$-isomorphic to an $\mathcal{O}^*$-algebra in the uniform topology τ_u defined in (5.2.4).*

We present an example of an $A\mathcal{O}^*$-algebra. For this we need the result that follows; for the topologies involved, see (5.2.1) and (5.2.4).

Theorem 6.3.9 (Lassner) *Let π : $\mathcal{A}[\tau]$ $\to$ $\mathcal{L}^\dagger(\mathcal{D})[\tau_w]$ be a continuous $*$-representation of a barrelled locally convex $*$-algebra $\mathcal{A}[\tau]$, when $\mathcal{L}^\dagger(\mathcal{D})$ carries the weak topology τ_w. Then, π : $\mathcal{A}[\tau]$ $\to$ $\mathcal{L}^\dagger(\mathcal{D})[\tau_u]$ is also continuous, when $\mathcal{L}^\dagger(\mathcal{D})$ carries the uniform topology τ_u.*

Proof Let V_0 be a τ_u-neighbourhood of $0 \in \pi(\mathcal{A})$ (for the notation see the discussion after (5.2.4)). Then, there exists a t_π-bounded subset M of $\mathcal{D}$ and $\varepsilon > 0$, such that

$$\big\{\pi(x) \in \pi(\mathcal{A}) : p_M(\pi(x)) := \sup_{\xi,\eta \in M} |\langle \xi, \pi(x)\eta\rangle| \leq \varepsilon\big\} \subseteq V_0.$$

Let

$$U_{M,\varepsilon} = \big\{x \in \mathcal{A} : p_M(\pi(x)) \leq \varepsilon\big\} = \bigcap_{\xi,\eta \in M} \big\{x \in \mathcal{A} : |\langle \xi, \pi(x)\eta\rangle| \leq \varepsilon\big\}.$$

It is clear that the function $x \mapsto \langle \xi, \pi(x)\eta\rangle$ is continuous, for all $\xi, \eta \in \mathcal{D}$, by hypothesis. Hence, the set $\{x \in \mathcal{A} : |\langle \xi, \pi(x)\eta\rangle| \leq \varepsilon\}$ is absolutely convex and closed, for all $\xi, \eta \in \mathcal{D}$. Consequently, $U_{M,\varepsilon}$ is absolutely convex and closed. Evidently, $U_{M,\varepsilon}$ is also balanced.

We show that $U_{M,\varepsilon}$ is absorbing. Let $x \in \mathcal{A}$ and suppose that $|\alpha| \geq \frac{1}{\varepsilon} p_M(\pi(x))$, $\alpha \in \mathbb{C}$. Then, $p_M(\pi(x)) \leq |\alpha|\varepsilon$, hence $\frac{1}{\alpha}x \in U_{M,\varepsilon}$, i.e., $x \in \alpha U_{M,\varepsilon}$; hence, $U_{M,\varepsilon}$ is absorbing.

It follows that $U_{M,\varepsilon}$ is a barrel of $\mathcal{A}[\tau]$ and since $\mathcal{A}[\tau]$ is barrelled, we have that $U_{M,\varepsilon}$ is a neighbourhood of $0 \in \mathcal{A}$. Since also $\pi(U_{M,\varepsilon}) \subseteq V_0$, we obtain the conclusion. $\square$

Note that Example 6.3.10(1) is due to Lassner, while Example 6.3.10(2) is due to Schmüdgen.

Examples 6.3.10

(1) We show that *every barrelled pro- C*-algebra* $\mathcal{A}[\tau]$ (hence, every $\sigma - C^*$-algebra; see Definition 3.1.2) *is an* $A\mathcal{O}^*$-*algebra*. Let $\{p_\nu\}_{\nu \in \Lambda}$ denote a family of C^*-seminorms on $\mathcal{A}$ defining the topology τ. If $N_\nu := \{x \in \mathcal{A} : p_\nu(x) = 0\}$, $\nu \in \Lambda$, it is easily verified that N_ν is a closed $*$-ideal of $\mathcal{A}$, for all $\nu \in \Lambda$. Let $A_\nu \equiv \mathcal{A}/N_\nu$, $\nu \in \Lambda$. Put

$$\|x + N_\nu\|_\nu := p_\nu(x), \ \forall x \in \mathcal{A}, \ \lambda \in \Lambda.$$

Then, $\mathcal{A}_\nu[\| \cdot \|_\nu]$ is a C*-algebra [60, p. 130, 10.(2)], for all $\nu \in \Lambda$. Therefore, $\mathcal{A}_\nu$ is topologically $*$-isomorphic to a C^*-subalgebra $\mathcal{M}_\nu$ of bounded linear operators on some Hilbert space $\mathcal{H}_\nu$, with norm $\| \cdot \|_\nu$. Denote this topological $*$-isomorphism by ϕ_ν. Then, $\|x + N_\nu\|_\nu = \|\phi_\nu(x + N_\nu)\|_\nu$, for all $x \in \mathcal{A}$ and $\nu \in \Lambda$. Let $q_\nu : A \to A_\nu$, $\nu \in \Lambda$, be the corresponding quotient map. Then, $\psi_\nu := \phi_\nu \circ q_\nu$ is a $*$-homomorphism from $\mathcal{A}$ to $\mathcal{M}_\nu$, with

$$\|\psi_\nu(x)\|_\nu = \|\phi_\nu(x + N_\nu)\|_\nu = \|x + N_\nu\|_\nu = p_\nu(x), \tag{6.3.5}$$

for all $x \in \mathcal{A}$, $\nu \in \Lambda$. Let $\mathcal{H}$ denote the Hilbert space direct sum of all $\mathcal{H}_\nu$, and let $\mathcal{D}$ denote the algebraic direct sum of all $\mathcal{H}_\nu$. Observe that $\mathcal{D}$ is a dense subspace of $\mathcal{H}$. For every $x \in \mathcal{A}$ and $\xi \in \mathcal{H}$, let $\pi(x)\xi = \sum_\nu \psi_\nu(x)\xi_\nu$, where $\xi = \sum_\nu \xi_\nu \in \mathcal{D}$; in both sums, only a finite number of summands are nonzero. Clearly, π is a $*$-homomorphism of $\mathcal{A}$ into $\mathcal{L}^\dagger(\mathcal{D})$.

We show that π is a topological $*$-isomorphism. For every $\nu \in \Lambda$, let U_ν denote the closed unit ball in $\mathcal{H}_\nu$. Let $\eta \in U_\nu$ and $\eta_1 = (\eta_\lambda^1)_\lambda \in \mathcal{D}$, where $\eta_\lambda^1 = 0$, for all $\lambda \neq \nu$ and $\eta_\lambda^1 = \eta$ if $\lambda = \nu$. Then, $\eta = \sum_\lambda \eta_\lambda^1 = \eta_1 \in \mathcal{D}$ and

$$\pi(x)\eta = \pi(x)\left(\sum_\lambda \eta_\lambda^1\right) = \sum_\lambda \psi_\lambda(x)\eta_\lambda^1 = \psi_\nu(x)\eta, \ \forall x \in \mathcal{A}.$$

Observe now that the set U_ν is a $t_{\mathcal{L}^\dagger(\mathcal{D})}$-bounded subset of $\mathcal{D}$, for all $\nu \in \Lambda$. Indeed, for $\xi \in U_\nu$ and $x \in \mathcal{A}$, we have

$$\|\xi\|_{\pi(x)} = \|\xi\| + \|\pi(x)\xi\| = \|\xi\| + \|\psi_\nu(x)\xi\| \leq \|\xi\|(1 + \|\psi_\nu(x)\|) \leq 1 + \|\psi_\nu(x)\|,$$

so U_ν is t_π- bounded (see notation before Example 5.1.3), for each $\nu \in \Lambda$, which equivalently means that U_ν is a $t_{\mathcal{L}^\dagger(\mathcal{D})}$-bounded set, for all $\nu \in \Lambda$; since $\pi(\mathcal{A})$ is a closed $\mathcal{O}^*$-algebra (see Definition 6.3.2) the latter follows from comments around (5.2.4). Thus, the seminorms $\{p_{U_\nu}\}_{\nu \in \Lambda}$, belong to the family of seminorms defining the topology τ_u. In particular, (see also (6.3.5))

$$p_{U_\nu}(\pi(x)) = \sup_{\xi,\eta \in U_\nu} |\langle \xi, \pi(x)\eta \rangle| = \sup_{\xi,\eta \in U_\nu} |\langle \xi, \psi_\nu(x)\eta \rangle|$$

$$= \|\psi_\nu(x)\|_\nu = p_\nu(x), \tag{6.3.6}$$

for all $x \in \mathcal{A}$ and $\nu \in \Lambda$. Therefore, if $\pi(x) = 0$, then $p_\nu(x) = 0$, for all $\nu \in \Lambda$, consequently $x = 0$. Furthermore, (6.3.6) shows, on the one hand that $\pi^{-1} : \pi(\mathcal{A})[\tau_u \upharpoonright_{\pi(\mathcal{A})}] \to \mathcal{A}[\tau]$ is continuous and on the other hand, that $\pi : \mathcal{A}[\tau] \to \mathcal{L}^\dagger(\mathcal{D})[\tau_w]$ is also continuous. The latter, according to Theorem 6.3.9, implies that $\pi : \mathcal{A}[\tau] \to \mathcal{L}^\dagger(\mathcal{D})[\tau_u]$ is continuous. Thus, π is a topological *-isomorphism from $\mathcal{A}[\tau]$ on $\pi(\mathcal{A})[\tau_u]$ (see notation after (5.2.4)). The result now follows from Definition 6.3.8.

(2) Let $\{\mathcal{A}_\nu\}_{\nu \in \Lambda}$ be a directed family of barrelled pro-C^*-algebras, each one of them having an identity element. Let $\mathcal{A}[\tau] := \prod_{\nu \in \Lambda} \mathcal{A}_\nu$ be their product endowed with the product topology τ and the usual algebraic operations and involution defined coordinatewise. Then, $\mathcal{A}[\tau]$ is a barrelled pro-C^*-algebra; see respectively [123, Proposition 4.2.5(i)] and [60, 7.6 Examples, (2)]. Hence, from the preceding Example 6.3.10(1), the product algebra $\mathcal{A}[\tau]$ is an $A\mathcal{O}^*$-algebra.

For further information on $A\mathcal{O}^*$-algebras, see [104, 105, 133].

The following result, connects GB^*-algebras with $A\mathcal{O}^*$-algebras, where both are generalizations of C^*-algebras. Namely, we may say that Theorem 6.3.11 gives in the context of normal Fréchet GB^*-algebras, a noncommutative Gelfand–Naimark type theorem.

Theorem 6.3.11 (Schmüdgen) *A Fréchet GB^*-algebra $\mathcal{A}[\tau]$ is an $A\mathcal{O}^*$-algebra, if and only if, the (proper, convex) cone $\mathcal{A}^+$ of the positive elements in $\mathcal{A}[\tau]$ is normal.*

In what follows, we give a proof of Theorem 6.3.11, and for this, we require the two lemmas and proposition that follow. *The given proof is inspired from the proof of a result of* K. Schmüdgen (see [133, Theorem 5.1]), *which reads as follows: a barrelled topological *-algebra $\mathcal{A}[\tau]$ with an identity element is an $A\mathcal{O}^*$-algebra, if and only if, the cone $\mathcal{A}^+$ of positive elements in $\mathcal{A}[\tau]$ is normal in the ordered topological vector space $H(\mathcal{A})[\tau]$.* The concept of normality is given just before the proof of Theorem 6.3.11.

Lemma 6.3.12 *Let M be an absolutely convex subset of a pre-Hilbert space $\mathcal{D}$. Then, for every $T = T^\dagger \in \mathcal{L}^\dagger(\mathcal{D})$,*

$$\sup_{\xi \in M} |\langle T\xi, \xi \rangle| \leq \sup_{\xi,\eta \in M} |\langle T\xi, \eta \rangle| \leq 2 \sup_{\xi \in M} |\langle T\xi, \xi \rangle|.$$

Proof The first inequality is obvious, and so we give the proof of the second one. Let $\gamma = \sup_{\xi \in M} |\langle T\xi, \xi \rangle|$ and $\xi, \eta \in M$. Choose $r \in \mathbb{R}$, such that $\langle T(e^{ir}\xi), \eta \rangle \in \mathbb{R}$. Let $2\xi_1 = e^{ir}\xi + \eta$ and $2\eta_1 = e^{ir}\xi - \eta$. Since M is absolutely convex, we obtain that $\xi_1, \eta_1 \in M$. Therefore,

$$
\begin{aligned}
4|\langle T\xi, \eta \rangle| &= 4|\langle T(e^{ir}\xi), \eta \rangle| \\
&= |2\langle T\eta, e^{ir}\xi \rangle + 2\langle T(e^{ir}\xi), \eta \rangle| \\
&= 4|\langle T\xi_1, \xi_1 \rangle + \langle T\eta_1, \eta_1 \rangle| \\
&\leq 4|\langle T\xi_1, \xi_1 \rangle| + 4|\langle T\eta_1, \eta_1 \rangle| \\
&\leq 8\gamma.
\end{aligned}
$$

Therefore $|\langle T\xi, \eta \rangle| \leq 2\gamma = 2\sup_{\xi \in M} |\langle T\xi, \xi \rangle|$. $\qquad\qquad\square$

Lemma 6.3.13 *Let M be an absolutely convex subset of a pre-Hilbert space $\mathcal{D}$. If $T \in \mathcal{L}^\dagger(\mathcal{D})$, then*

$$
\sup_{\xi \in M} |\langle T\xi, \xi \rangle| \leq \sup_{\xi, \eta \in M} |\langle T\xi, \eta \rangle| \leq 4\sup_{\xi \in M} |\langle T\xi, \xi \rangle|.
$$

Proof Let $T = T_1 + iT_2$, where $T_i = T_i^\dagger \in \mathcal{L}^\dagger(\mathcal{D})$ for $i = 1, 2$. By Lemma 6.3.12,

$$
\sup_{\xi \in M} |\langle T_i\xi, \xi \rangle| \leq \sup_{\xi, \eta \in M} |\langle T_i\xi, \eta \rangle| \leq 2\sup_{\xi \in M} |\langle T_i\xi, \xi \rangle|, \; i = 1, 2.
$$

Therefore,

$$
\begin{aligned}
\sup_{\xi, \eta \in M} |\langle T\xi, \eta \rangle| &= \sup_{\xi, \eta \in M} |\langle (T_1 + iT_2)\xi, \eta \rangle| \\
&= \sup_{\xi, \eta \in M} |\langle T_1\xi, \eta \rangle + i\langle T_2\xi, \eta \rangle| \\
&\leq 2\sup_{\xi \in M} |\langle T_1\xi, \xi \rangle| + 2\sup_{\xi \in M} |\langle T_2\xi, \xi \rangle| \\
&\leq 4\sup_{\xi \in M} |\langle T\xi, \xi \rangle|,
\end{aligned}
$$

since $\sup_{\xi \in M} |\langle T_i\xi, \xi \rangle| \leq \sup_{\xi \in M} |\langle T\xi, \xi \rangle|$, for $i = 1, 2$. $\qquad\qquad\square$

Proposition 6.3.14 *For an $\mathcal{O}^*$-algebra $\mathcal{M}$ on $\mathcal{D}$, the uniform topology τ_u on $\mathcal{M}$ can also be defined by the family of seminorms*

$$
p'_M(T) := \sup_{\xi \in M} |\langle T\xi, \xi \rangle|, \; T \in \mathcal{M},
$$

where M is a $t_\mathcal{M}$-bounded subset of $\mathcal{D}$ (for $t_\mathcal{M}$, see after (5.1.1)).

Proof If M_1 denotes the absolute convex hull of M, then M_1 is $t_{\mathcal{M}}$-bounded if M is $t_{\mathcal{M}}$-bounded. Observe that

$$\sup_{\xi,\eta\in M} |\langle T\xi,\eta\rangle| \leq \sup_{\xi,\eta\in M_1} |\langle T\xi,\eta\rangle|, \quad T\in\mathcal{M}. \tag{6.3.7}$$

By Lemma 6.3.13,

$$\sup_{\xi\in M_1} |\langle T\xi,\xi\rangle| \leq \sup_{\xi,\eta\in M_1} |\langle T\xi,\eta\rangle| \leq 4 \sup_{\xi\in M_1} |\langle T\xi,\xi\rangle|, \quad T\in\mathcal{M}. \tag{6.3.8}$$

Therefore, if (T_α) is a net in $\mathcal{M}$ with $p'_M(T_\alpha - T) \to 0$, for all $t_{\mathcal{M}}$-bounded subsets M of $\mathcal{D}$, we get from (6.3.8) that

$$\sup_{\xi,\eta\in M_1} |\langle (T_\alpha - T)\xi,\eta\rangle| \to 0.$$

By (6.3.7),

$$\sup_{\xi,\eta\in M} |\langle (T_\alpha - T)\xi,\eta\rangle| \to 0,$$

for all $t_{\mathcal{M}}$-bounded subsets M of $\mathcal{D}$.

Conversely, if

$$\sup_{\xi,\eta\in M} |\langle (T_\alpha - T)\xi,\eta\rangle| \to 0,$$

for all $t_{\mathcal{M}}$-bounded subsets M of $\mathcal{D}$, then it is clear that $p'_M(T_\alpha - T) \to 0$, for all $t_{\mathcal{M}}$-bounded subsets M of $\mathcal{D}$. $\qquad\square$

Let E be a (real or complex) vector space. A subset C of E is called a *cone*, if it is invariant under addition and scalar multiplication by strictly positive scalars and $C \cap (-C) = \{0\}$ [136, p. 20]. Let $E[\tau]$ be a real locally convex space, such that the topology τ is determined by a family $\Gamma = \{p\}$ of seminorms. A cone C in E is called *normal*, with respect to τ, if (see [131, p. 215, 3.1 (e)] and [133, p. 220, Section 4])

$$p(x) \leq p(x+y), \quad \forall\, x, y \in C \text{ and } \forall\, p \in \Gamma. \tag{6.3.9}$$

In [90, p. 86] a normal cone is called *a self-allied cone*. According to the second reference, just before (6.3.9), the seminorms $p \in \Gamma$, having the previous property, are called *monotone* or *normal seminorms for* τ. If C is a cone in a topological vector space E, the *dual cone* C' of C is given by the set

$$\{f \in E' : \mathrm{Re}\, f(x) \geq 0, \text{ for } x \in C\},$$

where $\mathrm{Re} f(x)$ denotes the real part of $f(x)$; namely, C' is the polar of $-C$ with respect to the dual pair (E, E') (see [131, p. 218, comments before Lemma 1]).

Recall that if $\mathcal{A}[\tau]$ is a Fréchet GB^*-algebra, the subset $\mathcal{A}^+$ of positive elements of $\mathcal{A}$ is a τ-closed proper (the latter meaning that $\mathcal{A}^+ \cap (-\mathcal{A}^+) = 0$) convex cone in $\mathcal{A}[\tau]$. Hence, $\mathcal{A}^+$ fulfills all the properties of a cone as given above, so that *in what follows we shall refer to $\mathcal{A}^+$ as a cone of $\mathcal{A}$*.

We are ready now to exhibit the proof of Theorem 6.3.11.

Proof First suppose that $\mathcal{A}[\tau]$ is an $A\mathcal{O}^*$-algebra. Then, $\mathcal{A}[\tau]$ is topologically $*$-isomorphic to an $\mathcal{O}^*$-algebra $\mathcal{M}$ on a pre-Hilbert space $\mathcal{D}$, where $\mathcal{M}$ is equipped with the uniform topology τ_u. By Proposition 6.3.14, the uniform topology τ_u can be defined by the family of seminorms

$$p'_M(T) = \sup_{\xi \in M} |\langle T\xi, \xi \rangle|, \ T \in \mathcal{M},$$

where M is a $t_{\mathcal{M}}$-bounded subset of $\mathcal{D}$. For all $T, S \in \mathcal{M}^+$ and for all $t_{\mathcal{M}}$-bounded subsets M of $\mathcal{D}$,

$$p'_M(T + S) = \sup_{\xi \in M}(\langle T\xi, \xi \rangle + \langle S\xi, \xi \rangle) \geq \sup_{\xi \in M}\langle T\xi, \xi \rangle = p'_M(T).$$

Concerning the inequality in the previous relation, observe that T, S are positive operators, so that $\langle T\xi, \xi \rangle \geq 0$ and $\langle S\xi, \xi \rangle \geq 0$, for every $\xi \in M$. Therefore, $\mathcal{M}^+$ is normal.

We prove now that $\mathcal{A}^+$ is normal in the ordered real locally convex space $H(A)[\tau]$. Let ϕ be the topological $*$-isomorphism from $\mathcal{A}[\tau]$ to $\mathcal{M}[\tau_u]$. Consider the family of $*$-seminorms $\{p'_M \circ \phi\}$ on $\mathcal{A}$, where M runs in the set of $t_{\mathcal{M}}$-bounded subsets of $\mathcal{D}$. Then, it is readily seen that the topology τ is equivalent to the topology induced on $\mathcal{A}$ by the preceding family of $*$-seminorms. Thus, for any $a, b \in \mathcal{A}^+$, we have

$$(p'_M \circ \phi)(a) = p'_M(\phi(a)) \leq p'_M(\phi(a) + \phi(b)) = (p'_M \circ \phi)(a + b), \ \forall \ M,$$

i.e., (6.3.9) is fulfilled and this shows that $\mathcal{A}^+$ is normal in $H(A)[\tau]$.

Conversely, suppose that $\mathcal{A}^+$ is normal in $H(A)[\tau]$. Recall that since $\mathcal{A}[\tau]$ is a Fréchet GB^*-algebra, *each positive linear functional on $\mathcal{A}[\tau]$ is continuous* (see Corollary 7.2.3, in Sect. 7.2). The *-representation, π (algebraic) direct sum of the family $\{\pi_f : f \in \mathcal{P}(A)\}$ (Definition 6.3.2), is a $*$-homomorphism of $\mathcal{A}$ onto $\pi(\mathcal{A})$. We prove that it is also injective. Let $\pi(x) = 0$, $x \in \mathcal{A}$. Then, $f(x^*x) = 0$, for all $f \in \mathcal{P}(A)$, therefore (by the Cauchy–Schwarz inequality) $f(x) = 0$, for all $f \in \mathcal{P}(A)$, hence $x = 0$; if $x \neq 0$, from Theorem 6.3.4 (see also discussion after it) we would have $f(x) \neq 0$, for some $f \in \mathcal{P}(A)$, a contradiction. So π is an algebraic $*$-isomorphism onto $\pi(\mathcal{A})$.

Furthermore, we show that $\pi : \mathcal{A}[\tau] \rightarrow \pi(\mathcal{A})[\tau_u]$ is a topological $*$-isomorphism, thereby proving that $\mathcal{A}[\tau]$ is an $A\mathcal{O}^*$algebra. Let

$$F_{\xi,\eta}(x) = \langle \pi(x)\xi, \eta \rangle, \ \forall \, x \in \mathcal{A}.$$

For every ξ, η in the domain of π, we observe that $F_{\xi,\eta}$ is a finite sum of linear functionals F_{ξ_f, η_f}, where ξ_f and η_f are of the form $\pi_f(y)\lambda_f(e)$ and $\pi_f(z)\lambda_f(e)$ respectively, with $y, z \in \mathcal{A}$ and f a positive linear functional on $\mathcal{A}$. Then,

$$F_{\xi_f, \eta_f}(x) = \langle \pi_f(z^*xy)\lambda_f(e), \lambda_f(e) \rangle = f(z^*xy).$$

Multiplication in $\mathcal{A}[\tau]$ is jointly continuous and since f is also continuous, it follows from the above that each F_{ξ_f, η_f} is continuous and consequently $F_{\xi,\eta}$ too. Therefore, π is weakly continuous. It follows now from Theorem 6.3.9 that $\pi : \mathcal{A}[\tau] \rightarrow \pi(\mathcal{A})[\tau_u]$ is continuous.

We are ready to prove that $\pi^{-1} : \pi(\mathcal{A})[\tau_u] \rightarrow \mathcal{A}[\tau]$ is continuous. Since $\pi(\mathcal{A})[\tau_u]$ can be identified with $H(\pi(\mathcal{A}))[\tau_u] \oplus iH(\pi(\mathcal{A}))[\tau_u]$ as locally convex spaces, it is sufficient to prove that $\pi^{-1} : H(\pi(\mathcal{A}))[\tau_u] \rightarrow H(\mathcal{A})[\tau]$ is continuous.

Employ the *GNS*-construction $(\pi_f, \lambda_f, \mathcal{H}_f)$, for each $f \in \mathcal{P}(\mathcal{A}) \subset \mathcal{A}'$ (see discussion after Proposition 6.3.1). Then, we have that

$$f(x) = \langle \pi_f(x)\lambda_f(e), \lambda_f(e) \rangle = \langle \pi(x)\xi, \xi \rangle, \ \forall \, x \in \mathcal{A}, \qquad (6.3.10)$$

where $\pi = \oplus_{f \in \mathcal{P}(\mathcal{A})} \pi_f$ and $\xi = (\xi_g)_{g \in \mathcal{P}(\mathcal{A}^+)}$, with $\xi_g = 0$, for $g \neq f$ and $\xi_f = \lambda_f(e)$.

Denote by M the subset of the domain $\mathcal{D}_\pi$ of π, consisting of all ξ defined as before, for a given fixed $f \in \mathcal{P}(\mathcal{A})$, every time. Consider the set

$$F \equiv \big\{ f \in \mathcal{P}(\mathcal{A}) : f(x) = \langle \pi(x)\xi, \xi \rangle \text{ with } x \in \mathcal{A} \text{ and } \xi \in M \big\}. \qquad (6.3.11)$$

Then, F is an equicontinuous subset of $\mathcal{A}'$. Indeed, let $V = \{\lambda \in \mathbb{C} : |\lambda| < \varepsilon\}$, $\varepsilon > 0$, be a 0-neighbourhood in $\mathbb{C}$. Let

$$U = \Big\{ x \in \mathcal{A} : \sup_{\xi \in M} p_M(\pi(x)) < \varepsilon \Big\},$$

where $p_M(\pi(x)) = |\langle \pi(x)\xi, \xi \rangle|$, $x \in \mathcal{A}$, and M a $t_\mathcal{M}$-bounded subset of $\mathcal{D}$. Now if $W = \{T \in \pi(\mathcal{A}) : p_M(T) < \varepsilon\}$ is a τ_u-neighbourhood of 0 in $\pi(\mathcal{A})$, we clearly have that $U = \pi^{-1}(W)$ and since π is $\tau - \tau_u$ continuous, U is a 0-neighbourhood in $\mathcal{A}[\tau]$. Moreover, for each $x \in U$ and for each $f \in F$, we obtain $|f(x)| = |\langle \pi(x)\xi, \xi \rangle| < \varepsilon$, $x \in \mathcal{A}$. Therefore, $f(U) \subseteq V$, for all $f \in F$.

To continue we have to justify that F is weakly bounded: $\mathcal{A}[\tau]$ being a Fréchet locally convex space is barrelled [74, p. 214, Corollary], therefore F as an equicontinuous subset of $\mathcal{A}'$ is weakly bounded [74, p. 212, Corollary].

By the weak boundedness of F, it follows now (see also (6.3.11)) that

$$\sup_{\xi \in M} \|\pi(x)\xi\|^2 = \sup_{\xi \in M} \langle \pi(x^*x)\xi, \xi \rangle = \sup_{f \in F} f(x^*x) < +\infty,$$

for all $x \in \mathcal{A}$. Hence, M is t_π-bounded.

Now let F_0 be an equicontinuous subset of $\mathcal{A}'$. From [131, p. 219, Corollary 1], the normality of $\mathcal{A}^+$ in $H(A)[\tau]$ (see (6.3.9)) is equivalent to the existence of an equicontinuous subset F_1 of $(\mathcal{A}^+)'$ (see comments before the proof of Theorem 6.3.11), such that $F_0 \subseteq F_1 - F_1$. So, we have (see (6.3.10) and Proposition 6.3.14)

$$p_{F_1}(x) := \sup_{f \in F_1} |f(x)| = \sup_{\xi \in M} |\langle \pi(x)\xi, \xi \rangle| = p'_M(\pi(x)), \; x \in \mathcal{A}^+.$$

Consequently,

$$p_{F_0}(x) \leq 2p_{F_1}(x) = 2p'_M(\pi(x)), \; \forall \, x \in \mathcal{A}^+. \tag{6.3.12}$$

Again the normality of $\mathcal{A}^+$ in $\mathcal{A}[\tau]$, from the preceding reference in [131], equivalently means that the locally convex topology τ on $\mathcal{A}$ is the topology of uniform convergence on the equicontinuous subsets of $(\mathcal{A}^+)'$. It is now clear from (6.3.12) and Proposition 6.2.3 that π^{-1} is $\tau_u - \tau$ continuous when restricted to $H(A)[\tau]$. This completes the proof. $\qquad\square$

Notice that in the complex locally convex space $\mathcal{A}[\tau]$, the cone $\mathcal{A}^+ + i\mathcal{A}^+$ is normal, if and only if, $\mathcal{A}^+$ is a normal cone in the real ordered locally convex space $H(\mathcal{A})[\tau \restriction_{H(\mathcal{A})}]$.

Corollary 6.3.15 *Let $\mathcal{A}[\tau]$ be a Fréchet GB*-algebra. The following statements are equivalent:*

(i) *the given algebra $\mathcal{A}[\tau]$ is an $A\mathcal{O}^*$-algebra;*
(ii) *the cone $\mathcal{A}^+$ is normal with respect to the topology τ;*
(iii) *the cone $\mathcal{A}^+$ is normal with respect to the weak topology $\sigma(\mathcal{A}, \mathcal{A}')$;*
(iv) *for every continuous, self-adjoint, linear functional on $\mathcal{A}[\tau]$, there are two continuous, positive, linear functionals f_1, f_2 on $\mathcal{A}[\tau]$, such that $f = f_1 - f_2$.*

Proof (i) $\Leftrightarrow$ (ii) follows from Theorem 6.3.11.

(ii) $\Rightarrow$ (iii) Every element $f \in \mathcal{A}'$, can be written in the form $f = g_1 + ig_2$, where g_1, g_2 are continuous self-adjoint linear functionals on $\mathcal{A}[\tau]$. Thus, we obtain $\sigma \equiv \sigma(\mathcal{A}, \mathcal{A}') \restriction_{H(\mathcal{A})} = \sigma(H(\mathcal{A}), H(\mathcal{A})')$. Hence, since every normal cone in $H(\mathcal{A})$ is also weakly normal by the second part of [131, p. 220, Corollary 3], the implication (ii) $\Rightarrow$ (iii) is proved.

(iii) $\Rightarrow$ (ii) It is known that the bounded sets of $H(\mathcal{A})$, with respect to the locally convex topologies $\tau \restriction_{H(\mathcal{A})}$ and $\sigma(H(\mathcal{A}), H(\mathcal{A})')$, compatible with the dual pair

$(H(\mathcal{A}), (H(\mathcal{A})')$, are the same (cf. [74, p. 209, Theorem 3 and p. 198, Definition 1]). Moreover, $H(\mathcal{A})[\tau \restriction_{H(\mathcal{A})}]$ is a real ordered metrizable locally convex space, so that the weakly normal cone $\mathcal{A}^+$ is also normal with respect to τ; see [90, Theorem 3.2.13 and the discusiion after it]. Thus, (iii) $\Rightarrow$ (ii).

(iii) $\Rightarrow$ (iv) From the first part of [131, p. 220, Corollary 3], we have that $\mathcal{A}' = (\mathcal{A}^+)' - (\mathcal{A}^+)'$, where $(\mathcal{A}^+)'$ is the dual cone of $\mathcal{A}^+$ (see comments after (6.3.9)), i.e.,

$$(\mathcal{A}^+)' = \{ f \in \mathcal{A}' : \mathrm{Re}\, f(x) \geq 0, \ \forall\, x \in \mathcal{A}^+ \}. \tag{6.3.13}$$

From the preceding expression of $\mathcal{A}'$, we conclude that each $f \in \mathcal{A}'$ is of the form

$$f = f_1 - f_2, \ \text{ with } \ f_i \in (\mathcal{A}^+)', \ i = 1, 2.$$

We shall show that $f_i \in \mathcal{P}(\mathcal{A})$, $i = 1, 2$, i.e., taking, for instance $i = 1$, we must prove that $f_1(x^*x) \geq 0$, for all $x \in \mathcal{A}$. From Definition 6.2.2, all elements x^*x, $x \in \mathcal{A}$, belong to $\mathcal{A}^+$, hence (6.3.13) implies that $\mathrm{Re}\, f_1(x^*x) \geq 0$, where it is readily seen that $\overline{f_1(x^*x)} = f_1(x^*x)$, that is $f_1(x^*x)$ is a real number. It follows now from (6.3.13) that $f_1 \in \mathcal{P}(\mathcal{A})$. Similarly, $f_2 \in \mathcal{P}(\mathcal{A})$ and this proves (iv).

(iv) $\Rightarrow$ (iii) Let $f \in \mathcal{A}'$. Then, $f = f_1 + i f_2$, with f_1, f_2 self-adjoint elements in $\mathcal{A}'$. Thus, from (iv), we conclude that each $f \in \mathcal{A}'$ is a linear combination of continuous, positive, linear functionals of $\mathcal{A}$. Let f_i ($i = 1, \cdots, n$) be one of them. It is then easily seen that

$$\mathrm{Re}\, f_i(x) \geq 0, \ \forall\, x \in \mathcal{A}^+; \ \text{ i.e., } \ f_i \in (\mathcal{A}^+)',$$

(see (6.3.13)). Finally we obtain that $\mathcal{A}' = (\mathcal{A}^+)' - (\mathcal{A}^+)'$, from which follows that the cone $\mathcal{A}^+$ is weakly normal (see the first part of [131, p. 220, Corollary 3]).

(iv) $\Leftrightarrow$ (i) From what we have shown before, (iv) $\Leftrightarrow$ (iii) $\Leftrightarrow$ (ii) $\Leftrightarrow$ (i) and this completes the proof.   □

▶ An immediate consequence of Corollary 6.3.15 is Corollary 6.3.16, just below, that provides *an analogue* of Theorem 4.2.5 *in the case of noncommutative GB*-algebras.*

For the *classical C*-algebra counterpart* of the latter, see [122, Theorem 3.2.5].

Corollary 6.3.16 *Let $\mathcal{A}[\tau]$ be a Fréchet GB*-algebra. The following are equivalent:*

(i) *$\mathcal{A}[\tau]$ is an $A\mathcal{O}^*$-algebra;*
(ii) *for every self-adjoint, continuous linear functional f on $\mathcal{A}[\tau]$ there are two continuous, positive, linear functionals f_1, f_2 on $\mathcal{A}[\tau]$, such that $f = f_1 - f_2$.*

Corollary 6.3.17 *Let $\mathcal{A}[\tau]$ be a Fréchet GB*-algebra, such that its cone $\mathcal{A}^+$ is normal. Let $\mathcal{B}[\tau\!\upharpoonright_\mathcal{B}]$ be a closed *-subalgebra of $\mathcal{A}[\tau]$ containing the identity e of $\mathcal{A}$. Then, $\mathcal{B}[\tau\!\upharpoonright_\mathcal{B}]$ is an $A\mathcal{O}^*$-subalgebra of $\mathcal{A}[\tau]$.*

Proof The closed *-subalgebra $\mathcal{B}[\tau\!\upharpoonright_\mathcal{B}]$ of $\mathcal{A}[\tau]$ is clearly Fréchet, but also a GB*-algebra from Proposition 3.3.19. Moreover, $\mathcal{B}^+ \subset \mathcal{A}^+$ and since $\mathcal{A}^+$ is normal, the same is true for $\mathcal{B}^+$. Hence, Corollary 6.3.15 yields that $\mathcal{B}[\tau\!\upharpoonright_\mathcal{B}]$ is an $A\mathcal{O}^*$-subalgebra of the $A\mathcal{O}^*$-algebra (Theorem 6.3.11)) $\mathcal{A}[\tau]$. □

Corollary 6.3.18 *The Arens algebra $L^\omega[0, 1]$ is an $A\mathcal{O}^*$-algebra.*

Proof From Examples 3.3.16(5) $\mathcal{A}[\tau] \equiv L^\omega[0, 1] := \bigcap_{1 \leq p < \infty} L^p[0, 1]$ is a Fréchet GB*-algebra, where τ is given by the L^p-norms $\|\cdot\|_p$, $p = 1, 2, \cdots$. Let $f \in \mathcal{A}^+$. Then, $f(t) \geq 0$, almost everywhere. Hence, for $f, g \in \mathcal{A}^+$, we shall have $\|f + g\|_p \geq \|f\|_p$, $p = 1, 2, \cdots$, which shows that the cone $\mathcal{A}^+$ of positive elements in $L^\omega[0, 1]$ is normal. The result now follows from Theorem 6.3.11. □

R. Arens proved in [9, p. 933, Theorem 2; p. 934, Corollary] that the topology τ on $L^\omega[0, 1]$ cannot be given by any norm, consequently by any C^*-norm too. For instance, let us take the C^*-norm $\|\cdot\|_\infty$ on $L^\infty[0, 1]$, where $\tau \prec \|\cdot\|_\infty$, and moreover,

$$L^\infty[0, 1] \subsetneqq L^\omega[0, 1]; \quad \overline{L^\infty[0, 1]}^\tau = L^\omega[0, 1].$$

Then, clearly $L^\omega[0, 1]$ cannot be a C^*-algebra under $\|\cdot\|_\infty$.

A completely analogous situation we have by looking at $L^\omega[0, 1]$ as a GB*-algebra. Indeed, consider the C^*-subalgebra $(L^\omega[0, 1])[B_0]$ of $L^\omega[0, 1]$, generated by the greatest member B_0 of the $\mathfrak{B}^*$-collection establishing the structure of $L^\omega[0, 1]$, where B_0 is the closed unit ball of the C^*-algebra $L^\infty[0, 1]$. The C^*-structure of $(L^\omega[0, 1])[B_0]$ is figured by the gauge function of B_0. At the same time, $(L^\omega[0, 1])[B_0]$ is τ-dense in $L^\omega[0, 1]$ (τ as above) according to Theorem 4.2.11. On the other hand, $(L^\omega[0, 1])[B_0]$ coincides with the C^*-algebra $L^\infty[0, 1]$ under its usual C^*-norm $\|\cdot\|_\infty$ and, of course, $\|\cdot\|_{B_0} = \|\cdot\|_\infty$. For more details on the Arens algebra $L^\omega[0, 1]$, see Examples 3.3.16(5).

We finish the present section with some information concerning *-representations of a GB*-algebra.

It is well known that if $\pi : \mathcal{A} \to \mathcal{B}(\mathcal{H})$ is a *-representation of a C^*-algebra $\mathcal{A}$, then $\pi(\mathcal{A})$ is a C^*-subalgebra of $\mathcal{B}(\mathcal{H})$. We now investigate *how this result extends to GB*-algebras*. For this, we require the following

Theorem 6.3.19 *If $\phi : \mathcal{A}_1 \to \mathcal{A}_2$ is an injective *-homomorphism of the GB*-algebra $\mathcal{A}_1[\tau_1]$ into the GB*-algebra $\mathcal{A}_2[\tau_2]$, then $\phi(B_0^1) = \phi(\mathcal{A}_1) \cap B_0^2$, where B_0^1, B_0^2, are the respective maximal members in $\mathfrak{B}^*_{\mathcal{A}_1}$ and $\mathfrak{B}^*_{\mathcal{A}_2}$.*

Proof We first show that $\phi(B_0^1) \subseteq B_0^2$. By Theorem 6.3.5, there is a faithful closed *-representation π of $\mathcal{A}_2$, such that $\pi(B_0^2) = U(\pi(\mathcal{A}_2))$. Then, $\pi \circ \phi$ is a closed *-representation of $\mathcal{A}_1$. It follows from Lemma 6.3.3 that

$$(\pi \circ \phi)(B_0^1) \subseteq U\big((\pi \circ \phi)(\mathcal{A}_1)\big) \subseteq U\big(\pi(\mathcal{A}_2)\big) = \pi(B_0^2).$$

Since π is faithful, $\phi(B_0^1) \subseteq B_0^2$, therefore $\phi(B_0^1) \subseteq \phi(\mathcal{A}_1) \cap B_0^2$. Hence, $\phi(\mathcal{A}_1[B_0^1]) \subseteq \mathcal{A}_2[B_0^2]$.

We now show that $\phi(\mathcal{A}_1) \cap B_0^2 \subseteq \phi(B_0^1)$. We first prove that if $h \in H(\mathcal{A}_1)$ and $\phi(h) \in B_0^2$, then $h \in B_0^1$. We break the proof up into two cases.

Case I: $\mathcal{A}_1$ and $\mathcal{A}_2$ are commutative. Then, from Theorem 3.4.9

$$\mathcal{A}_1[B_0^1] = C(\mathfrak{M}_0^1) \text{ and } \mathcal{A}_2[B_0^2] = C(\mathfrak{M}_0^2), \text{ algebraically,}$$

where $\mathfrak{M}_0^1$ and $\mathfrak{M}_0^2$ are the maximal ideal spaces of $\mathcal{A}_1[B_0^1]$ and $\mathcal{A}_2[B_0^2]$ respectively (see Sect. 2.5). By the fact that $\phi(B_0^1) \subseteq B_0^2$ proved above, we have that $\phi(\mathcal{A}_1[B_0^1]) \subseteq \mathcal{A}_2[B_0^2]$, hence $\phi \upharpoonright_{\mathcal{A}_1[B_0^1]}$ is an injection of $\mathcal{A}_1[B_0^1]$ into $\mathcal{A}_2[B_0^2]$, which induces a one to one correspondence ϕ' between $\mathfrak{M}_0^2$ and $\mathfrak{M}_0^1$ given by

$$(\phi'(\chi))(x) = \chi(\phi(x)), \text{ with } \chi \in \mathfrak{M}_0^2 \text{ and } x \in \mathcal{A}_1[B_0^1].$$

The multiplicative linear functional $\phi'(\chi)$ on $\mathcal{A}_1[B_0^1]$ extends to a $\mathbb{C}^*$-valued partial homomorphism on the entirety of $\mathcal{A}_1$ (by Proposition 3.4.1). This extension coincides with the extension given by

$$(\phi'(\chi))(x) = \chi'(\phi(x)),$$

where $\chi' \in \mathfrak{M}_0^2$ is the extension of χ to $\mathcal{A}_2$ (Definition 3.4.7). Now, since $\phi(h) \in B_0^2$,

$$\sup\big\{|\chi'(\phi(h))| : \chi' \in \mathfrak{M}_0^2\big\} = \|\widehat{\phi(h)}\|_\infty = \|\phi(h)\|_{B_0^2} \leq 1.$$

However,

$$\sup\big\{|\phi'(\chi)(h)| : \chi \in \mathfrak{M}_0^2\big\} = \|\widehat{h}\|_\infty,$$

since ϕ' is surjective. Therefore $\|h\|_{B_0^1} = \|\widehat{h}\|_\infty \leq 1$, i.e., $h \in B_0^1$.

Case II: $\mathcal{A}_1$ and $\mathcal{A}_2$ are in general noncommutative. Since $h \in \mathcal{A}_1$ and $\phi(h) \in \mathcal{A}_2$, we can consider the maximal commutative *-subalgebras $\mathcal{B}_1$ and $\mathcal{B}_2$ of $\mathcal{A}_1$ and $\mathcal{A}_2$ respectively, generated by h and $\phi(h)$ respectively. Since $\mathcal{B}_1$ and $\mathcal{B}_2$ are closed, it follows from Proposition 3.3.19 that $\mathcal{B}_1$ and $\mathcal{B}_2$ are commutative GB^*-algebras over $B_0^1 \cap \mathcal{B}_1$ and $B_0^2 \cap \mathcal{B}_2$ respectively. We apply Case I to obtain that $h \in B_0^1 \cap \mathcal{B}_1 \subset B_0^1$.

This concludes the proof of the fact that if $h \in H(\mathcal{A}_1)$ and $\phi(h) \subset B_0^2$, then $h \in B_0^1$. If $x \in \mathcal{A}_1$ with $\phi(x) \in \phi(\mathcal{A}_1) \cap B_0^2$, then we can write $x = h + ik$, where $h, k \in H(\mathcal{A}_1)$, hence $\phi(h)$ and $\phi(k)$ are in B_0^2, implying that h, k are in B_0^1. Therefore, $x \in 2B_0^1$, so that $x \in \mathcal{A}_1[B_0^1]$. Now, since $\phi \upharpoonright_{\mathcal{A}_1[B_0^1]}$ is an injective *-homomorphism into $\mathcal{A}_2[B_0^2]$, and $\phi(x) \in B_0^2$, we obtain that

$$\|x\|_{B_0^1} \le \|\phi(x)\|_{B_0^2} \le 1,$$

consequently $x \in B_0^1$. $\qquad\square$

Corollary 6.3.20 *If $\phi : \mathcal{A}_1 \to \mathcal{A}_2$ is an algebraic *-isomorphism of two GB^*-algebras $\mathcal{A}_1[\tau_1]$ and $\mathcal{A}_2[\tau_2]$, then $\phi(\mathcal{A}_1[B_0^1]) = \mathcal{A}_2[B_0^2]$.*

Theorem 6.3.5 says that there exists a faithful closed *-representation π of a GB^*-algebra $\mathcal{A}[\tau]$ such that

$$\pi(\mathcal{A}) \text{ is an } EC^* - \text{algebra and } U\big(\pi(\mathcal{A})\big) = \pi(B_0).$$

Now, using Theorem 6.3.19, we shall show that every faithful closed *-representation of a GB^*-algebra $\mathcal{A}[\tau]$ has this property.

Theorem 6.3.21 *If π is a faithful closed *-representation of a GB^*-algebra $\mathcal{A}[\tau]$, then $\pi(B_0) = U\big(\pi(\mathcal{A})\big)$ and $\pi(\mathcal{A})$ is an EC^*-algebra.*

Proof We have from Lemma 6.3.3 that $\pi(B_0) \subseteq U\big(\pi(\mathcal{A})\big)$. Thus, it remains to show that $U\big(\pi(\mathcal{A})\big) \subseteq \pi(B_0)$. Let $\mathcal{A}^0 = \pi^{-1}(\overline{\pi(\mathcal{A})} \cap \mathcal{B}(\mathcal{H}_\pi))$, where $\overline{\pi(\mathcal{A})}$ denotes the set $\{\overline{\pi(x)} : x \in \mathcal{A}\}$. Clearly, $\mathcal{A}[B_0] \subseteq \mathcal{A}^0 \subseteq \mathcal{A}$. If $\mathfrak{B}^*_{\mathcal{A}}$ denotes the collection of sets as in Definition 3.3.1, then $\mathfrak{B}^*_{\mathcal{A}^0} := \{B \cap \mathcal{A}^0 : B \in \mathfrak{B}^*_{\mathcal{A}}\}$ also satisfies the corresponding properties of $\mathfrak{B}^*_{\mathcal{A}}$, therefore $\mathcal{A}^0$ is a GB^*-algebra with respect to the relative topology on $\mathcal{A}^0$ coming from τ. Observe that the restriction of π to $\mathcal{A}^0$ is an injective *-homomorphism into $\mathcal{B}(\mathcal{H}_\pi)$, where the latter being a C^*-algebra under the operator norm, is a GB^*-algebra. It follows from Theorem 6.3.19 that $U\big(\pi(\mathcal{A})\big) \subseteq U\big(\pi(\mathcal{A}^0)\big) = \pi(B_0)$. It is now immediate that $\pi(\mathcal{A})$ is an EC^*-algebra. $\qquad\square$

6.4   Ogasawara's Commutativity Condition in GB*-Algebras

In 1955, T. Ogasawara proved that if $\mathcal{A}[\|\cdot\|]$ is a C^*-algebra and for any positive elements x, y in $\mathcal{A}$ (i.e., x, y in $\mathcal{A}^+$), one has that

$$x \leq y \text{ always implies } x^2 \leq y^2, \tag{6.4.14}$$

then $\mathcal{A}[\|\cdot\|]$ is commutative (see [116]). Note that the double inequality $0 \leq x \leq y$ does not imply, in general, that $x^2 \leq y^2$; see [73, p. 369, Exercise 12*]. Recently, M. Oudadess showed in [118] that if condition (6.4.14) is true in a GB^*-algebra $\mathcal{A}[\tau]$, for any x, y in $\mathcal{A}^+$, then $\mathcal{A}[\tau]$ is commutative too. Namely, using results of Sect. 6.2, as well of Chap. 3, one obtains the following,

Theorem 6.4.1 (Oudadess) *Every GB^*-algebra $\mathcal{A}[\tau]$ that satisfies condition (6.4.14) is commutative.*

Proof At the beginning of the proof of Theorem 6.2.5, it is shown that $(\mathcal{A}[B_0])^+ = \mathcal{A}[B_0] \cap \mathcal{A}^+$. So, it is obvious that the C^*-algebra $\mathcal{A}[B_0]$ also fulfills Ogasawara's condition (6.4.14), therefore it is commutative by Ogasawara's theorem. Let now x be an arbitrary element in $\mathcal{A}$. Since $\mathcal{A}[B_0]$ is dense in $\mathcal{A}[\tau]$ (see Theorem 4.2.11), it follows from the separate continuity of multiplication of $\mathcal{A}[\tau]$ that $xy = yx$, for all $y \in \mathcal{A}[B_0]$. Again from the separate continuity of multiplication and the density of $\mathcal{A}[B_0]$ in $\mathcal{A}[\tau]$, we obtain that $xz = zx$, for all $z \in \mathcal{A}$. Therefore, $\mathcal{A}$ is commutative.   □

Corollary 6.4.2 *Every pro-C^*-algebra with identity and every C^*-like locally convex $*$-algebra with identity, that fulfills Ogasawara's condition (6.4.14), is commutative.*

Let now $\mathcal{A}[\tau]$ be a pseudo-complete locally convex algebra and $\mathcal{A}_0$ the set of its bounded elements. In general, $\mathcal{A}_0$ is not even a subspace of $\mathcal{A}$. If $\mathcal{A}[\tau]$, as before has also a continuous involution, then x belongs to $\mathcal{A}_0$, if and only if, x^* belongs to $\mathcal{A}_0$. In particular, *in every commutative pseudo-complete locally convex $*$-algebra $\mathcal{A}[\tau]$, $\mathcal{A}_0$ is a $*$-subalgebra of $\mathcal{A}$* (Corollary 2.2.11). The assumption that $\mathcal{A}_0$ is a subspace of $\mathcal{A}$, is very important, as the next result of M. Oudadess shows; at the same time, this result constitutes a more general version of Ogasawara's theorem.

Theorem 6.4.3 (Oudadess) *Let $\mathcal{A}[\tau]$ be a pseudo-complete locally convex $*$-algebra with an identity element e that satisfies the properties:*

(α) *there is a C^*-subalgebra C of $\mathcal{A}$ contained in $\mathcal{A}_0$;*
(β) *the subset of all self-adjoint elements of $\mathcal{A}_0$ is contained in C;*
(γ) *the subset $\mathcal{A}_0$ is dense in $\mathcal{A}[\tau]$;*
(δ) *$\mathcal{A}[\tau]$ fulfills the Ogasawara's condition (6.4.14).*

Then, the following are equivalent:

(i) $\mathcal{A}[\tau]$ *is commutative;*
(ii) *the subset $\mathcal{A}_0$ of $\mathcal{A}[\tau]$ is a $*$-subalgebra;*
(iii) *the subset $\mathcal{A}_0$ of $\mathcal{A}[\tau]$ is a subspace.*

Proof (i) $\Rightarrow$ (ii) follows from Corollary 2.2.11.

(ii) $\Rightarrow$ (iii) Obvious.

(iii) $\Rightarrow$ (i) First we show that $\mathcal{C} = \mathcal{A}_0$. By (α), it is sufficient to show that $\mathcal{A}_0 \subset \mathcal{C}$. Let $x = h + ik$ be in $\mathcal{A}_0$, with h, k in $H(\mathcal{A})$. Then, $x^* = h - ik$ belongs also to $\mathcal{A}_0$ and because of (iii), $2h = x + x^*$ is in $\mathcal{A}_0$ too. Thus, the self-adjoint elements h, k sit in $\mathcal{A}_0$ and because of (β) they belong to $\mathcal{C}$, therefore x lies in $\mathcal{C}$.

Hence, $\mathcal{A}_0$ becomes a $*$-algebra, therefore we may define the spectrum of an element x in $\mathcal{A}_0$. Since $x \in \mathcal{A}_0$, $\infty \notin \sigma_{\mathcal{A}_0}(x)$ (see Definition 2.3.1); at the same time, it is easily seen that

$$\sigma_{\mathcal{A}}(x) = \sigma_{\mathcal{A}_0}(x) = \sigma_{\mathcal{C}}(x), \ \forall\, x \in \mathcal{A}_0 = \mathcal{C}.$$

Therefore, by Proposition 6.2.1, an element x is positive in $\mathcal{C}$, if and only if, it is positive in $\mathcal{A}$. In conclusion, by (δ), the C^*-algebra $\mathcal{C}$ fulfills Ogasawara's condition, therefore it is commutative by Ogasawara's theorem. Finally, since by (γ) $\mathcal{A}_0$ is dense in $\mathcal{A}$, using the separate continuity of multiplication in $\mathcal{A}$ we first prove that every element x in $\mathcal{A}_0$ commutes with every element y in $\mathcal{A}$; applying this and again the separate continuity of multiplication in $\mathcal{A}$ we show that $\mathcal{A}$ is commutative and this completes the proof. $\qquad\square$

In Lemma 3.3.7(ii), it is shown that in a commutative GB^*-algebra $\mathcal{A}[\tau]$ one has that $\mathcal{A}_0 = A[B_0]$ (as we said before, the set $\mathcal{A}_0$, in general, is not even a subspace). The next corollary shows that *if in a GB^*-algebra $\mathcal{A}[\tau]$, the set $\mathcal{A}_0$ is a subspace, then we have $\mathcal{A}_0 = A[B_0]$* (without any commutativity condition). On the other hand, *using this fact, together with the assumption that C^*-algebra $A[B_0]$ satisfies the Ogasawara's condition*, Corollary 6.4.5, below, *shows that the GB^*-algebra $\mathcal{A}[\tau]$ is commutative*!

Corollary 6.4.4 *Let $\mathcal{A}[\tau]$ be a GB^*-algebra, such that $\mathcal{A}_0$ is subspace of A. Then, $\mathcal{A}_0 = A[B_0]$.*

Proof Since $\tau \prec \|\cdot\|_{B_0}$ on $A[B_0]$ and every element in $A[B_0]$ is (Allan-)bounded with respect to $\|\cdot\|_{B_0}$ (see Definition 2.2.1), it is also (Allan-)bounded with respect to τ. Hence, $A[B_0] \subseteq \mathcal{A}_0$. Thus, condition (α) of Theorem 6.4.3 is fulfilled, with $A[B_0]$ in the place of $\mathcal{C}$. Condition (β) of the same Theorem also holds, according to Corollary 3.3.8. Arguing now as at the beginning of proof in (iii) $\Rightarrow$ (i) of Theorem 6.4.3, we conclude that $\mathcal{A}_0 \subseteq A[B_0]$ and this completes the proof. $\qquad\square$

Corollary 6.4.5 *Let $\mathcal{A}[\tau]$ be a GB*-algebra, such that C*-subalgebra $\mathcal{A}[B_0]$ of $\mathcal{A}[\tau]$ fulfills the Ogasawara's condition (6.4.14). Then, the following statements hold:*

(i) *$\mathcal{A}[\tau]$ is commutative;*
(ii) *the subset $\mathcal{A}_0$ of $\mathcal{A}[\tau]$ is a *-subalgebra that coincides with the C*-algebra $\mathcal{A}[B_0]$;*
(iii) *the subset $\mathcal{A}_0$ of $\mathcal{A}[\tau]$ is a subspace.*

Proof

(i) As in the proof of Theorem 6.4.1, we conclude that $\mathcal{A}[\tau]$ is commutative.
(ii) Since $\mathcal{A}[\tau]$ is commutative, (ii) follows from Corollary 2.2.11 and Lemma 3.3.7(ii), respectively..
(iii) Clearly, (iii) is an immediate consequence of the first part of (ii).

$\square$

Notes A locally convex *-algebra $\mathcal{A}[\tau]$ is called an $A\widehat{\mathcal{O}}^*$-*algebra* if it is algebraically and topologically isomorphic to an $\mathcal{O}^*$-algebra $\mathcal{M}[\tau_u]$. If $\mathcal{A}[\tau]$ is moreover complete, then it is an $A\mathcal{O}^*$-*algebra* (Definition 6.3.8). Theorem 6.3.9, as well Example 6.3.10(1) are owed to G. Lassner and the reader can find them in [103, Theorem 4.1] and [104, p. 162, Theorem 1], respectively. Example 6.3.10(2) is due to K. Schmúdgen and it is Corollary 6.6 in [133]. Furthermore, K. Schmüdgen [133, Corollary 6.1] has investigated when a non-complete topological *-algebra is an $A\widehat{\mathcal{O}}^*$-algebra and he obtained that *every metrizable barrelled topological *-algebra is an $A\widehat{\mathcal{O}}^*$-algebra, if and only if, every continuous self-adjoint linear functional on $\mathcal{A}$ is the difference of two positive linear functionals of $\mathcal{A}$* (an analogous result in our case is given by Corollary 6.3.16). The results of Sects. 6.1 and 6.3, up to Corollary 6.3.7 are owed to P.G. Dixon and can be found in [50]. Theorem 6.3.11, Lemmas 6.3.12 and 6.3.13, as well Proposition 6.3.14 are due to K. Schmüdgen and are included in [133]. Finally, Theorems 6.4.1, 6.4.3 in Sect. 6.4 are due to M. Oudadess; see respectively [118, Propositions III.1 and V.1].

Chapter 7
Applications I: Miscellanea

In the present chapter, we give various applications, using the theory of GB^*-algebras developed in the previous 5 chapters. For instance, in Sect. 7.1, starting with a classical C^*-algebra $\mathcal{A}[\|\cdot\|]$ endowed with a locally convex $*$-algebra topology τ coarser than the $\|\cdot\|$-topology of $\mathcal{A}$ and making the multiplication of $\mathcal{A}$ jointly continuous, we prove that $\widetilde{\mathcal{A}}[\tau]$ (the completion of the C^*-algebra $\mathcal{A}[\|\cdot\|]$ with respect to τ) is a GB^*-algebra (Theorem 7.1.2). Furthermore, continuity of positive linear functionals on GB^*-algebras is explored (see Theorem 7.2.1 and its corollaries), while a bounded approximate identity is confirmed on a GB^*-algebra (Theorem 7.3.1) and new GB^*-algebras are constructed (cf. Corollary 7.4.7 and Theorem 7.4.8). Finally, Sect. 7.5 exhibits various types of the celebrated Vidav–Palmer theorem, in the context of locally convex $*$-algebras (included GB^*-algebras); see, e.g., Theorems 7.5.43, 7.5.49, and 7.5.51.

7.1 The Completion of a C*-Algebra Under a Locally Convex Topology

Let $\mathcal{A}[\|\cdot\|]$ be a C^*-algebra with identity e and τ a locally convex $*$-algebra topology on $\mathcal{A}$, such that $\tau \prec \|\cdot\|$. For the structure of the completion $\widetilde{\mathcal{A}}[\tau]$ of the C^*-algebra $\mathcal{A}[\|\cdot\|]$, with respect to τ, the following cases arise:

- **Case 1:** If the multiplication in $\mathcal{A}$ is jointly continuous, with respect to the topology τ, then $\widetilde{\mathcal{A}}[\tau]$ is a complete locally convex $*$-algebra containing the C^*-algebra $\mathcal{A}[\|\cdot\|]$ as a dense subalgebra.
- **Case 2:** If the multiplication on $\mathcal{A}$ is not jointly continuous with respect to τ, then $\widetilde{\mathcal{A}}[\tau]$ is not necessarily a locally convex $*$-algebra. In this case, $\widetilde{\mathcal{A}}[\tau]$ has the structure of a locally convex quasi $*$-algebra (see [66, Chapter 6 and Chapter 7]), an important class of locally convex partial $*$-algebras [7]. For the corresponding results of Case 2 the reader is referred to [12, 62]; in this regard, see also [13,

© The Author(s), under exclusive license to Springer Nature Switzerland AG 2022   157
M. Fragoulopoulou et al., *Generalized B*-Algebras and Applications*, Lecture Notes in Mathematics 2298, https://doi.org/10.1007/978-3-030-96433-7_7

14, 146]. Note that topological quasi $*$-algebras were initiated by G. Lassner (see [106, 107]), in order to give solutions to some problems, in quantum statistics and quantum dynamics.

Here we do not treat Case 2. We only investigate the structure and the representation theory of the complete locally convex $*$-algebra $\widetilde{\mathcal{A}}[\tau]$, in Case 1. In Theorem 7.1.2, we prove that $\widetilde{\mathcal{A}}[\tau]$ is a GB^*-algebra (see [12, 62] and Corollary 7.1.4). Concerning the second stage of our investigation, we give conditions under which $\widetilde{\mathcal{A}}[\tau]$ attains a faithful (continuous or not) unbounded $*$-representation. For instance, such a condition is that $\widetilde{\mathcal{A}}[\tau]^+ \cap (-\widetilde{\mathcal{A}}[\tau]^+) = \{0\}$ (Lemma 7.1.5). In this aspect, see also Lemma 7.1.6 and Theorem 7.1.7.

▶ Denote by B_τ *the τ-closure of the closed unit ball* $U(\mathcal{A}) := \{x \in \mathcal{A} : \|x\| \leq 1\}$ *of the C^*-algebra* $\mathcal{A}[\|\cdot\|]$ and recall that $\mathcal{B}^*_{\widetilde{\mathcal{A}}}$ is as in Definition 3.3.1. In this regard, we have the following

Lemma 7.1.1 *The set B_τ belongs to $\mathcal{B}^*_{\widetilde{\mathcal{A}}}$ and $\widetilde{\mathcal{A}}[B_\tau]$ is a Banach $*$-algebra under the norm $\|\cdot\|_{B_\tau}$, satisfying the following properties:*

(i) *the elements $(e + x^*x)^{-1}$, $x(e + x^*x)^{-1}$ and $(e + x^*x)^{-1}x$ belong to B_τ, for every $x \in \widetilde{\mathcal{A}}[\tau]$;*

(ii) *one has that $\mathcal{A} \subseteq \widetilde{\mathcal{A}}[B_\tau]$ and $\|x\| = \|x\|_{B_\tau}$, for all $x \in \mathcal{A}$. Hence, $U(\mathcal{A}) = B_\tau \cap \mathcal{A}$ and $\mathcal{A}$ is a closed $*$-subalgebra of the Banach $*$-algebra $\widetilde{\mathcal{A}}[B_\tau]$;*

(iii) *the Banach $*$-algebra $\widetilde{\mathcal{A}}[B_\tau]$ is $\|\cdot\|_B$-dense in $\widetilde{\mathcal{A}}[B]$, for every $B \in \mathcal{B}^*_{\widetilde{\mathcal{A}}}$ containing $U(\mathcal{A})$.*

Proof It is clear that $B_\tau \in \mathcal{B}^*_{\widetilde{\mathcal{A}}}$ and tha $\widetilde{\mathcal{A}}[B_\tau]$ is a Banach $*$-algebra since $\widetilde{\mathcal{A}}[\tau]$ is complete.

(i) Take an arbitrary $x \in \widetilde{\mathcal{A}}[\tau]$ and (x_α) a net in $\mathcal{A}$, such that $\tau - \lim_\alpha x_\alpha = x$. Then, since $\mathcal{A}[\|\cdot\|]$ is a C^*-algebra, from the functional calculus in $\mathcal{A}[\|\cdot\|]$, it follows that $(e + x_\alpha^*x_\alpha)^{-1} \in U(\mathcal{A})$, for every α and from the joint continuity of multiplication in $\mathcal{A}$, with respect to τ, for any τ-continuous seminorm p, we can find $\gamma > 0$ and some τ-continuous seminorm q, such that

$$p((e + x_\alpha^*x_\alpha)^{-1} - (e + x_\beta^*x_\beta)^{-1})$$
$$= p((e + x_\alpha^*x_\alpha)^{-1}(x_\beta^*x_\beta - x_\alpha^*x_\alpha)(e + x_\beta^*x_\beta)^{-1})$$
$$\leq q((e + x_\alpha^*x_\alpha)^{-1})q((e + x_\beta^*x_\beta)^{-1})q(x_\beta^*x_\beta - x_\alpha^*x_\alpha)$$
$$\leq \gamma \|(e + x_\alpha^*x_\alpha)^{-1}\|\|(e + x_\beta^*x_\beta)^{-1}\|q(x_\beta^*x_\beta - x_\alpha^*x_\alpha)$$
$$\leq \gamma q(x_\beta^*x_\beta - x_\alpha^*x_\alpha),$$

Thus, $\left((e + x_\alpha^* x_\alpha)^{-1}\right)$ is a Cauchy net in $\widetilde{\mathcal{A}}[\tau]$ and $y \equiv \lim_\alpha (e + x_\alpha^* x_\alpha)^{-1}$ exists in $\widetilde{\mathcal{A}}[\tau]$. Since,

$$e = (e + x_\alpha^* x_\alpha)(e + x_\alpha^* x_\alpha)^{-1} = (e + x_\alpha^* x_\alpha)^{-1}(e + x_\alpha^* x_\alpha), \quad \forall\, \alpha,$$

it follows that $(e + x^* x)^{-1} \in \widetilde{\mathcal{A}}[\tau]$ and $y = (e + x^* x)^{-1}$. Also, $(e + x^* x)^{-1} \in B_\tau$ and in a similar way we have that

$$x(e + x^* x)^{-1} \text{ and } (e + x^* x)^{-1} x \text{ belong to } B_\tau.$$

(ii) Since $U(\mathcal{A}) \subseteq B_\tau$, it follows that $\mathcal{A} \subseteq \widetilde{\mathcal{A}}[B_\tau]$ and $\|x\|_{B_\tau} \leq \|x\|$, for each $x \in \mathcal{A}$. From the theory of C*-algebras (see, for example, [144, Proposition I.5.3]), we have $\|x\| \leq \|x\|_{B_\tau}$, for every $x \in \mathcal{A}$. Hence, $\|x\| = \|x\|_{B_\tau}$, for each $x \in \mathcal{A}$, which implies that $U(\mathcal{A}) = B_\tau \cap \mathcal{A}$ and $\mathcal{A}$ is a closed $*$-subalgebra of $\widetilde{\mathcal{A}}[B_\tau]$.

(iii) Take an arbitrary $B \in \mathfrak{B}^*_{\widetilde{\mathcal{A}}}$ containing $U(\mathcal{A})$. Since B is τ-closed, it follows that $B_\tau \subseteq B$, therefore $\widetilde{\mathcal{A}}[B_\tau] \subseteq \widetilde{\mathcal{A}}[B]$ and $\|x\|_B \leq \|x\|_{B_\tau}$, for each $x \in \widetilde{\mathcal{A}}[B_\tau]$. Let $x \in \widetilde{\mathcal{A}}[B]$. By (i) we have

$$x \left(e + \frac{1}{n} x^* x \right)^{-1} \in \widetilde{\mathcal{A}}[B_\tau], \quad \forall\, n \in \mathbb{N} \text{ and}$$

$$
\begin{aligned}
\lim_{n \to \infty} \left\| x \left(e + \frac{1}{n} x^* x \right)^{-1} - x \right\|_B &= \lim_{n \to \infty} \frac{1}{n} \left\| x x^* x \left(e + \frac{1}{n} x^* x \right)^{-1} \right\|_B \\
&\leq \lim_{n \to \infty} \frac{1}{n} \|x x^* x\|_B \left\| \left(e + \frac{1}{n} x^* x \right)^{-1} \right\|_B \\
&\leq \lim_{n \to \infty} \frac{1}{n} \|x x^* x\|_B \left\| \left(e + \frac{1}{n} x^* x \right)^{-1} \right\|_{B_\tau} \\
&\leq \lim_{n \to \infty} \frac{1}{n} \|x x^* x\|_B = 0.
\end{aligned}
$$

Hence, $\widetilde{\mathcal{A}}[B_\tau]$ is $\|\cdot\|_B$-dense in $\widetilde{\mathcal{A}}[B]$. This completes the proof. $\qquad \square$

For another aspect of Lemma 7.1.1, see Lemmas 3.4.2(i) and 7.3.2.

Theorem 7.1.2 *Let $\mathcal{A}[\|\cdot\|]$ be a C*-algebra with identity e and τ a locally convex topology on $\mathcal{A}$, such that $\tau \prec \|\cdot\|$ and $\mathcal{A}[\tau]$ is a locally convex $*$-algebra with jointly continuous multiplication. Then, the algebra $\widetilde{\mathcal{A}}[\tau]$ is a GB*-algebra over $B_0 = B_\tau$.*

Proof Since $\widetilde{A}[\tau]$ is a complete locally convex $*$-algebra, it suffices to show that B_τ is a greatest member in $\mathfrak{B}^*_{\widetilde{A}}$ and that $(e + x^*x)^{-1} \in \widetilde{A}[B_\tau]$, for every $x \in \widetilde{A}[\tau]$. The latter follows from Lemma 7.1.1(i).

- A crucial point for showing that B_τ is greatest in $\mathfrak{B}^*_{\widetilde{A}}$, is to show that $\widetilde{A}[B_\tau]$ is a C^*-algebra.

For this consider the C^*-algebra $\mathfrak{A}$ of all $\|\cdot\|$-bounded nets $(x_\lambda)_{\lambda \in \Lambda}$ in $A[\|\cdot\|]$, where the index set Λ consists of a 0-neighbourhood basis for τ, and the C^*-norm is given by $\|(x_\lambda)_{\lambda \in \Lambda}\|_\infty := \sup_{\lambda \in \Lambda} \|x_\lambda\|$, $(x_\lambda)_{\lambda \in \Lambda} \in \mathfrak{A}$. Take now,

$$\mathfrak{A}_c := \Big\{ (x_\lambda)_{\lambda \in \Lambda} \in \mathfrak{A} : (x_\lambda)_{\lambda \in \Lambda} \text{ is a } \tau - \text{Cauchy net} \Big\} \text{ and}$$

$$\mathfrak{A}_0 := \Big\{ (x_\lambda)_{\lambda \in \Lambda} \in \mathfrak{A}_c : \tau - \lim_\lambda x_\lambda = 0 \Big\}.$$

It is easily checked that $\mathfrak{A}_c$ is a closed $*$-subalgebra of the C^*-algebra $\mathfrak{A}[\|\cdot\|_\infty]$, hence a C^*-algebra, and that $\mathfrak{A}_0$ is a closed ideal (hence a $*$-ideal) in $\mathfrak{A}_c$. Therefore, the quotient $\mathfrak{A}_c/\mathfrak{A}_0$ is a C^*-algebra. Now, an element $a \in B_\tau$ is of the form $\tau - \lim_\lambda x_\lambda$, where $(x_\lambda)_{\lambda \in \Lambda}$ is a net in A with $\|x_\lambda\| \leq 1$, for all $\lambda \in \Lambda$. In other words, $a = \tau - \lim_\lambda x_\lambda$, with $(x_\lambda)_{\lambda \in \Lambda}$ in $\mathfrak{A}_c$. Thus, the following correspondence

$$\Theta : \mathfrak{A}_c \;\to\; \widetilde{A}[B_\tau] : (x_\lambda)_{\lambda \in \Lambda} \mapsto \tau - \lim_\lambda x_\lambda,$$

is a well-defined continuous $*$-homomorphism, such that $\ker \Theta = \mathfrak{A}_0$ and

$$\|\Theta((x_\lambda)_\lambda)\| \leq \sup_\lambda \|x_\lambda\| = \|(x_\lambda)_\lambda\|_\infty, \; \forall \, (x_\lambda)_\lambda \in \mathfrak{A}_c.$$

Observe that Θ induces an isometric $*$-isomorphism $\widetilde{\Theta}$ from $\mathfrak{A}_c/\mathfrak{A}_0$ onto $\widetilde{A}[B_\tau]$. Indeed, since $\widetilde{\Theta}^{-1}$ is a $*$-homomorphism from a Banach $*$-algebra onto a C^*-algebra, we have that

$$\|(x_\lambda)_\lambda + \mathfrak{A}_0\|_q \leq \|\widetilde{\Theta}((x_\lambda)_\lambda + \mathfrak{A}_0)\|, \; \forall \, (x_\lambda)_\lambda + \mathfrak{A}_0 \in \mathfrak{A}_c/\mathfrak{A}_0,$$

where $\|\cdot\|_q$ is the quotient C^*-norm on $\mathfrak{A}_c/\mathfrak{A}_0$. On the other hand, from the very definitions we have

$$\|\widetilde{\Theta}((x_\lambda)_\lambda + \mathfrak{A}_0)\| \leq \|(x_\lambda)_\lambda\|_\infty \leq \|(x_\lambda)_\lambda + a_0\|_\infty + \|a_0\|_\infty,$$

with $a_0 \in \mathfrak{A}_0$. Hence,

$$\|\widetilde{\Theta}((x_\lambda)_\lambda + \mathfrak{A}_0)\| \leq \|(x_\lambda)_\lambda + \mathfrak{A}_0\|_q, \; \forall \, (x_\lambda)_\lambda \in \mathfrak{A}_c.$$

This completes the proof of our claim.

Consequently, the Banach $*$-algebra $\widetilde{A}[B_\tau]$ is a C^*-algebra.

- We show now that B_τ is the greatest member in $\mathfrak{B}^*_{\widetilde{A}}$.

Take an arbitrary $B \in \mathfrak{B}^*_{\widetilde{\mathcal{A}}}$ and $h^* = h \in B$. Let $\mathcal{C}$ be a maximal commutative $*$-subalgebra of $\widetilde{\mathcal{A}}[\tau]$ containing h. Then, $\mathcal{C}$ is complete with respect to the topology induced by τ. We denote by $\mathfrak{B}^*_{\mathcal{C}}$ the collection of all closed, bounded, absolutely convex subsets B_1 of $\mathcal{C}$ satisfying: $e \in B_1$, $B_1^* = B_1$ and $B_1^2 \subseteq B_1$. Then, $\mathfrak{B}^*_{\mathcal{C}} = \{B_2 \cap \mathcal{C} : B_2 \in \mathfrak{B}^*_{\widetilde{\mathcal{A}}}\}$. We show that $B \cap \mathcal{C} \subseteq B_\tau \cap \mathcal{C}$. Since $\mathcal{C}$ is commutative and complete, it follows from Theorem 2.2.10, that $\mathfrak{B}^*_{\mathcal{C}}$ is directed, so there exists $B_1 \in \mathfrak{B}^*_{\mathcal{C}}$, such that $(B \cap \mathcal{C}) \cup (B_\tau \cap \mathcal{C}) \subseteq B_1$. So, since $B_\tau \cap \mathcal{C} \subseteq B_1$, we obtain

$$\|x\|_{B_1} \leq \|x\|_{B_\tau \cap \mathcal{C}} = \|x\|_{B_\tau}, \ \forall \, x \in \widetilde{\mathcal{A}}[B_\tau] \cap \mathcal{C}.$$

On the other hand, since the C^*-algebra $\mathcal{C}[B_\tau \cap \mathcal{C}] = \widetilde{\mathcal{A}}[B_\tau] \cap \mathcal{C}$ is contained in the Banach $*$-algebra $\mathcal{C}[B_1]$, it follows from [144, Proposition I.5.3] that

$$\|x\|_{B_\tau} = \|x\|_{B_\tau \cap \mathcal{C}} \leq \|x\|_{B_1}, \ \forall \, x \in \widetilde{\mathcal{A}}[B_\tau] \cap \mathcal{C}.$$

Thus, we have

$$\|x\|_{B_1} = \|x\|_{B_\tau}, \ \forall \, x \in \widetilde{\mathcal{A}}[B_\tau] \cap \mathcal{C}. \tag{7.1.1}$$

Next we show that $B_\tau \cap \mathcal{C} = B_1$. Take an arbitrary $x \in \mathcal{C}[B_1]$ and $n \in \mathbb{N}$. By Lemma 7.1.1(i), we obtain

$$x\left(e + \frac{1}{n}x^*x\right)^{-1} \in \widetilde{\mathcal{A}}[B_\tau], \ \text{so that} \ x\left(e + \frac{1}{n}x^*x\right)^{-1} \in \widetilde{\mathcal{A}}[B_\tau] \cap \mathcal{C}.$$

The estimate

$$\left\| x\left(e + \frac{1}{n}x^*x\right)^{-1} - x \right\|_{B_1} = \frac{1}{n}\left\| xx^*x\left(e + \frac{1}{n}x^*x\right)^{-1} \right\|_{B_1}$$

$$\leq \frac{1}{n}\|xx^*x\|_{B_1}\left\|\left(e + \frac{1}{n}x^*x\right)^{-1}\right\|_{B_1}$$

$$= \frac{1}{n}\|xx^*x\|_{B_1}\left\|\left(e + \frac{1}{n}x^*x\right)^{-1}\right\|_{B_\tau}$$

$$\leq \frac{1}{n}\|xx^*x\|_{B_1}$$

implies now that $\widetilde{\mathcal{A}}[B_\tau] \cap \mathcal{C}$ is $\|\cdot\|_{B_1}$-dense in $\mathcal{C}[B_1]$. Therefore, by (7.1.1) we conclude that

$$\widetilde{\mathcal{A}}[B_\tau] \cap \mathcal{C} = \mathcal{C}[B_\tau \cap \mathcal{C}] = \mathcal{C}[B_1],$$

and thus $B_\tau \cap \mathcal{C} = B_1$.

This implies that $h \in B \cap C \subseteq B_\tau \cap C$. Consequently, we have shown that $h \in B_\tau$, for each $h = h^* \in B$. So,

$$\widetilde{\mathcal{A}}[B] \subseteq \widetilde{\mathcal{A}}[B_\tau] \quad \text{and} \quad \|x\|_{B_\tau}^2 = \|x^*x\|_{B_\tau} \le 1, \ \forall \, x \in B.$$

Hence, $B \subseteq B_\tau$ and B_τ is a greatest member in $\mathfrak{B}_{\mathcal{A}}^*$. This completes the proof. $\square$

The C^*-normed algebra $\mathcal{A}[\| \cdot \|]$ that determines the locally convex $*$-algebra $\widetilde{\mathcal{A}}[\tau]$ is not unique. For this reason, we denote by $C^*(\mathcal{A}, \tau)$ the set of all C^*-normed algebras $\mathcal{A}'[\| \cdot \|']$, such that

$$\mathcal{A} \subseteq \mathcal{A}' \subseteq \widetilde{\mathcal{A}}[\tau], \ \tau \prec \| \cdot \|' \quad \text{and} \quad \|x\|' = \|x\|, \ \forall \, x \in \mathcal{A}.$$

Then $C^*(\mathcal{A}, \tau)$ is an ordered set under the following order:

$$\mathcal{A}_1[\| \cdot \|_1] \prec \mathcal{A}_2[\| \cdot \|_2], \ \text{if and only if,} \ \mathcal{A}_1 \subseteq \mathcal{A}_2 \ \text{and} \ \|x\|_1 = \|x\|_2, \ \forall \, x \in \mathcal{A}_1.$$

Theorem 7.1.2 shows that $\widetilde{\mathcal{A}}[B_\tau]$ is the largest element in $C^*(\mathcal{A}, \tau)$. It also implies the following

Corollary 7.1.3 *Let $\mathcal{A}[\| \cdot \|]$ be a C^*-algebra with identity e and τ a locally convex $*$-algebra topology on $\mathcal{A}$, such that $\tau \prec \| \cdot \|$. Then, the following statements are equivalent:*

(i) $\widetilde{\mathcal{A}}[\tau]$ *is a GB*-algebra over the closed unit ball $U(\mathcal{A}) := \{x \in \mathcal{A} : \|x\| \le 1\}$ of the C^*-algebra $\mathcal{A}[\| \cdot \|]$.*
(ii) $U(\mathcal{A})$ *is τ-closed.*

Let now $\mathcal{D}$ be a dense subspace in a Hilbert space $\mathcal{H}$, such that $\mathcal{L}^\dagger(\mathcal{D})$ is closed. Let τ_u be the uniform topology on $\mathcal{L}^\dagger(\mathcal{D})$ (see discussion after (5.2.4)). Suppose that $\mathcal{M}$ is a C^*-algebra on $\mathcal{H}$, such that $\mathcal{MD} \subseteq \mathcal{D}$ and let the multiplication on $\mathcal{M}$ be jointly continuous, with respect to the topology τ_u. Furthermore, suppose that the $\mathcal{O}^*$-algebra $\widetilde{\mathcal{M}}[\tau_u]$ is closed. If this does not happen, we simply consider the τ_u-closure of $\widetilde{\mathcal{M}}[\tau_u]$ in $\mathcal{L}^\dagger(\mathcal{D})[\tau_u]$. Then, by Lemma 5.3.3 and Corollary 7.1.3 we obtain the following

Corollary 7.1.4 $\widetilde{\mathcal{M}}[\tau_u]$ *is a GB*-algebra over $U(\mathcal{M})$.*

Further, we investigate the representation theory of the complete locally convex $*$-algebra $\widetilde{\mathcal{A}}[\tau]$ as in Corollary 7.1.3. Let $\widetilde{\mathcal{A}}[\tau]^+$ denote the τ-closure of the positive cone $\mathcal{A}^+$ in the C^*-algebra $\mathcal{A}[\| \cdot \|]$. Then, $\widetilde{\mathcal{A}}[\tau]^+$ is a *wedge*, in the sense that,

$$x + y \ \text{and} \ \lambda x \in \widetilde{\mathcal{A}}[\tau]^+, \ \forall \ x, y \in \widetilde{\mathcal{A}}[\tau]^+ \ \text{and} \ \forall \, \lambda \ge 0.$$

Moreover, $\widetilde{\mathcal{A}}[\tau]^+$ is identical to the τ-closure of the algebraic wedge

$$\mathcal{P}(\widetilde{\mathcal{A}}[\tau]) := \left\{ \sum_{k=1}^{n} x_k^* x_k : x_k \in \widetilde{\mathcal{A}}[\tau], k = 1, 2, \ldots, n, \ n \in \mathbb{N} \right\}.$$

A linear functional f on $\widetilde{\mathcal{A}}[\tau]$ is said to be *strongly positive* if $f(x) \geq 0$, for all $x \in \widetilde{\mathcal{A}}[\tau]^+$.

Furthermore, note that $\mathcal{L}^\dagger(\mathcal{D})$ is a locally convex $*$-algebra equipped with the weak topology τ_w (resp. strong* topology τ_{s*}) defined by the family $\{ p_{\xi,\eta}(\cdot) : \xi, \eta \in \mathcal{D} \}$ of seminorms with $p_{\xi,\eta}(T) := |\langle T\xi, \eta \rangle|$, $T \in \mathcal{L}^\dagger(\mathcal{D})$ (resp. the family $\{ p_\xi^*(\cdot) : \xi \in \mathcal{D} \}$ of seminorms with $p_\xi^*(T) := \|T\xi\| + \|T^*\xi\|$, $T \in \mathcal{L}^\dagger(\mathcal{D}))$.

We shall say that a $*$-representation π of a locally convex $*$-algebra $\mathcal{A}[\tau]$ is $(\tau - \tau_w)$-*continuous* (reps. $(\tau - \tau_{s*})$-*continuous*) if it is continuous from $\mathcal{A}[\tau]$ to $\pi(\mathcal{A})[\tau_w]$ (resp. to $\pi(\mathcal{A})[\tau_{s*}]$).

In this regard, we have (for Pták function, see Definition 3.2.8)

Lemma 7.1.5 *The following statements are equivalent:*

(i) $\widetilde{\mathcal{A}}[\tau]^+ \cap \left(- \widetilde{\mathcal{A}}[\tau]^+ \right) = \{0\}$.

(ii) $\widetilde{\mathcal{A}}[B_\tau]^+ \cap \left(- \widetilde{\mathcal{A}}[B_\tau]^+ \right) = \{0\}$.

(iii) *The Pták function* $p_{\widetilde{\mathcal{A}}[B_\tau]}$ *on the Banach $*$-algebra* $\widetilde{\mathcal{A}}[B_\tau]$ *is a C^*-norm.*

(iv) *There exists a faithful $*$-representation of* $\widetilde{\mathcal{A}}[\tau]$.

(v) *There exists a faithful* $(\tau - \tau_{s*})$-*continuous $*$-representation of* $\widetilde{\mathcal{A}}[\tau]$.

Proof (i) $\Rightarrow$ (v) Let $\mathcal{F}$ be the set of all τ-continuous strongly positive linear functionals on $\widetilde{\mathcal{A}}[\tau]$. Let $(\pi_f, \lambda_f, \mathcal{H}_f)$ be the *GNS*-construction for $f \in \mathcal{F}$. We put

$$\mathcal{D}(\pi) := \left\{ (\lambda_f(x_f)) \in \bigoplus_{f \in \mathcal{F}} \mathcal{H}_f : \lambda_f(x_f) = 0, \ \forall f \in \mathcal{F}, \ \text{except for a finite number} \right\},$$

$$\pi(a)(\lambda_f(x_f)) := (\lambda_f(ax_f)), \ \forall a \in \widetilde{\mathcal{A}}[\tau], \ (\lambda_f(x_f)) \in \mathcal{D}(\pi).$$

Then, it is easily shown that π is a $(\tau - \tau_{s*})$-continuous $*$-representation of $\widetilde{\mathcal{A}}[\tau]$. We show that π is faithful. Take $0 \neq a \in H(\widetilde{\mathcal{A}}[\tau])$ (i.e., $a^* = a$). Assume that $a \in \widetilde{\mathcal{A}}[\tau]^+$. Since $\widetilde{\mathcal{A}}[\tau]^+ \cap (-\widetilde{\mathcal{A}}[\tau]^+) = \{0\}$, we have $\widetilde{\mathcal{A}}[\tau]^+ \cap \{-a\} = \emptyset$. Then, it follows from [35, Chap. II, §5, Proposition 4] that there exists a τ-continuous strongly positive linear functional f on $\widetilde{\mathcal{A}}[\tau]$, such that $f(a) > 0$. Now assume that $a \notin \widetilde{\mathcal{A}}[\tau]^+$. Since $\widetilde{\mathcal{A}}[\tau]^+ \cap \{a\} = \emptyset$, we can show in a similar way that there exists a τ-continuous strongly positive linear functional f on $\widetilde{\mathcal{A}}[\tau]$, such that $f(a) < 0$. Since $\langle \pi_f(a)\lambda_f(e) | \lambda_f(e) \rangle = f(a) \neq 0$, this implies that $\pi_f(a) \neq 0$, and so $\pi(a) \neq 0$. Similarly, for any $0 \neq a \in \widetilde{\mathcal{A}}[\tau]$ we have $\pi(a) \neq 0$ by considering $a = a_1 + ia_2$ $(a_1, a_2 \in H(\widetilde{\mathcal{A}}[\tau]))$.

(v) ⇒ (iv) This is trivial.

(iv) ⇒ (iii) Let π be a faithful $*$-representation of $\widetilde{\mathcal{A}}[\tau]$. By Lemma 7.1.1(i), $\widetilde{\mathcal{A}}[B_\tau]$ is a symmetric Banach $*$-algebra, hence hermitian (see discussion after Definition 3.1.2). Therefore, the Pták function $p_{\widetilde{\mathcal{A}}[B_\tau]}(x) := r_{\widetilde{\mathcal{A}}[B_\tau]}(x^*x)^{1/2}$, $x \in \widetilde{\mathcal{A}}[B_\tau]$, is a C^*-seminorm (cf. [52, (33.1) Theorem]). In particular (Raikov criterion for symmetry [127, (4.7.21) Theorem]),

$$p_{\widetilde{\mathcal{A}}[B_\tau]}(x) = \sup_{\rho \in Rep(\widetilde{\mathcal{A}}[B_\tau])} \|\rho(x)\|, \ x \in \mathcal{A}[B_\tau],$$

where $Rep(\widetilde{\mathcal{A}}[B_\tau])$ denotes the set of all $*$-representations of $\widetilde{\mathcal{A}}[B_\tau]$. Suppose $p_{\widetilde{\mathcal{A}}[B_\tau]}(x) = 0$. Since $\pi \upharpoonright_{\widetilde{\mathcal{A}}[B_\tau]} \in Rep(\widetilde{\mathcal{A}}[B_\tau])$, we have $\pi(x) = 0$, and so $x = 0$. Thus, $p_{\widetilde{\mathcal{A}}[B_\tau]}$ is a C^*-norm.

(iii) ⇒ (ii) We first show that

$$sp_{\widetilde{\mathcal{A}}[B_\tau]}(x) \subseteq \mathbb{R}_+ \equiv \{\lambda \in \mathbb{R} : \lambda \geq 0\}, \ \forall\, x \in \widetilde{\mathcal{A}}[B_\tau]^+. \tag{7.1.2}$$

In fact, take an arbitrary $x \in \widetilde{\mathcal{A}}[B_\tau]^+$ and a net (x_α) in $\mathcal{A}^+$ that converges to x, with respect to τ. Since $\widetilde{\mathcal{A}}[B_\tau]$ is hermitian we have that $sp_{\widetilde{\mathcal{A}}[B_\tau]}(x) \subseteq \mathbb{R}$. Let $\lambda < 0$. Notice that $\lambda(\lambda e - x_\alpha)^{-1} \in U(\mathcal{A})$, for every α. Then, for any τ-continuous seminorm p on $\widetilde{\mathcal{A}}[\tau]$

$$
\begin{aligned}
p\big(\lambda(\lambda e - x_\alpha)^{-1} &- \lambda(\lambda e - x_\beta)^{-1}\big) \\
&= |\lambda|\, p\big((\lambda e - x_\alpha)^{-1}(x_\alpha - x_\beta)(\lambda e - x_\beta)^{-1}\big) \\
&\leq |\lambda|\, q\big((\lambda e - x_\alpha)^{-1}\big) q(x_\alpha - x_\beta) q\big((\lambda e - x_\beta)^{-1}\big) \\
&\leq \frac{1}{|\lambda|}\gamma \|\lambda(\lambda e - x_\alpha)^{-1}\| \|\lambda(\lambda e - x_\beta)^{-1}\| q(x_\alpha - x_\beta) \\
&\leq \frac{\gamma}{|\lambda|} q(x_\alpha - x_\beta),
\end{aligned}
$$

for some constant $\gamma > 0$ and a τ-continuous seminorm q on $\widetilde{\mathcal{A}}[\tau]$. It follows that $\lambda(\lambda e - x_\alpha)^{-1}$ converges to an element y of B_τ with respect to τ, which implies that $\lambda(\lambda e - x)^{-1}$ exists and equals y. Hence, $\lambda \notin sp_{\widetilde{\mathcal{A}}[B_\tau]}(x)$. Thus, we have $sp_{\widetilde{\mathcal{A}}[B_\tau]}(x) \subseteq \mathbb{R}_+$. Take an arbitrary $x \in \widetilde{\mathcal{A}}[B_\tau]^+ \cap (-\widetilde{\mathcal{A}}[B_\tau]^+)$. Then, from (7.1.2), it follows that $sp_{\widetilde{\mathcal{A}}[B_\tau]}(x) = \{0\}$, therefore $p_{\widetilde{\mathcal{A}}[B_\tau]}(x) = r_{\widetilde{\mathcal{A}}[B_\tau]}(x) = 0$. Since $p_{\widetilde{\mathcal{A}}[B_\tau]}$ is a norm, we have $x = 0$.

(ii) ⇒ (i) Take an arbitrary $a \in \widetilde{\mathcal{A}}[\tau]^+ \cap (-\widetilde{\mathcal{A}}[\tau]^+)$. Then, from Lemma 7.1.1(i) it follows that $a(e + a^2)^{-1} \in \widetilde{\mathcal{A}}[B_\tau]^+ \cap (-\widetilde{\mathcal{A}}[B_\tau]^+) = \{0\}$, which implies $a = 0$. This completes the proof.   □

Let now π be a $*$-representation of $\widetilde{\mathcal{A}}[\tau]$. We define

$$\mathcal{D}(r_\pi) = \widetilde{\mathcal{A}}[\tau]_b^\pi := \left\{ x \in \widetilde{\mathcal{A}}[\tau] : \overline{\pi(x)} \in \mathcal{B}(\mathcal{H}_\pi) \right\}$$

$$r_\pi(x) := \|\overline{\pi(x)}\|, \ \forall \ x \in \mathcal{D}(r_\pi),$$

where $\mathcal{H}_\pi$ is a Hilbert space. Then, r_π is an unbounded C^*-seminorm (see discussion before Definition 3.5.1) on the $*$-subalgebra $\mathcal{D}(r_\pi)$ of $\widetilde{\mathcal{A}}[\tau]$, induced by π.

Lemma 7.1.6 *Let π be a faithful $*$-representation of $\widetilde{\mathcal{A}}[\tau]$ and B any element of $\mathcal{B}^*_{\widetilde{\mathcal{A}}}$ containing $U(\mathcal{A})$. Then, the following statements hold:*

(1) $\mathcal{A} \subseteq \widetilde{\mathcal{A}}[B_\tau] \subseteq \widetilde{\mathcal{A}}[B] \subseteq \mathcal{D}(r_\pi) = \widetilde{\mathcal{A}}[\tau]_b^\pi$ *and* $\|\pi(x)\| \le \|x\|_B$, *for all* $x \in \widetilde{\mathcal{A}}[B]$. *In particular,* $\|\pi(x)\| = \|x\|_{B_\tau} = \|x\|$, *for all* $x \in \mathcal{A}$.
(2) $\pi\big(\widetilde{\mathcal{A}}[B]\big)$ *is* τ_s^*-*dense in* $\pi\big(\widetilde{\mathcal{A}}[\tau]\big)$, *and it is also uniformly dense in* $\pi\big(\widetilde{\mathcal{A}}[\tau]_b^\pi\big)$.
(3) *Suppose that* π *is* $(\tau - \tau_w)$-*continuous. Then,*

$$\pi\big(\widetilde{\mathcal{A}}[\tau]^+\big) \subseteq \mathcal{L}^\dagger\big(\mathcal{D}(\pi)\big)^+ := \big\{ T \in \mathcal{L}^\dagger\big(\mathcal{D}(\pi)\big) : T \ge 0 \big\}.$$

Proof Statement (1) is easily checked.

(2) Take an arbitrary $a \in \widetilde{\mathcal{A}}[\tau]$. Then, it follows that

$$(e + \varepsilon a^* a)^{-1} a = \frac{1}{\sqrt{\varepsilon}} \big(e + (\sqrt{\varepsilon}a)^*(\sqrt{\varepsilon}a)\big)^{-1}(\sqrt{\varepsilon}a) \in \widetilde{\mathcal{A}}[B_\tau], \ \forall \ \varepsilon > 0$$

and for each $\xi \in \mathcal{D}(\pi)$

$$\begin{aligned}
\big\|\pi\big((e + \varepsilon a^* a)^{-1} a\big)\xi - \pi(a)\xi\big\| &= \varepsilon\big\|\pi\big((e + \varepsilon a^* a)^{-1}\big)\pi(a^* a^2)\xi\big\| \\
&\le \varepsilon\big\|\pi\big((e + \varepsilon a^* a)^{-1}\big)\big\|\big\|\pi(a^* a^2)\xi\big\| \\
&= \varepsilon\big\|(e + \varepsilon a^* a)^{-1}\big\|_{B_\tau}\big\|\pi(a^* a^2)\xi\big\| \\
&\le \varepsilon\big\|\pi(a^* a^2)\xi\big\| \xrightarrow[\varepsilon\downarrow 0]{} 0,
\end{aligned}$$

so that $\pi\big(\widetilde{\mathcal{A}}[B_\tau]\big)$ is τ_{s^*}-dense in $\pi\big(\widetilde{\mathcal{A}}[\tau]\big)$. Take an arbitrary $a \in \widetilde{\mathcal{A}}[\tau]_b^\pi$. Then, since

$$\big\|\pi\big((e + \varepsilon a^* a)^{-1} a\big)\xi - \pi(a)\xi\big\| \le \varepsilon\big\|\pi(a^* a^2)\big\|\|\xi\|, \ \forall \ \xi \in \mathcal{D}(\pi),$$

it follows that $\lim_{\varepsilon\downarrow 0} \pi\big((e + \varepsilon a^* a)^{-1} a\big) = \pi(a)$ uniformly, which implies that $\pi\big(\widetilde{\mathcal{A}}[B_\tau]\big)$ is uniformly dense in $\pi\big(\widetilde{\mathcal{A}}[\tau]_b^\pi\big)$. Since $\widetilde{\mathcal{A}}[B_\tau] \subseteq \widetilde{\mathcal{A}}[B]$, we conclude (2).

(3) This follows from $(\tau - \tau_w)$-continuity of π and the fact that $\pi(\mathcal{A}^+) \subseteq \mathcal{L}^\dagger(\mathcal{D}(\pi))^+$. This completes the proof. $\square$

Theorem 7.1.7 *The following statements are equivalent:*

(i) $\widetilde{\mathcal{A}}[\tau]$ *is a* GB^**-algebra.*
(ii) *There exists a faithful* $(\tau - \tau_{s^*})$*-continuous* $*$*-representation* π *of* $\widetilde{\mathcal{A}}[\tau]$, *such that* $\tau \prec r_\pi$.

Proof (i) $\Rightarrow$ (ii) Suppose $\widetilde{\mathcal{A}}[\tau]$ is a GB^*-algebra over B_0. Since

$$\widetilde{\mathcal{A}}[B_\tau]^+ \cap \left(-\widetilde{\mathcal{A}}[B_\tau]^+\right) \subseteq \widetilde{\mathcal{A}}[B_0]^+ \cap \left(-\widetilde{\mathcal{A}}[B_0]^+\right) = \{0\},$$

Lemma 7.1.5 implies the existence of a faithful $(\tau - \tau_{s^*})$-continuous $*$-representation of $\widetilde{\mathcal{A}}[\tau]$. Furthermore, since $\pi(\widetilde{\mathcal{A}}[B_0])$ is a C^*-algebra, Lemma 7.1.6(2) yields that

$$\pi\left(\widetilde{\mathcal{A}}[B_0]\right) = \pi\left(\widetilde{\mathcal{A}}[\tau]_b^\pi\right) \text{ and } r_\pi(x) = \left\|\overline{\pi(x)}\right\| = \|x\|_{B_0}, \quad \forall\, x \in \mathcal{D}(r_\pi),$$

which implies $\tau \prec r_\pi$.

(ii) $\Rightarrow$ (i) Since $\tau \prec r_\pi$ and π is $(\tau - \tau_{s^*})$-continuous, it follows that τ and r_π are compatible, so one obtains that the completion $\mathcal{A}_{r_\pi}$ of $\mathcal{D}(r_\pi)[r_\pi]$ is embedded in $\widetilde{\mathcal{A}}[\tau]$. We denote by B_0 the τ-closure of the closed unit ball $U(\mathcal{A}_{r_\pi})$ of the C^*-algebra $\mathcal{A}_{r_\pi}$. Then, $B_0 \in \mathfrak{B}^*_{\widetilde{\mathcal{A}}}$ and from Lemma 7.1.6(1) we have

$$B \subseteq U\left(\widetilde{\mathcal{A}}[\tau]_b^\pi\right) \subseteq B_0, \ \forall\, B \in \mathfrak{B}^*_{\widetilde{\mathcal{A}}},$$

which implies that $B_0 = U\left(\widetilde{\mathcal{A}}[\tau]_b^\pi\right)$, is a greatest member in $\mathfrak{B}^*_{\widetilde{\mathcal{A}}}$. Thus, from Theorem 7.1.2, we conclude that $\widetilde{\mathcal{A}}[\tau]$ is a GB^*-algebra over $U\left(\widetilde{\mathcal{A}}[\tau]_b^\pi\right)$ and this completes the proof. $\square$

Corollary 7.1.8 *If* $\widetilde{\mathcal{A}}[\tau]$ *is a Fréchet* $*$*-algebra, the following are equivalent:*

(i) $\widetilde{\mathcal{A}}[\tau]$ *is a* GB^**-algebra.*
(ii) *There exists a faithful* $*$*-representation* π *of* $\widetilde{\mathcal{A}}[\tau]$ *such that* $\tau \prec r_\pi$.

Proof Every $*$-representation of the Fréchet locally convex $*$-algebra $\widetilde{\mathcal{A}}[\tau]$ is $(\tau - \tau_s^*)$-continuous. Indeed, take an arbitrary $\xi \in \mathcal{D}(\pi)$ and put $f_\xi(x) := \langle \pi(x)\xi, \xi \rangle$, $x \subset \mathcal{A}$. Then, f_ξ is a positive linear functional on the Fréchet $*$-algebra $\mathcal{A}[\tau]$, which is continuous by [51, Theorem 4.3] (see also [60, Theorem 15.5]). Furthermore, since the multiplication of a Fréchet $*$-algebra is jointly continuous, it follows that π is $(\tau - \tau_{s^*})$-continuous. Hence, the assumption follows from Theorem 7.1.7. $\square$

Notes Almost for all the results of this section, the reader is referred to [12] and [62]. In particular, Lemmas 7.1.1, 7.1.5, 7.1.6, Theorem 7.1.7 and Corollary 7.1.8 can be found in [12, Section 2], while Theorem 7.1.2 comes from [62, Theorem 2.1] (see also [12, Theorem 2.2]). Related studies can be found in [22–25]. For the investigation of Case 2 (where the multiplication is considered separately continuous, see beginning of this section), we refer the reader to [12–14, 62].

7.2 Continuity of Positive Linear Functionals of GB*-Algebras

In this section the presented results concern GB^*-algebras of Dixon (Definition 3.3.5).

It is well known that all positive linear functionals of a C^*-algebra are continuous and so the question is if one can extend this result to GB^*-algebras. We shall see in Corollary 7.2.2 below that at least some GB^*-algebras have this property. We also give examples of GB^*-algebras, which admit at least one discontinuous positive linear functional, thereby showing that the conditions in Theorem 7.2.1 cannot be dropped.

Theorem 7.2.1 (Dixon) *If $\mathcal{A}[\tau]$ is a sequentially complete GB^*-algebra satisfying property [A] of Section 6.2, then every positive linear functional on $\mathcal{A}$ is bounded.*

Proof Let f be a positive linear functional of $\mathcal{A}$. We first show that if f is bounded on all bounded sequences of $\mathcal{A}$, then f is bounded. So suppose that f is bounded on all bounded sequences of $\mathcal{A}$. If f is not bounded, then there exists a bounded subset C of $\mathcal{A}$, such that for every $n \in \mathbb{N}$, there exists $y_n \in C$, such that $|f(y_n)| > n$. This implies that f is not bounded on the (y_n), which contradicts our claim.

Therefore, it suffices to prove that f is bounded on all bounded sequences of $\mathcal{A}$. Let (x_n) be a bounded sequence of positive elements of $\mathcal{A}$. Also, let (λ_n) be a sequence in l^1 with $\lambda_n \geq 0$, for all $n \in \mathbb{N}$. For every absolutely convex 0-neighbourhood U in $\mathcal{A}$, there exists $k > 0$, such that $x_n \in kU$, for all $n \in \mathbb{N}$. Since $(\lambda_n) \in l^1$, there exists $N \in \mathbb{N}$, such that $\sum_{i=p}^{q} \lambda_i < \frac{1}{k}$ for all $p, q \geq N$. Therefore,

$$\sum_{i=p}^{q} \lambda_i x_i \in U, \ \forall \, p, q \geq N.$$

Since $\mathcal{A}$ is sequentially complete, it follows that $\sum_{i=1}^{\infty} \lambda_i x_i$ converges to $x \in \mathcal{A}$, say. Since $\mathcal{A}$ satisfies property [A], it follows from Theorem 6.2.11 that $\mathcal{A}^+$ is closed, hence $x \in \mathcal{A}^+$. Consequently,

$$\sum_{i=n+1}^{\infty} \lambda_i x_i = x - \sum_{i=1}^{n} \lambda_i x_i \geq 0, \ \forall \, n \in \mathbb{N}.$$

Thus,

$$\sum_{i=1}^{n} \lambda_i f(x_i) = f\left(\sum_{i=1}^{n} \lambda_i x_i\right) = f(x) - f\left(\sum_{i=n+1}^{\infty} \lambda_i x_i\right) \leq f(x).$$

Therefore, by the monotone convergence theorem for sequences of real numbers, $\sum_{n=1}^{\infty} \lambda_n f(x_n)$ is convergent for all (λ_n) in l^1. Hence, $(f(x_n))$, $n \in \mathbb{N}$, is a bounded sequence.

Now let (x_n) be a bounded sequence of self-adjoint elements of $\mathcal{A}$. By taking the Gelfand representation of x_n, we obtain that

$$-\left(e + 2x_n^2(e + x_n^2)^{-1/2}\right) \leq x_n \leq \left(e + 2x_n^2(e + x_n^2)^{-1/2}\right);$$

hence,

$$|f(x_n)| \leq f(e) + 2f\left(x_n^2(e + x_n^2)^{-1/2}\right). \tag{7.2.3}$$

Using again the Gelfand representation of x_n, we obtain that $x_n(e + x_n^2)^{-1/2} \in B_0$. Since $\mathcal{A}$ satisfies property [A], then for any 0-neighbourhood U in $\mathcal{A}$, there exists a 0-neighbourhood V, such that $B_0 V \subset U$. Due to the fact that (x_n) is a bounded sequence, there exists $m \in \mathbb{N}$, such that $x_n \in mV$ for all $n \in \mathbb{N}$. Therefore,

$$x_n^2(e + x_n^2)^{-1/2} = x_n(e + x_n^2)^{-1/2}x_n \in B_0 m V \subset mU,$$

for all $m \in \mathbb{N}$, implying that $\left(x_n^2(e + x_n^2)^{-1/2}\right)$ is a bounded sequence of positive elements in $\mathcal{A}$. From the above, it follows now that $f\left(x_n^2(e + x_n^2)^{-1/2}\right)$ is bounded. Hence, by (7.2.3), $f(x_n)$, $n \in \mathbb{N}$, is a bounded sequence.

Finally, let (x_n) be any bounded sequence in $\mathcal{A}$. Then,

$$y_n := \frac{1}{2}(x_n + x_n^*) \text{ and } z_n := \frac{1}{2i}(x_n - x_n^*),$$

define bounded sequences in $\mathcal{A}$ by continuity of involution. Since $f(x_n) = f(y_n) + if(z_n)$, for all $n \in \mathbb{N}$, it follows that the sequence $f(x_n)$, $n \in \mathbb{N}$, is bounded and this completes the proof. □

Corollary 7.2.2 *Every positive linear functional on a sequentially complete, bornological GB*- algebra is continuous.*

Proof Let $\mathcal{A}[\tau]$ be a GB^*-algebra and f a positive linear functional on $\mathcal{A}$. A sequentially complete bornological space is barrelled, and therefore $\mathcal{A}$ has hypocontinuous multiplication, therefore it has property [A]. By Theorem 7.2.1, it follows that f is bounded, and since $\mathcal{A}$ is bornological, f is continuous. □

Since every Fréchet locally convex algebra is sequentially complete and bornological, we obtain the following result.

Corollary 7.2.3 *Every positive linear functional on a Fréchet GB*-algebra is continuous.*

In 1959, D-S. Sya proved that the above corollary holds for any Fréchet locally convex ∗-algebra with an identity element, i.e., he proved that *every positive linear functional of a Fréchet locally convex ∗-algebra with identity is continuous* [143] (see also [51, Theorem 4.3] and/or [60, Theorem 15.5]). Later, in 1978, G. Dales, with instruction from Dixon, proved that *every positive linear functional of a Fréchet topological algebra, with identity and continuous involution, is continuous* [42, Theorem 11.1].

In the next two examples, we will show that *neither assumptions of sequential completeness nor the algebra being bornological can be dropped from* Corollary 7.2.2, thereby giving *examples of GB*-algebras admitting at least one discontinuous positive linear functional.*

Example 7.2.4 ([48, Example 8.4]) Let $\mathcal{A} = L^{\infty}[0, 1]$ be equipped with the topology defined by the family of seminorms being the L_p-norms $\| \cdot \|_p$, $p \in \mathbb{N}$. Since $L^{\omega}[0, 1]$ is a *GB**-algebra over the *C**-algebra $L^{\infty}[0, 1]$ (see Example 3.3.16(5)), then

$$\{B \cap L^{\infty}[0, 1] : B \in \mathfrak{B}_{L^{\omega}[0,1]}\}$$

has a largest element B_0 with respect to set inclusion, which is the closed unit ball of $L^{\infty}[0, 1]$. Furthermore, the bounded part of $L^{\infty}[0, 1]$ is a *C**-algebra, namely itself, and symmetric. Therefore, $\mathcal{A} = L^{\infty}[0, 1]$ is a *GB**-algebra of Dixon (i.e., a locally convex *GB**-algebra, in the sense of Definition 3.3.5). We define a positive linear functional F on $\mathcal{A}$ by

$$F(f) = \int_0^1 \frac{f(t)}{t(\log(t) - 1)^2} dt, \ \forall f \in \mathcal{A}.$$

Let $N \in \mathbb{N}$. We define

$$f_{q,N}(t) = \min\{(4t)^{-1/2q}, N\}, \ \forall q = 1, 2, 3, \cdots.$$

It easily follows that

$$\|f_{q,N}\|_q \leq \left(\int_0^1 (4t)^{-1/2} dt\right)^{1/q} dt = \left(\lim_{s \to 0^+} \int_s^1 (4t)^{-1/2} dt\right)^{1/q} dt = 1.$$

Now,

$$F(f_{q,N}) \geq \int_0^{\eta} \frac{N}{t(\log(t) - 1)^2} dt = \frac{N}{2q\log(N) + \log(4) + 1},$$

where $\eta = \frac{1}{4N^{2q}}$. The last expression goes to infinity as $N \to \infty$, for every q. If F is continuous, then there exists $p \in \mathbb{N}$ and a constant $K > 0$, such that $|F(f)| \leq K\|f\|_p$ for all $f \in \mathcal{A}$. This gives a contradiction for $f = f_{p,N}$, upon letting $N \to \infty$.

The topology defined above is metrizable, hence bornological and therefore, in light of Corollary 7.2.2, $\mathcal{A}$ cannot be sequentially complete.

Example 7.2.5 ([48, Example 8.3]) Suppose that X is a countably compact, non-compact, completely regular space, being either locally compact or first countable. Let $\mathcal{A} \equiv \mathcal{C}_c(X)$ be the commutative pro-C^*-algebra of all continuous $\mathbb{C}$-valued functions on X endowed with the compact-open topology 'c' [112, p. 52, Proposition 12.2, a)]. As such, $\mathcal{A}$ is a GB^*-algebra (see Examples 3.3.16(1)), that also has property [A] according to the observation in ▶ after Corollary 6.2.9. Therefore, Corollary 7.2.2 yields that *every positive linear functional on $\mathcal{A}$ is continuous*. But, each character of $\mathcal{A}$ is the point evaluation corresponding to an element $x \in \beta X$, the Stone–Čech compactification of X. On the other hand, from a known result of Gelfand–Kolmogorov (cf. [68, Theorem 1]), one concludes that the character of $\mathcal{A}$ corresponding to an element $x \in \beta X \setminus X$ is discontinuous, and since every character of $\mathcal{A}$ is $*$-preserving (see [60, Corollary 9.3(1)] (Michael)) is clearly a positive linear functional. So, $\mathcal{A}$ accepts discontinuous positive linear functionals.

Since $\mathcal{A}$ is complete, and therefore sequentially complete, this is attributed to the fact that $\mathcal{A}$ cannot be bornological.

The above example also illustrates that $*$-representations of GB^*-algebras are not continuous, in general. Recall, in this regard, that all $*$-representations of a C^*-algebra are continuous. Moreover, notice that R.M. Brooks [37, p. 17, Example 6.1] has given an example of a discontinuous $*$-representation on a pro-C^*-algebra, hence a GB^*-algebra (see also [60, Sect. 17, pp. 196–197]).

Notes All the results of this Section are due to P.G. Dixon and the reader can find them in [48, 50].

7.3 Every GB*-Algebra has a Bounded Approximate Identity

It is well known that every C^*-algebra and every pro-C^*-algebra (Inoue) [60, p. 137, Theorem 11.5] has a bounded approximate identity. Below, we extend this result to GB^*-algebras.

If $\mathcal{A}[\tau]$ is a topological $*$-algebra, we recall that a bounded approximate identity of $\mathcal{A}$ is a bounded net (e_α) in $\mathcal{A}$, such that

$$\lim_\alpha e_\alpha x = x = \lim_\alpha x e_\alpha, \ \forall \, x \in \mathcal{A}.$$

For the definition of *hypocontinuity*, used in the statement of Theorem 7.3.1, see comments right after Remark 6.2.7.

The following Theorem 7.3.1 is valid for every GB^*-algebra in the sense of Dixon (Definition 3.3.3).

Theorem 7.3.1 (Bhatt) *Let $\mathcal{A}[\tau]$ be a GB^*-algebra. The following hold:*

(i) *if the multiplication of $\mathcal{A}[\tau]$ is hypocontinuous and I is a closed two-sided ideal of $\mathcal{A}[\tau]$, then every bounded approximate identity of $I \cap \mathcal{A}[B_0]$ is also an approximate identity of I;*

(ii) *every GB^*-algebra $\mathcal{A}[\tau]$ admits a bounded approximate identity.*

Proof

(i) Since $\mathcal{A}[\tau]$ has hypocontinuous multiplication and I is a two-sided ideal of $\mathcal{A}$, it follows that I also has hypocontinuous multiplication with respect to the relative topology inherited from the topology τ on $\mathcal{A}$.

Let (e_α) be a bounded approximate identity of $I \cap \mathcal{A}[B_0]$. By the continuity of addition and hypocontinuity of multiplication on I, it follows that for any given 0-neighbourhood U in I, we can choose 0-neighbourhoods V and V_1 in I, such that $V_1 B_0 + V + V \subset U$. By Proposition 3.3.19, I is a GB^*-algebra over the C^*-algebra $I \cap \mathcal{A}[B_0]$.

Let $x \in I$, and let $x_n = x(e + \frac{1}{n}x^*x)^{-1}$, for all $n \in \mathbb{N}$. From Theorem 6.3.5 $\mathcal{A}[\tau]$ is represented as a unital $*$-algebra of closed operators sitting in $\pi(\mathcal{A})$ and having a common domain, dense in some Hilbert space, $\mathcal{H}_\pi$. Besides, $\mathcal{A}[B_0]$ coincides with the $*$-algebra of bounded operators of $\pi(\mathcal{A})$. Thus, the element $x \in \mathcal{A}$ can be identified with a closed densely defined operator T from $\pi(\mathcal{A})$. By Rudin [129, Theorem 13.13(a)], the operator $T(I + T^*T)^{-1}$ is bounded, therefore the elements x_n, $n \in \mathbb{N}$, corresponding to the bounded operators $T(I + \frac{1}{n}T^*T)^{-1}$, $n \in \mathbb{N}$, belong to $\mathcal{A}[B_0]$. Hence, finally we have that $x_n \in I \cap \mathcal{A}[B_0]$, for all $n \in \mathbb{N}$. Moreover, $x_n \xrightarrow{\tau} x$, as follows from a similar argument as in the proof of Lemma 7.1.1(iii); consequently there exists $n_0 \in \mathbb{N}$, such that $x - x_n \in V_1$, for all $n \geq n_0$. Similarly, there exists $n_1 \in \mathbb{N}$, such that $x_n - x \in V$, for all $n \geq n_1$. Let $m = \max\{n_0, n_1\}$. Since $x_n \in I$, for all $n \in \mathbb{N}$, and (e_α) is a bounded approximate identity of $I \cap \mathcal{A}[B_0]$, it follows that $x_n e_\alpha \to x_n$, for all $n \in \mathbb{N}$. Therefore, by taking $n \geq m$, there exists an α_0, such that for all $\alpha \geq \alpha_0$, we have

$$x e_\alpha - x = (x - x_n)e_\alpha + (x_n e_\alpha - x_n) + (x_n - x)$$
$$\in V_1 B_0 + V + V$$
$$\subset U.$$

Hence, $x e_\alpha \to x$, for all $x \in I$. Similarly, $e_\alpha x \to x$, so we conclude that (e_α) is a bounded approximate identity for I.

(ii) Let $\mathcal{A}[\tau]$ be a GB^*-algebra. We recall that $\mathcal{A}$ can be equipped with the topology T, as defined just before Theorem 6.2.6. By the latter theorem, $\mathcal{A}[T]$ is a barrelled GB^*-algebra, and therefore $\mathcal{A}[T]$ has hypocontinuous multiplication (see [131, p. 89, 5.2] and/or [74, p. 360 Theorem 2]). Furthermore, Theorem 6.2.6 also informs us that $B_0(T) = B_0(\tau)$. Since $\mathcal{A}[B_0]$ is a C^*-algebra, it admits a bounded approximate identity (e_α). By applying (i) to the case where $I = \mathcal{A}$, it follows that (e_α) is also a bounded approximate identity for $\mathcal{A}[T]$. Since the topology T on $\mathcal{A}$ is stronger than the given topology τ, we get that (e_α) is a bounded approximate identity of $\mathcal{A}[\tau]$ too.

□

Note that from the third paragraph of the proof of Theorem 7.3.1(i), thanks to Theorem 6.3.5 and to [129, Theorem 13.13(a)], we conclude the following result, which extends in the noncommutative case Lemma 3.4.2(i); recall that in the latter reference $\mathcal{A}_0 = \mathcal{A}[B_0]$, since $\mathcal{A}[\tau]$ is commutative (cf. Lemma 3.3.7(ii)). In this regard, the reader may also look at Lemma 7.1.1, for another aspect of the content of the lemma that follows.

Lemma 7.3.2 *Let $\mathcal{A}[\tau]$ be a GB^*-algebra. Then, for every $x \in \mathcal{A}$, one has that the element $x(e + x^*x)^{-1}$ belongs to the C^*-algebra $\mathcal{A}[B_0]$.*

Notes Theorem 7.3.1 is due to S.J. Bhatt, and can be found in [18].

7.4   Inverse Limits and Quotients of GB*-Algebras

▶ In this section, unless stated otherwise, a GB^*-algebra will mean a GB^*-algebra in the sense of *Definition 3.3.3* (Dixon).

A GB^*-algebra $\mathcal{A}[\tau]$, in the sense of Definition 3.3.3 (Dixon), has the property that the corresponding family of sets $\mathcal{B}_\mathcal{A}$, defined just before Definition 3.3.3, has a largest element B_0 with respect to set inclusion, which gives rise to a C^*-algebra $\mathcal{A}[B_0]$ contained in $\mathcal{A}$ (see also Remarks 3.3.4 and 3.3.10). In this section, we show that we can avoid having to make use of the set $\mathcal{B}_0$ in that we can characterize GB^*-algebras by using more algebraic techniques (see Theorem 7.4.5). This will lead to some information of inverse limits of GB^*-algebras.

In order to prove Theorem 7.4.5 below, we require Proposition 7.4.1 and Lemma 7.4.3 that we first prove.

Proposition 7.4.1 *Let $\mathcal{A}[\| \cdot \|]$ be a normed $*$-algebra with identity element e and let $\mathcal{B}[\| \cdot \|']$ be a C^*-algebra, which is also a $*$-subalgebra of $\mathcal{A}$ with $e \in \mathcal{B}$. If*

$(e + x^*x)^{-1} \in \mathcal{B}$, *for all* $x \in \mathcal{A}$, *then* $\mathcal{A} = \mathcal{B}$ *as algebras* (and not necessarily as normed algebras).

Proof Let $\widetilde{\mathcal{A}}[\|\cdot\|]$ denote the completion of $\mathcal{A}[\|\cdot\|]$. Taking into consideration that $e \in \mathcal{B} \subseteq \widetilde{\mathcal{A}}$, it follows immediately from [127, Theorem 4.8.5] that $sp_{\widetilde{\mathcal{A}}}(x) = sp_{\mathcal{B}}(x)$ for all $x \in \mathcal{B}$. It follows from this and the hypothesis that

$$0 \notin sp_{\widetilde{\mathcal{A}}}\big((e + x^*x)^{-1}\big) = sp_{\mathcal{B}}\big((e + x^*x)^{-1}\big)$$

for all $x \in \mathcal{A}$. Therefore $e + x^*x = \big((e + x^*x)^{-1}\big)^{-1} \in \mathcal{B}$, hence $x^*x \in \mathcal{B}$ for all $x \in \mathcal{A}$. Therefore, if $x \in \mathcal{A}$, then, since

$$x = \frac{1}{4}\Big((x+e)^*(x+e) - (x-e)^*(x-e) + i(x+ie)^*(x+ie) - i(x-ie)^*(x-ie)\Big),$$

it follows that $x \in \mathcal{B}$, i.e., $\mathcal{A} = \mathcal{B}$.   □

Corollary 7.4.2 *Let* $\mathcal{A}[\|\cdot\|]$ *be a normed* *-*algebra, which is also a* GB^*-*algebra, such that the topology on* $\mathcal{A}[B_0]$ *defined by* $\|\cdot\|$ *is weaker than the* $\|\cdot\|_{B_0}$-*topology. Then, there exists a norm* $\|\cdot\|'$ *equivalent to* $\|\cdot\|$, *such that* $\mathcal{A}[\|\cdot\|']$ *is a* C^*-*algebra.*

Proof Since $\mathcal{A}[\|\cdot\|]$ is a GB^*-algebra, we have that $(e + x^*x)^{-1} \in \mathcal{A}[B_0]$, for all $x \in \mathcal{A}$, and that $\mathcal{A}[B_0]$ is a C^*-algebra with respect to the Minkowski functional $\|\cdot\|_{B_0}$ on B_0 (see Theorem 3.3.9). Take, $\|\cdot\|' := \|\cdot\|_{B_0}$. By Proposition 7.4.1, it follows that $\mathcal{A} = \mathcal{A}[B_0]$. Hence, by [144, Proposition I.5.3], $\|x\|_{B_0} \leq \|x\|$, for all $x \in \mathcal{A}$. Since $\mathcal{A}[\|\cdot\|]$ is a GB^*-algebra, it follows that the $\|\cdot\|$-topology is weaker than the $\|\cdot\|_{B_0}$-topology on $\mathcal{A}[B_0] = \mathcal{A}$. Therefore, the norms $\|\cdot\|$ and $\|\cdot\|'$ are equivalent on $\mathcal{A}$.   □

The following lemma is Corollary 4.8.4 of [127].

Lemma 7.4.3 *If* $\mathcal{A}[\|\cdot\|]$ *is a* C^*-*algebra and* $\|\cdot\|'$ *is another norm on* $\mathcal{A}$ *with respect to which* $\mathcal{A}$ *is a* *-*normed algebra, then* $\|x\|^2 \leq \|x^*\|'\|x\|'$, *for all* $x \in \mathcal{A}$.

We let

$$[-e, e] = \big\{h \in H(\mathcal{A}) : -e \leq h \leq e\big\},$$

and E the absolute convex hull of $[-e, e]$. When the natural ordering is Archimedean, then the vector space $\mathcal{E}$ generated by E becomes a normed *-space when equipped with the Minkowski functional $\|\cdot\|' \equiv \|\cdot\|_E$ of E.

For the proof of the next theorem, we still need the following

Lemma 7.4.4 *Let* $\mathcal{A}[\tau]$ *be a* GB^*-*algebra. Then, the* C^*-*algebra* $\mathcal{A}[B_0]$ *is the linear hull of* $[-e, e]$.

Proof Let $h \in \mathcal{A}[B_0]$ be self-adjoint. Take the commutative C^*-subalgebra of $\mathcal{A}[B_0]$, say $\mathcal{C}$, generated by $\{h, e\}$. Then, $\mathcal{C} \cong C(X)$, where X is the maximal ideal space of $\mathcal{C}$. We can therefore identify h with $\widehat{h} \in C(X)$. Since $\widehat{h}$ is bounded, there exists $k > 0$, such that $|\widehat{h}(\varphi)| \leq k$, for each character $\varphi \in X$. By the functional calculus in C^*-algebras, it follows that $-ke \leq h \leq ke$. Since now every $x \in \mathcal{A}[B_0]$ is a linear combination of self-adjoint elements, we conclude that $\mathcal{A}[B_0]$ is the linear hull of $[-e, e]$. $\square$

The following theorem is the promised characterization of GB^*-algebras using more algebraic techniques.

Theorem 7.4.5 *Let $\mathcal{A}[\tau]$ be a topological $*$-algebra with identity e. The following conditions are equivalent:*

(i) *$\mathcal{A}[\tau]$ is a GB^*-algebra with underlying C^*-algebra $\mathcal{A}[B_0]$;*

(ii) *$\mathcal{A}[\tau]$ has a $*$-subalgebra $\mathcal{B}$, which is a C^*-algebra with respect to some norm $\| \cdot \|$, the element $(e + x^*x)^{-1} \in \mathcal{B}$, for all $x \in \mathcal{A}$, and the closed unit ball $U(\mathcal{B})$ of $\mathcal{B}$ is τ-bounded;*

(iii) *$\mathcal{E}$ is, with respect to some norm $\| \cdot \|'$, a C^*-algebra, such that E is τ-bounded, and $(e + x^*x)^{-1} \in \mathcal{E}$, for all $x \in \mathcal{A}$.*

If (i), (ii) *and* (iii) *hold, then $\mathcal{A}[B_0] = \mathcal{E} = \mathcal{B}$.*

Proof (i) $\Rightarrow$ (iii) We shall show that $\mathcal{E} = \mathcal{A}[B_0]$. By Lemma 7.4.4, $\mathcal{A}[B_0]$ is the linear hull of $[-e, e]$, therefore is contained in the linear hull of E (the latter being the absolutely convex hull of $[-e, e]$), which yields that $\mathcal{A}[B_0] \subseteq \mathcal{E}$.

We now show that $\mathcal{E} \subseteq \mathcal{A}[B_0]$. Let $h \in [-e, e]$. Consider the maximal commutative GB^*-subalgebra $\mathcal{C}$ of $\mathcal{A}[\tau]$ containing h. By Remark 3.4.3(3), $\mathcal{C}$ is algebraically $*$-isomorphic to the $*$-algebra $\widehat{\mathcal{C}}$ of $\mathbb{C}^*$-valued functions on the compact Hausdorff space X corresponding to the Gelfand space of the commutative C^*-algebra $\mathcal{C}_0 = \mathcal{C}[B_0 \cap \mathcal{C}]$ with an identity element (see Lemma 6.1.1, which will be used again right after). Under this isomorphism, h is identified with an element $\widehat{h} \in \widehat{\mathcal{C}}$, such that $-1 \leq \widehat{h}(\varphi) \leq 1$, for every character $\varphi \in X$. By Remark 3.3.10(a), $h \in \mathcal{C}_0 = \mathcal{C}[B_0 \cap \mathcal{C}] \subset \mathcal{C}$. It follows that

$$\|h\|_{B_0} = \|\widehat{h}\|_\infty \leq 1,$$

which yields $h \in B_0$. This proves $\mathcal{E} \subseteq \mathcal{A}[B_0]$ and so $\mathcal{E} = \mathcal{A}[B_0]$.

Therefore, $(e + x^*x)^{-1} \in \mathcal{A}[B_0] = \mathcal{E}$, for all $x \in \mathcal{A}$ (see Theorem 3.3.9). Lastly, since $\mathcal{E} = \mathcal{A}[B_0]$ and B_0 is τ-bounded, it follows that E is τ-bounded.

(iii) $\Rightarrow$ (ii) This follows by taking $\mathcal{B} = \mathcal{E}$.

(ii) $\Rightarrow$ (i) It remains to show that $U(\mathcal{B})$ is the largest element of $\mathfrak{B}_{\mathcal{A}}$, with respect to set inclusion.

Let $B \in \mathfrak{B}_{\mathcal{A}}$, $h \in B \cap H(\mathcal{A})$, and $\mathcal{C}$ (as above) the maximal commutative $*$-subalgebra of $\mathcal{A}$ containing h (with h now in $B \cap H(\mathcal{A})$). Then, $C \equiv \mathcal{C} \cap \mathcal{B}$ is a commutative C^*-subalgebra of $\mathcal{B}[\| \cdot \|]$, while $\mathcal{C}$ is a τ-closed $*$-subalgebra of $\mathcal{A}[\tau]$. Take the closed unit ball $U(C)$ of $C[\| \cdot \|]$ and if $H \equiv \{e, h^n : n \in \mathbb{N}\}$, denote by Y the absolutely convex hull of $H \cdot U(C)$ in $\mathcal{C}$.

We prove that Y is τ-bounded. Let L_n denote the mapping $x \mapsto h^n x$ of $C[\| \cdot \|]$ into $\mathcal{A}[\tau]$, and let $\mathcal{L} = \{L_n : n \in \mathbb{N}\}$. It follows from the continuity of multiplication and the τ-boundedness of $H \subseteq B$, that $\mathcal{L}$ is pointwise bounded, i.e., $\{L_n(x) : x \in C, n \in \mathbb{N}\}$ is τ-bounded. Since $C[\| \cdot \|]$ is of second category, it follows from a generalization of the uniform boundedness theorem [131, Theorem, p. 83] that $\mathcal{L}$ is uniformly bounded. In particular, for every 0-neighbourhood V in $\mathcal{A}[\tau]$, there exists $\lambda > 0$, such that $H \cdot U(C) \subseteq \lambda V$. So, $H \cdot U(C)$ is τ-bounded and since Y is the absolutely convex hull of $H \cdot U(C)$, it follows easily that Y is τ-bounded.

Furthermore, simple calculations show that $Y^2 \subseteq Y$, $Y^* = Y$ and $e \in Y$. If $\mathcal{Y}$ denotes the vector space generated by Y, then we show that $\mathcal{Y}$ is a normed $*$-algebra, when equipped with the Minkowski functional $\| \cdot \|_Y$ of Y.

We already know that $\| \cdot \|_Y$ is a $*$-seminorm on $\mathcal{Y}$, therefore it remains to show that $\| \cdot \|_Y$ is a norm. Let $\mathcal{U}$ denote a τ-neighbourhood base of $0 \in \mathcal{A}$ and assume that $\|x\|_Y = 0$. Since Y is τ-bounded, for every $U \in \mathcal{U}$, there exists a scalar $\mu(U) > 0$, such that $Y \subseteq \mu(U)U$. Therefore, $x \in \lambda Y \subseteq (\lambda \cdot \mu(U))U$, for all $\lambda > 0$ and for all $U \in \mathcal{U}$. Since τ is Hausdorff, it follows that $x = 0$.

Observe that $U(C) \subseteq Y \subseteq \mathcal{Y}$, hence $C \subseteq \mathcal{Y}$. Moreover, by our assumption (ii), $(e + y^*y)^{-1} \in B$, for all $y \in \mathcal{Y}$. On the other hand, by the construction of $\mathcal{Y}$, it is evident that $\mathcal{Y} \subseteq C$ and C as a C^*-algebra is symmetric, so that $(e+y^*y)^{-1} \in C$, for each $y \in C$. Thus finally, $(e+y^*y)^{-1} \in B \cap C = C$, for all $y \in \mathcal{Y}$. Proposition 7.4.1 implies now that $\mathcal{Y} = C$. Therefore, by Lemma 7.4.3, $\|x\|^2 \leq \|x^*\|_Y\|x\|_Y$, for all $x \in Y$. Consequently, since $Y^* = Y$, we obtain that $\|x\|_Y \leq 1$, for all $x \in Y$. Hence, $\|x\|^2 \leq 1$, for all $x \in Y$. Therefore, $Y \subseteq U(C)$ and so $h \in Y \subseteq U(C) \subseteq U(\mathcal{B})$, i.e., $B \cap H(\mathcal{A}) \subseteq U(\mathcal{B})$.

If $x \in B$, then by the above, $x^*x \in B \cap H(\mathcal{A}) \subseteq U(\mathcal{B})$. Since $\mathcal{B}[\| \cdot \|]$ is a C^*-algebra, it follows that $\|x\| = \|x^*x\|^{\frac{1}{2}} \leq 1$, i.e., $x \in U(\mathcal{B})$. This proves that $U(\mathcal{B})$ is the largest element of $\mathfrak{B}_A$, with respect to set inclusion. $\qquad \square$

From Theorem 7.4.5, we obtain the following results.

Corollary 7.4.6 Let $\mathcal{A}_\alpha[\tau_\alpha]$, $\alpha \in J$, be a family of GB*-algebras, where J is a directed index set. Then, the product $\mathcal{A} = \Pi_{\alpha \in J}\mathcal{A}_\alpha$, with respect to the product topology and algebraic operations and involution defined coordinatewise, is a GB*-algebra. Furthermore,

$$\mathcal{A}[B_0] = \left\{(x_\alpha)_{\alpha \in J} \in \mathcal{A} : x_\alpha \in \mathcal{A}_\alpha[(B_0)_\alpha] \text{ and } \sup_{\alpha \in J} \|x_\alpha\|_\alpha < \infty\right\}.$$

Proof Since $(B_0)_\alpha$ is τ_α-bounded for all $\alpha \in J$, it follows easily that $B_0 = \Pi_{\alpha \in J}((B_0)_\alpha)$ is bounded in $\mathcal{A}$ with respect to the product topology. On the other hand, it is clear that

$$\mathcal{B} = \left\{(x_\alpha)_{\alpha \in J} \in \mathcal{A} : x_\alpha \in \mathcal{A}_\alpha[(B_0)_\alpha] \text{ and } \sup_{\alpha \in J} \|x_\alpha\|_\alpha < \infty\right\}$$

is a C^*-algebra with respect to the norm

$$\|x\| = \sup_{\alpha} \|x_\alpha\|_\alpha, \ x = (x_\alpha)_\alpha \in \mathcal{A}.$$

Let e_α be the identity element of $\mathcal{A}_\alpha$, for all $\alpha \in J$, and let $x = (x_\alpha)_\alpha \in \mathcal{A}$. Clearly, $e = (e_\alpha)_\alpha$ is the identity element of $\mathcal{A}$. Since $\mathcal{A}_\alpha[\tau_\alpha]$ is a GB^*-algebra over the C^*-algebra $\mathcal{A}_\alpha[(B_0)_\alpha]$, for all $\alpha \in J$, it follows that $(e_\alpha + x_\alpha^* x_\alpha)^{-1} \in (B_0)_\alpha \subset \mathcal{A}_\alpha[(B_0)_\alpha]$, for all $\alpha \in J$. Consequently, $(e + x^*x)^{-1} \in \mathcal{B}$ for all $x \in \mathcal{A}$. Therefore, by Theorem 7.4.5, $\mathcal{A}[\tau]$ is a GB^*-algebra and $\mathcal{A}[B_0] = \mathcal{B}$. $\square$

By similar reasoning as in the proof of Corollary 7.4.6, one obtains the following

Corollary 7.4.7 (Kunze) *Let $\mathcal{A}_\alpha[\tau_\alpha]$, $\alpha \in J$, denote a family of GB^*-algebras, for which J is a directed index set. Let $h_{\alpha\beta} : \mathcal{A}_\beta[\tau_\beta] \to \mathcal{A}_\alpha[\tau_\alpha]$ denote a family of continuous $*$-homomorphisms with $h_{\alpha\beta}(e_\beta) = e_\alpha$, for $\alpha \leq \beta$, where e_α is the identity element of $\mathcal{A}_\alpha$, for each $\alpha \in J$, such that $(\mathcal{A}_\alpha[\tau_\alpha], h_{\alpha\beta})_{(\alpha \leq \beta \ in \ J)}$ is an inverse system of GB^*-algebras. Then, the inverse limit*

$$\mathcal{A}[\tau] = \lim_{\leftarrow} h_{\alpha\beta}\big(\mathcal{A}_\beta[\tau_\beta]\big)$$

is a GB^*-algebra with

$$\mathcal{A}[B_0] = \left\{ (x_\alpha)_{\alpha \in J} \in \mathcal{A} : x_\alpha \in \mathcal{A}_\alpha[(B_0)_\alpha] \ and \ \sup_{\alpha \in J} \|x_\alpha\|_\alpha < \infty \right\}.$$

Every pro-C^*-algebra $\mathcal{A}[\tau]$ with identity is an inverse limit of C^*-algebras. Therefore, by Corollary 7.4.7, we retrieve the result that every pro-C^*-algebra is a GB^*-algebra, as shown in Example 3.3.16(1).

Recall that the quotient algebra of a closed ideal in a C^*-algebra is also a C^*-algebra. The following result shows that this is true for all GB^*-algebras too.

Theorem 7.4.8 (Kunze) *Let $\mathcal{A}[\tau]$ be a GB^*-algebra and I a closed two-sided ideal in $\mathcal{A}[\tau]$. Then, I is a $*$-ideal and the quotient algebra $\mathcal{A}/I$, equipped with the quotient topology τ_q, is a GB^*-algebra with underlying C^*-algebra $\mathcal{A}[B_0]/(I \cap \mathcal{A}[B_0])$.*

Proof Observe that, since $\mathcal{A}[B_0]$ is norm complete and I is τ-closed in $\mathcal{A}$, $I_0 \equiv I \cap \mathcal{A}[B_0]$ is a norm closed two-sided ideal of the C^*-algebra $\mathcal{A}[B_0]$. Therefore, I_0 is a $*$-ideal of $\mathcal{A}[B_0]$, and $\mathcal{A}[B_0]/I_0$ is a C^*-algebra. Let $x \in I$. Then, by Remark 3.3.10(b), we obtain that $x(e + x^*x)^{-1} \in I$. On the other hand, arguing as in the proof of Theorem 7.3.1(i), we have that $x(e + x^*x)^{-1} \in \mathcal{A}[B_0]$ and finally $x(e + x^*x)^{-1} \in I_0$. Then,

$$(e + x^*x)^{-1}x^* = \big(x(e + x^*x)^{-1}\big)^* \in I_0.$$

Therefore, $x^* = (e + x^*x)(e + x^*x)^{-1}x^* \in I$. It follows that I is a $*$-ideal of $\mathcal{A}$ and that $\mathcal{A}/I$ is a topological $*$-algebra with respect to the quotient topology τ_q. Let ϕ denote the quotient $*$-homomorphism of $\mathcal{A}$ onto $\mathcal{A}/I$. Then, $\phi \lceil_{\mathcal{A}[B_0]}$ is a $*$-homomorphism of $\mathcal{A}[B_0]$ into $\mathcal{A}/I$, with the kernel being the norm closed two-sided ideal I_0 of $\mathcal{A}[B_0]$. Consequently, $\phi(\mathcal{A}[B_0])$, equipped with the quotient norm

$$\|\phi(x)\|_q = \inf_{y \in I_0} \|x + y\|, \ x \in \mathcal{A}[B_0],$$

is $*$-isomorphic to $\mathcal{A}[B_0]/I_0$, therefore a C^*-algebra.

Since $x(e + x^*x)^{-1} \in \mathcal{A}[B_0]$, for all $x \in \mathcal{A}$, as we noticed above, it follows that

$$(x + I)\big((e + I) + (x + I)^*(x + I)\big)^{-1} = x(e + x^*x)^{-1} + I \in \phi(\mathcal{A}[B_0])$$

for all $x \in \mathcal{A}$. Since the relative topology on $\phi(\mathcal{A}[B_0])$ of the topology τ_q is weaker than the $\|\cdot\|$-topology on $\phi(\mathcal{A}[B_0])$, and since the norm closed unit ball of $\phi(\mathcal{A}[B_0])$ is bounded with respect to the norm topology on $\phi(\mathcal{A}[B_0])$, it is immediate that the norm closed unit ball of $\phi(\mathcal{A}[B_0])$ is τ_q-bounded. Hence, by Theorem 7.4.5(ii), $(\mathcal{A}/I)[\tau_q]$ is a GB^*-algebra over the C^*-algebra $\mathcal{A}[B_0]/(I \cap \mathcal{A}[B_0])$. □

An interesting application of Theorem 7.4.8 is the following, which clearly extends Corollary 3.3.11(i).

Theorem 7.4.9 (Kunze) *If $\mathcal{A}[\tau]$ is a complete m^*-convex algebra with an identity element e, which is also a GB^*-algebra, then $\mathcal{A}[\tau]$ is a pro-C^*-algebra.*

Proof For each $p \in \Gamma$, the set $N_p = \{x \in A : p(x) = 0\}$ is a closed $*$-ideal of $\mathcal{A}$, therefore by Theorem 7.4.8, we obtain that $(\mathcal{A}/N_p)[\tau_q]$ is a GB^*-algebra (with respect to the quotient topology τ_q). On the other hand, $\mathcal{A}/N_p$ is a normed $*$-algebra with respect to the topology defined by the norm $\|x + N_p\|_p := p(x)$, $x \in \mathcal{A}$. Since $(\mathcal{A}/N_p)[\tau_q]$ is a GB^*-algebra, it follows that $(\mathcal{A}/N_p)[B_0]$ is a C^*-algebra with respect to $\|\cdot\|_{B_0}$, where B_0 is now the greatest member in $\mathfrak{B}^*_{(\mathcal{A}/N_p)[\tau_q]}$. But, $(\mathcal{A}/N_p)[\|\cdot\|_p]$ is also a $*$-normed algebra and moreover "$\mathcal{A}/N_p$ is Allan symmetric", being a GB^*-algebra. So, it follows from Proposition 7.4.1 that $\mathcal{A}/N_p = (\mathcal{A}/N_p)[B_0]$, as algebras. Let $\mathcal{B} = (\mathcal{A}/N_p)[B_0]$. Then $U(\mathcal{B}) = B_0$ is $\|\cdot\|_p$-bounded. This is due to the fact that B_0 is τ_q-bounded, hence $\|\cdot\|_p$-bounded, because $\|\cdot\|_p \prec \tau_q$. Therefore, by Theorem 7.4.5, $(\mathcal{A}/N_p)[\|\cdot\|_p]$ is a GB^*-algebra.

Hence, by Corollary 7.4.2, $\mathcal{A}/N_p$ is a C^*-algebra with respect to some norm $\|\cdot\|'_p$ equivalent to $\|\cdot\|_p$. It follows that the topology τ can be defined by a family $\Gamma' = \{p'\}$ of C^*-seminorms with $p'(x) := \|x + N_p\|'_p, x \in \mathcal{A}$, so that $\mathcal{A}$ becomes a pro-C^*-algebra, under the equivalent to τ topology τ' determined by the family Γ' of C^*-seminorms as before. □

Notes All results in this section are due to W. Kunze and can be found in [98].

7.5 A Vidav-Palmer Theorem for GB*-Algebras

The well known Vidav–Palmer theorem provides us with a concrete characterization
of C^*-algebras as those complex Banach algebras with identity, which are spanned,
as linear spaces, by elements with real numerical range (for the definition of the
numerical range of an element in a normed algebra see Remark 7.5.5). In this section
we present various types of Vidav–Palmer Theorem in the context of locally convex
$*$-algebras (among them being GB^*-algebras), all of them credited to A.W. Wood
[156].

Theorem 7.5.1 (Vidav–Palmer) *Let* $\mathcal{A}[\| \cdot \|]$ *be a Banach algebra with identity,
such that* $\mathcal{A} = \mathfrak{H} + i\mathfrak{H}$, *where* $\mathfrak{H}$ *denotes the set of elements with real numerical
range. Then* $\mathcal{A}[\| \cdot \|]$ *is isometrically isomorphic with a* C^*-*algebra.*

In [156], A.W. Wood explores a possible generalization of the aforementioned
result in the realm of GB^*-algebras. In order to do so, he defines a natural
generalization of the notion of the numerical range, in the setting of a locally convex
algebra (see Definition 7.5.2).

Recall that given a locally convex space $E[\tau]$, the symbol E' denotes its
topological dual and $\Gamma = \{p\}$ a family of seminorms defining the topology τ. In
the sequel, Γ' stands for the family of seminorms

$$q_B(f) = \sup \{|f(x)| : x \in B\}, \ B \in \mathbb{B}, \ f \in E',$$

where $\mathbb{B}$ denotes the collection of τ-bounded subsets of E, of the form

$$\{x \in E : p(x) \leq M_p, \ p \in \Gamma, \ M_p > 0\}.$$

The family $\{q_B\}_{B \in \mathbb{B}}$ defines the topology $\beta(E', E)$ on E', i.e., *the topology of
uniform convergence on weakly bounded subsets of E*. The latter implication
follows from the fact that the τ-bounded and the weakly bounded subsets of E
coincide (see [128, p. 67, Theorem 1]).

▶ In the rest of this section, *we shall use the notation* (E, Γ), $(\mathcal{A}, \Gamma)$, for a
given locally convex space $E[\tau]$, respectively a given locally convex algebra
$\mathcal{A}[\tau]$, whose topology τ is defined by a family of seminorms $\Gamma = \{p\}$.

Definition 7.5.2 Let (E, Γ) be a locally convex space. Then, for every $p \in \Gamma$, let

$$\Pi_p := \{(x, f) \in E \times E' : p(x) = f(x) = 1, \ |f(y)| \leq p(y), \ \forall\, y \in E\}.$$

For $T \in L(E)$, where $L(E)$ *is the set of all linear continuous operators on the space* E, let

$$W_p^1(T) := \left\{ f(Tx) : (x, f) \in \Pi_p \right\} \text{ and } W^1(T) := \bigcup_{p \in \Gamma} W_p^1(T).$$

Likewise, if E'' *is the double topological dual of* (E, Γ), *let* $q \equiv q_B \in \Gamma'$ *and*

$$\Pi_q := \left\{ (f, x'') \in E' \times E'' : x''(f) = 1 = q(f), \ |x''(g)| \leq q(g), \ \forall g \in E' \right\}$$

and

$$W_q^1(T') := \left\{ x''(T'f) : (f, x'') \in \Pi_q \right\}, \quad W^2(T) := \bigcup_{q \in \Gamma'} W_q^1(T'),$$

where T' *denotes the transpose of* T. *Then,*

$$W_\Gamma(T) := W^1(T) \bigcup W^2(T) \text{ is called the } numerical\ range \text{ of } T.$$

For simplicity we shall use the notation $W(T)$.

Remark 7.5.3

(1) The previous definition is a generalization of the respective definition of the spatial numerical range of an operator $T \in L(E)$ for a normed space E. We recall that in the latter case, the spatial numerical range of $T \in L(E)$ is denoted by $V(T)$ and is the set

$$V(T) := \left\{ f(Tx) : (x, f) \in \Pi \right\}, \quad \text{where}$$
$$\Pi := \{ (x, f) \in E \times E' : \|x\| = |f(x)| = \|f\| = 1 \}.$$

Clearly, $V(T)$ coincides with $W^1(T)$ of Definition 7.5.2. In addition, when E is a Banach space, by [31, Corollary 17.3] we have that $V(T) \subset V(T') \subset \overline{V(T)}$. Therefore, since $V(T')$ coincides with $W^2(T)$ of Definition 7.5.2, we have that for a Banach space E and $T \in L(E)$, the sets $W^1(T)$ and $W^2(T)$ have the same closure.

(2) It is apparent from Definition 7.5.2 that the numerical range of $T \in L(E)$ depends on the family Γ of seminorms defining the topology of the locally convex space E.

Based on Definition 7.5.2, the following definition describes the numerical range of an element in an abstract locally convex algebra.

Definition 7.5.4 Let $(\mathcal{A}, \Gamma)$ be a locally convex algebra and let $a \in \mathcal{A}$. The *numerical range* of a, denoted by $W_\Gamma(a)$ (or simply by $W(a)$), is the set $W_\Gamma(T_a)$

(or simply the set $W(T_a)$ as noticed above), where $T_a \in L(\mathcal{A})$ is the linear operator $T_a x := ax$, $x \in \mathcal{A}$. An element $h \in \mathcal{A}$, such that $W_\Gamma(h) \subset \mathbb{R}$, is said to be *hermitian*. The set of all hermitian elements of $\mathcal{A}$ is denoted by $\mathfrak{H}_\Gamma$ (or by $\mathfrak{H}$ if no confusion arises).

Remark 7.5.5 In case $\mathcal{A}$ is a Banach algebra with identity, the classical definition of the numerical range of an element $a \in \mathcal{A}$ is that given by the set

$$V(\mathcal{A}, a) := \{ f(ax) : x \in S(\mathcal{A}), \ f \in D(\mathcal{A}, x) \},$$

where $S(\mathcal{A}) := \{ x \in \mathcal{A} : \|x\| = 1 \}$ and

$$D(\mathcal{A}, x) = \{ f \in \mathcal{A}' : f(x) = 1 = \|f\| \}$$

(more generally, this is the definition given for the numerical range of an element of a normed algebra with identity; see [30, Definition 2.1]). We note that $V(\mathcal{A}, a)$ coincides with $W^1(T_a)$ of Definition 7.5.4. So,

$$V(\mathcal{A}, a) = W^1(T_a) \subset W(a).$$

On the other hand, for $\lambda \in W^1(T_a')$, we have that $\lambda \in \overline{W^1(T_a')} = \overline{W^1(T_a)}$, where the equality is based on [31, Corollary 17.3]. Thus, since $W^1(T_a)$ is closed (see [30, §2, Theorem 3]), we have that $\lambda \in W^1(T_a)$, i.e., $W^1(T_a') \subset V(\mathcal{A}, a)$. Therefore, we conclude that in the case of a Banach algebra, the definition of the numerical range of an element of the algebra, as it is given in Definition 7.5.4, coincides with the classical definition.

For a hypocontinuous locally convex algebra, the set $\mathfrak{H}$ of hermitian elements is closed, as it is implied by Proposition 7.5.6, below. We recall that *a locally convex algebra $\mathcal{A}[\tau]$ is hypocontinuous*, if for every bounded subset B of $\mathcal{A}[\tau]$ and each 0-neighbourhood U there is a 0-neighbourhood V, such that $BV \subset U$ and $VB \subset U$. Note that *in a hypocontinuous topological algebra the product of two bounded subsets is bounded*. Indeed, let $\mathcal{A}[\tau]$ be a hypocontinuous locally convex algebra and let B_1, B_2 be two bounded subsets of $\mathcal{A}[\tau]$. Let U be a 0-neighbourhood in $\mathcal{A}[\tau]$. Then, there is a 0-neighbourhood V and a positive number λ, such that $B_1 \subset \lambda V$ and $V B_2 \subset U$, $B_2 V \subset U$. Hence, $B_1 B_2 \subset \lambda V B_2 \subset \lambda U$. Therefore, $B_1 B_2$ is bounded.

Proposition 7.5.6 *Let $(\mathcal{A}, \Gamma)$ be a hypocontinuous locally convex algebra and let $(a_\lambda)_{\lambda \in \Lambda}$ be a net in $\mathcal{A}$, such that $a_\lambda \to a \in \mathcal{A}$. If K is a closed subset of $\mathbb{C}$, such that $W(a_\lambda) \subset K$, for all $\lambda \in \Lambda$, then $W(a) \subset K$.*

Proof Let $p \in \Gamma$ and $(x, f) \in \Pi_p$. Then $f(a_\lambda x) \in W(a_\lambda)$ and so $f(a_\lambda x) \in K$, for all $\lambda \in \Lambda$. Since the multiplication in $\mathcal{A}$ is separately continuous and K is closed, we have that $f(ax) \in K$. Thus, $W^1(T_a) \subset K$.

Next, we are going to show that $W^2(T_a) \subset K$. Let $(f, x'') \in \Pi_q$, for some $q \equiv q_B \in \Gamma'$. We show that $T_{a_\lambda}' f \to T_a' f$ with respect to $\beta(\mathcal{A}', \mathcal{A})$: let $\varepsilon > 0$

and B a $\sigma(\mathcal{A}, \mathcal{A}')$-bounded subset of $\mathcal{A}$. By the comments before Definition 7.5.2, it suffices to take $B \equiv \{x \in \mathcal{A} : p(x) \leq M_p\}$, $p \in \Gamma$, where $\{M_p : p \in \Gamma\}$ is any family of positive numbers. Clearly the set B is then a bounded subset of the locally convex algebra $(\mathcal{A}, \Gamma)$. By continuity of $f \in \mathcal{A}'$, there is a seminorm $p \in \Gamma$, such that $|f(x)| \leq p(x)$, for all $x \in \mathcal{A}$. If we consider the 0-neighbourhood $U = \{y \in \mathcal{A} : p(y) < \frac{\varepsilon}{2}\}$ of $\mathcal{A}$, due to hypocontinuity of $\mathcal{A}$ there is a 0-neighbourhood V, such that $BV \subset U$ and $VB \subset U$. Since $a_\lambda \to a$, there is $\lambda_0 \in \Lambda$, with $a_\lambda - a$ in V, for all $\lambda \geq \lambda_0$. Therefore,

$$|f((a_\lambda - a)x)| \leq p((a_\lambda - a)x) < \frac{\varepsilon}{2}, \ \forall \, \lambda \geq \lambda_0 \text{ and } x \in B.$$

Observe that, there is some $x_0 \in B$, such that

$$\sup \{|f((a_\lambda - a)x)| : x \in B\} < |f((a_\lambda - a)x_0)| + \frac{\varepsilon}{2}.$$

Hence, $\sup \{|f((a_\lambda - a)x)| : x \in B\} < \varepsilon$, for all $\lambda \geq \lambda_0$. So,

$$\sup \{|(T'_{a_\lambda} - T'_a)f(x)| : x \in B\} \to 0, \ \text{ i.e., } \ T'_{a_\lambda} f \to T'_a f, \ \text{ with respect to } \ \beta(\mathcal{A}', \mathcal{A}).$$

Consequently, $x''(T'_{a_\lambda} f) \to x''(T'_a f)$, from which it follows that $x''(T'_\alpha f) \in K$. Hence, we derive the desired inclusion $W^2(T_a) \subset K$ and that

$$W(a) = W^1(T_a) \cup W^2(T_a) \subset K. \qquad \square$$

Definitions 7.5.2 and 7.5.4 suggest that an element of a locally convex algebra $\mathcal{A}[\tau]$ can interchangeably be viewed, in terms of its numerical range, as a continuous operator in the algebra $L(\mathcal{A})$. The next definition distinguishes a particular bounded subset of $L(E)$, where E is a general locally convex space, which holds an important role in the course of obtaining a generalized Vidav–Palmer theorem for locally convex algebras (in this respect, see, for instance, Theorem 7.5.42).

Definition 7.5.7 Let (E, Γ) be a locally convex space. The set of Γ-contractions, denoted by C_Γ, is the set

$$C_\Gamma = \{T \in L(E) : p(Tx) \leq p(x), \ x \in E, \ p \in \Gamma\}.$$

It is clear that C_Γ is an absolutely convex, bounded, closed subset of $L(E)$ with respect to the topology of uniform convergence on bounded subsets of E. Moreover it is straightforward that $C_\Gamma^2 \subset C_\Gamma$ and $Id_E \in C_\Gamma$, where Id_E denotes the identity map on E.

Then, $\mathcal{A}[C_\Gamma]$, the subalgebra of $L(E)$ generated by C_Γ, equals to $\{\lambda T : \lambda \in \mathbb{C}, \ T \in C_\Gamma\}$ and it is a normed algebra, with respect to the norm

$$\|T\|_\Gamma = \inf \{\mu > 0 : T \in \mu C_\Gamma\}, \ T \in \mathcal{A}[C_\Gamma].$$

We note that

$$\|T\|_\Gamma = \sup\{p(Tx) : p \in \Gamma,\ x \in E,\ p(x) \leq 1\}.$$

Indeed, if $T \in \lambda C_\Gamma$, for some $\lambda > 0$, then $T = \lambda Q,\ Q \in C_\Gamma$. Hence,

$$\sup\{p(Tx) : p \in \Gamma,\ x \in E,\ p(x) \leq 1\} \leq$$
$$\lambda \sup\{p(Qx) : p \in \Gamma,\ x \in E,\ p(x) \leq 1\} \leq \lambda,$$

thus

$$\sup\{p(Tx) : p \in \Gamma,\ x \in E,\ p(x) \leq 1\} \leq \inf\{\lambda > 0 : T \in \lambda C_\Gamma\}.$$

On the other hand, if $\sup\{p(Tx) : p \in \Gamma,\ x \in E,\ p(x) \leq 1\} \equiv s$, then for every $\varepsilon > 0$ we have that

$$p\left(\left(\frac{1}{s+\varepsilon}T\right)x\right) = \frac{1}{s+\varepsilon}p(Tx) \leq p(x).$$

The previous inequality is certainly true for $x \in E$ with $p(x) \neq 0$. For $x \in E$ with $p(x) = 0$, if $T = \mu Q$, where $\mu \in \mathbb{C},\ Q \in C_\Gamma$, we have that

$$p(Tx) = |\mu|p(Qx) \leq |\mu|p(x) = 0,$$

thus the aforementioned inequality is still valid. So, by $p\left(\left(\frac{1}{s+\varepsilon}T\right)x\right) \leq p(x)$, we have that $\frac{1}{s+\varepsilon}T \in C_\Gamma$, thus $\inf\{\lambda > 0 : T \in \lambda C_\Gamma\} \leq s+\varepsilon$. Since $\varepsilon > 0$ is arbitrary we obtain $\inf\{\lambda > 0 : T \in \lambda C_\Gamma\} \leq s$.

Furthermore, by considerations similar to those of the paragraph preceding Definition 2.2.3, the topology of uniform convergence on bounded subsets of E, which $\mathcal{A}[C_\Gamma]$ carries as a subalgebra of $L(E)$, is weaker than the norm-topology $\|\cdot\|_\Gamma$.

If (E, Γ) is a locally convex space and $T \in \mathcal{A}[C_\Gamma]$, the numerical range of T regarded as an element of the normed algebra $\mathcal{A}[C_\Gamma]$ with identity, is denoted by $V_\Gamma(\mathcal{A}[C_\Gamma], T)$ (or $V(\mathcal{A}[C_\Gamma], T)$). By [30, Lemma 2.2], the numerical range $V_\Gamma(\mathcal{A}[C_\Gamma], T)$ of an element $T \in \mathcal{A}[C_\Gamma]$, coincides with the set $\{f(T) : f \in D(\mathcal{A}[C_\Gamma], Id)\}$, where $D(\mathcal{A}[C_\Gamma], Id)$ stands for the set of all continuous linear functionals $f : \mathcal{A}[C_\Gamma] \to \mathbb{C}$, such that $\|f\| = 1 = f(Id)$ and Id denotes the identity map on $\mathcal{A}[C_\Gamma]$. For an element $T \in \mathcal{A}[C_\Gamma]$, the relation between the sets $V(\mathcal{A}[C_\Gamma], T)$ and $W(T)$ is given by the following result.

Proposition 7.5.8 *Let (E, Γ) be a locally convex space and let $T \in \mathcal{A}[C_\Gamma]$. Then,*

$$W(T) \subset V(\mathcal{A}[C_\Gamma], T).$$

Proof Let $\mu \in W^1(T)$. Then, $\mu \in W_p^1(T)$, for some $p \in \Gamma$, i.e., $\mu = f(Tx)$, for some $(x, f) \in \Pi_p$. Consider the map

$$g : \mathcal{A}[C_\Gamma] \to \mathbb{C} : S \mapsto g(S) := f(Sx), \ (x, f) \in \Pi_p.$$

For every $S \in \mathcal{A}[C_\Gamma]$, we have that

$$|g(S)| = |f(Sx)| \leq p(Sx) \leq \|S\|_\Gamma \, p(x) = \|S\|_\Gamma.$$

So, $\|g\| \leq 1$ and since $g(Id) = 1 = f(x)$, we have that $g \in D(\mathcal{A}[C_\Gamma], Id)$. Therefore, $f(Tx) = g(T) \in V(\mathcal{A}[C_\Gamma], T)$, which proves the inclusion of $W^1(T)$ in $V(\mathcal{A}[C_\Gamma], T)$.

In order to obtain the inclusion $W^2(T) \subset V(\mathcal{A}[C_\Gamma], T)$, we first note that $\|T\|_\Gamma = \|T'\|_{\Gamma'}$, for every $T \in \mathcal{A}[C_\Gamma]$. The latter relation holds true, since for $F \in C_\Gamma$, for a bounded set $B = \{x \in E : p(x) \leq M_p, \ p \in \Gamma\}$, where $M_p > 0$, $p \in \Gamma$, and for every $f \in E'$, we have that

$$q_B(F'f) = \sup\{|(F'f)(x)| : x \in B\} = \sup\{|f(Fx)| : x \in B\}$$
$$\leq \sup\{|f(x)| : x \in B\} = q_B(f),$$

where the inequality is based on the inclusion $F(B) \subset B$, as can be easily seen by the fact that $F \in C_\Gamma$. Hence, $F' \in C_{\Gamma'}$. Therefore, for $T \in \mathcal{A}[C_\Gamma]$, $\frac{T}{\|T\|_\Gamma} \in C_\Gamma$ and thus $\frac{T'}{\|T\|_\Gamma} \in C_{\Gamma'}$, so $\|T'\|_{\Gamma'} \leq \|T\|_\Gamma$. By symmetrical arguments, $\|T\|_\Gamma \leq \|T'\|_{\Gamma'}$.

Let $\mu \in W^2(T)$. Then, $\mu \in W_q^1(T')$ for some $q \in \Gamma'$. Thus, $\mu = x''(T'f)$, for some $(f, x'') \in \Pi_q$. Consider the map $h : \mathcal{A}[C_\Gamma] \to \mathbb{C} : S \mapsto h(S) := x''(S'f)$. Then,

$$|h(S)| = |x''(S'f)| \leq q(S'f) \leq \|S'\|_{\Gamma'} q(f) = \|S\|_\Gamma q(f).$$

So, $\|h\| = 1 = h(Id)$, hence, $\mu = x''(T'f) = h(T) \in V_\Gamma(\mathcal{A}[C_\Gamma], T)$.

Therefore, we conclude that $W(T) = W^1(T) \cup W^2(T) \subset V(\mathcal{A}[C_\Gamma], T)$.   □

Proposition 7.5.9 *Let $E[\tau]$ be a locally convex space. If B is a subset of $L(E)$, such that $B^2 \subset B$, then B is equicontinuous, if and only if, there is a defining family of seminorms $\widetilde{\Gamma}$ for τ, equivalent to the given one Γ, such that $B \subset C_{\widetilde{\Gamma}}$.*

Proof $\Leftarrow$ Let $\widetilde{\Gamma}$ be a defining family of seminorms for τ, equivalent to the respective given family Γ (for τ), such that $B \subset C_{\widetilde{\Gamma}}$. Then, it is straightforward that $T(V) \subset V$, for every 0-neighbourhood V, with respect to $\widetilde{\Gamma}$ and every $T \in B$. Hence, B is equicontinuous, with respect to $\widetilde{\Gamma}$, hence with respect to Γ, too.

$\Rightarrow$ Let B be equicontinuous, with respect to $\Gamma = \{p\}$. Consider,

$$\widetilde{p}(x) := \sup\{p(Tx) : T \in B\}, \ x \in E.$$

Then, the seminorm $\widetilde{p}$ is well-defined and this is a straightforward consequence of the equicontinuity of B, with respect to Γ. The family of seminorms $\widetilde{\Gamma} = \{\widetilde{p} : p \in \Gamma\}$ defines the same topology on E as the topology defined on E by the family Γ. For this, note that on the one hand, we can assume without loss of generality, that $Id_E \in B$. Then, if $(x_\lambda)_{\lambda \in \Lambda}$ is a net in E, such that $x_\lambda \underset{\{\widetilde{p}\}}{\to} 0$, it follows immediately that $x_\lambda \underset{\{p\}}{\to} 0$. On the other hand, if $x_\lambda \underset{\{p\}}{\to} 0$, then for every $T \in B$, $Tx_\lambda \underset{\{p\}}{\to} 0$. So, for $\varepsilon > 0$ there exist $\lambda_0 \in \Lambda$ and $T_0 \in B$, such that

$$\sup\left\{p(Tx_\lambda) : T \in B\right\} < \frac{\varepsilon}{2} + p(T_0 x_\lambda) < \varepsilon, \ \forall \lambda \geq \lambda_0; \ \text{hence,} \ x_\lambda \underset{\{\widetilde{p}\}}{\to} 0.$$

Moreover, for $T \in B$ we have that

$$\widetilde{p}(Tx) = \sup\left\{p\big(S(Tx)\big) : S \in B\right\} \leq \sup\left\{p(Rx) : R \in B\right\} = \widetilde{p}(x),$$

for every $x \in E$, $p \in \Gamma$, where the inequality is due to the property $B^2 \subset B$. Therefore, $T \in C_{\widetilde{\Gamma}}$, thus $B \subset C_{\widetilde{\Gamma}}$.   □

Lemma 7.5.10 *Let $\mathcal{A}[\tau]$ be a hypocontinuous locally convex algebra and B a bounded subset of $\mathcal{A}$, such that $B^2 \subset B$. Then, there is a defining family of seminorms $\widetilde{\Gamma}$ for τ, equivalent to the given one Γ, such that $\{T_a : a \in B\} \subset C_{\widetilde{\Gamma}}$ (for T_a, $a \in \mathcal{A}$, see Definition 7.5.4).*

Proof Let $B' = \{T_a : a \in B\}$. For every $b, c \in B$, $T_b T_c = T_{bc}$, where $bc \in B^2 \subset B$. So, $(B')^2 \subset B'$. Moreover, by hypocontinuity of $\mathcal{A}[\tau]$ and the assumption for B being bounded, we have that for any given 0-neighbourhood V in $\mathcal{A}[\tau]$, there is a 0-neighbourhood U in $\mathcal{A}[\tau]$, such that $BU \subset V$. The previous inclusion implies that B' is equicontinuous. Then, by Proposition 7.5.9, there is a defining family of seminorms $\widetilde{\Gamma}$ for τ, equivalent to the given one Γ, such that $B' \subset C_{\widetilde{\Gamma}}$.   □

Remark 7.5.11 Recall that the map $\mathcal{A} \to L(\mathcal{A}) : a \mapsto T_a$, with $T_a(x) := ax$, for all x in $\mathcal{A}$, is injective. So, *in what follows, for simplicity of notation, we find it convenient to identify an element of a locally convex algebra $\mathcal{A}[\tau]$ with its left regular representation $a \in \mathcal{A} \mapsto T_a \in L(\mathcal{A})$ and write a' instead of T_a'.* Moreover, we *choose to write $\|a\|_\Gamma$ in place of $\|T_a\|_\Gamma$, for $a \in \mathcal{A}$ with $T_a \in \mathcal{A}[C_\Gamma]$.* Observe that *by this natural identification of a with T_a there are two relevant spectra of $a \in \mathcal{A}$.* Namely,

$$\sigma_{\mathcal{A}}(a) = \left\{\lambda \in \mathbb{C} : \lambda e - a \ \text{has no inverse in} \ \mathcal{A}_0\right\} \cup \left\{\infty, \ \text{if and only if,} \ a \notin \mathcal{A}_0\right\}$$

and $\sigma_{C_\Gamma}(a) = \left\{\lambda \in \mathbb{C} : \lambda Id_{\mathcal{A}} - T_a \ \text{has no inverse in} \ \mathcal{A}[C_\Gamma]\right\} \cup$

$$\left\{\infty, \ \text{if and only if,} \ T_a \notin \mathcal{A}[C_\Gamma]\right\}.$$

The following lemma shows that these two sets coincide for a hypocontinuous locally convex algebra. Hence, *for a hypocontinuous locally convex algebra, the* *aforementioned identifications of a with T_a would not lead to any ambiguity with respect to the spectrum of a.*

Lemma 7.5.12 *Let $(\mathcal{A}, \Gamma)$ be a hypocontinuous locally convex algebra with identity e and a $\in \mathcal{A}$. Then, $\sigma_{\mathcal{A}}(a) = \sigma_{C_\Gamma}(a)$.*

Proof We begin by establishing that $\infty \in \sigma_{C_\Gamma}(a)$ if and only if $\infty \in \sigma_{\mathcal{A}}(a)$. Let $\infty \in \sigma_{\mathcal{A}}(a)$. Then, $a \notin \mathcal{A}_0$. With the aim to reach a contradiction suppose that $\infty \notin \sigma_{C_\Gamma}(a)$. Hence, $T_a \in \mathcal{A}[C_\Gamma]$. So, there exists $k > 0$, such that $p(T_a x) \leq kp(x)$, for every $x \in \mathcal{A}, p \in \Gamma$. Therefore, by induction we have that $p((\frac{1}{k}a)^n) \leq 1$, i.e., $a \in \mathcal{A}_0$, a contradiction.

On the other hand let $\infty \in \sigma_{C_\Gamma}(a)$ and suppose that $\infty \notin \sigma_{\mathcal{A}}(a)$. Then, there is $\lambda > 0$ such that the set $B = \{(\lambda a)^n : n \in \mathbb{N}\}$ is bounded. Consider $F = B \cup \{e\}$. Clearly, F is bounded and $F^2 \subset F$. By Lemma 7.5.10 there is a defining family of seminorms for the topology of $\mathcal{A}$, say $\widetilde{\Gamma}$, such that $\{T_b : b \in F\} \subset C_{\widetilde{\Gamma}}$. We recall that by the proof of Proposition 7.5.9 the family $\widetilde{\Gamma}$ is given by the seminorms

$$\widetilde{p}(x) = \sup\{p(T_b x) : b \in F\}, \ p \in \Gamma.$$

Hence, by the very definitions, the form of $\widetilde{p}$ and the fact that $T_{\lambda a} \in C_{\widetilde{\Gamma}}$, we deduce $\|T_{\lambda a}\|_{\Gamma} \leq 1$, so $T_a \in \mathcal{A}[C_\Gamma]$. This last fact contradicts our assumption that $\infty \in \sigma_{C_\Gamma}(a)$ in the beginning of this paragraph.

Let now $\lambda \in \sigma_{\mathcal{A}}(a) \cap \mathbb{C}$. We want to show that $\lambda \in \sigma_{C_\Gamma}(a)$. Supposing that $\lambda \notin \sigma_{C_\Gamma}(a)$ we have

$$(\lambda Id - T_a)^{-1} = T^{-1}_{\lambda e - a} \in \mathcal{A}[C_\Gamma], \text{ so that } T_{\lambda e - a} T^{-1}_{\lambda e - a} = T^{-1}_{\lambda e - a} T_{\lambda e - a} = Id.$$

By this last equality we conclude that $(\lambda e - a)^{-1}$ exists and that $T_{(\lambda e - a)^{-1}} = T^{-1}_{\lambda e - a} \in \mathcal{A}[C_\Gamma]$. So, $\infty \notin \sigma_{C_\Gamma}((\lambda e - a)^{-1})$. Therefore, $\infty \notin \sigma_{\mathcal{A}}((\lambda e - a)^{-1})$, i.e., $(\lambda e - a)^{-1} \in \mathcal{A}_0$, a contradiction.

Finally, by similar to the considerations of the previous paragraph one can show that

$$\sigma_{C_\Gamma}(a) \cap \mathbb{C} \subset \sigma_{\mathcal{A}}(a), \ a \in \mathcal{A}.$$

$\square$

It should be noted that the spectrum $\sigma_{C_\Gamma}(a)$, a in $\mathcal{A}$, as this is defined in Remark 7.5.11, is a slight generalization of the spectrum denoted by $\sigma_{\mathcal{A}[C_\Gamma]}(a)$ in Proposition 2.3.12.

▶ *For simplicity of notation, in what follows throughout the section, we write $\sigma_\Gamma(a)$ for $\sigma_{C_\Gamma}(a)$. Theorem 7.5.14 reveals a relation between the spectrum $\sigma_\Gamma(a)$ and the numerical range $W(a)$ of an element a of a complete*

(continued)

locally convex algebra. The following result can be seen as a step towards
Theorem 7.5.14.

Theorem 7.5.13 *Let* (E, Γ) *be a complete locally convex space and* $T \in L(E)$. *If* $\lambda \in \mathbb{C} \setminus \overline{W(T)}$, *then* $(\lambda Id_E - T)^{-1}$ *exists and*

$$\left\| (\lambda Id_E - T)^{-1} \right\|_\Gamma \leq \frac{1}{k}, \quad \text{with } k := \inf \left\{ |\lambda - \mu| : \mu \in W(T) \right\}.$$

Proof Let $(x, f) \in \Pi_p$, $p \in \Gamma$. Then (see Definition 7.5.2),

$$p\big((\lambda Id_E - T)x\big) \geq \big|f(\lambda Id_E - T)x\big| = \big|\lambda f(x) - f(Tx)\big| = \big|\lambda - f(Tx)\big| \geq k.$$

Therefore, we have that

$$p\big((\lambda Id_E - T)x\big) \geq kp(x), \text{ for all } p \in \Gamma, \ x \in E. \tag{7.5.4}$$

Moreover, for $(h, x'') \in \Pi_q$, $q \in \Gamma'$ (ibid.), we have that

$$q\big((\lambda Id_E - T)'h\big) \geq \big|x''\big((\lambda Id_E - T)'h\big)\big| = \big|x''(\lambda Id_{E'} - T')h\big|$$
$$= \big|\lambda x''(h) - x''(T'h)\big| = \big|\lambda - x''(T'h)\big| \geq k.$$

Therefore,

$$q\big((\lambda Id_E - T)'h\big) \geq kq(h), \text{ for all } q \in \Gamma', \ h \in E'. \tag{7.5.5}$$

By (7.5.4) we have that $\lambda Id_E - T$ is injective and that $(\lambda Id_E - T)E$ is closed in E. Furthermore, by (7.5.5), $(\lambda Id_E - T)E$ is dense in E: indeed if $\overline{(\lambda Id_E - T)E} \neq E$, then by the Hahn–Banach theorem, there is an $f \in E'$, such that $f \neq 0$ and $f \restriction_{(\lambda Id_E - T)E} = 0$. Therefore, $(\lambda Id_E - T)'f(E) = 0$ with $f \neq 0$, which contradicts (7.5.5).

Hence, $\lambda Id_E - T$ is onto E. Hence, $(\lambda Id_E - T)^{-1}$ exists and applying (7.5.4) for the element $(\lambda Id_E - T)^{-1}x$, in the place of x we conclude that

$$p\big((\lambda Id_E - T)^{-1}x\big) \leq \frac{1}{k}p(x), \ \forall \ p \in \Gamma, \ x \in E.$$

Consequently, $(\lambda Id_E - T)^{-1} \in \mathcal{A}[C_\Gamma]$ and $\left\| (\lambda Id_E - T)^{-1} \right\|_\Gamma \leq \frac{1}{k}$. $\qquad \square$

Theorem 7.5.14 *Let* $(\mathcal{A}, \Gamma)$ *be a complete locally convex algebra and let* $\overline{W(a)}^*$ *denote the closure of* $W(a)$ *in* $\mathbb{C}^*$. *Then,* $\sigma_\Gamma(a) \subset \overline{W(a)}^*$. *Moreover, if* $\lambda \in \mathbb{C}$ *and* $k := \inf \left\{ |\lambda - \mu| : \mu \in W(a) \right\} > 0$, *then* $\|(\lambda e - a)^{-1}\|_\Gamma \leq \frac{1}{k}$.

Proof For $\lambda \in \mathbb{C}$, the result is a direct consequence of Theorem 7.5.13 and the comments just before it. Hence, what is left to be shown is that for $\infty \in \sigma_\Gamma(a)$; we have $\infty \in \overline{W(a)}^*$. Equivalently, it suffices to show that if $W(a)$ is bounded, then $a \in \mathcal{A}[C_\Gamma]$. Towards this direction, let $M > 0$, such that $|\mu| \leq M$, for all $\mu \in W(a)$. Then, for every $|\lambda| > M$, one has $\lambda \notin \overline{W(a)}$. So, by Theorem 7.5.13, $(\lambda e - a)^{-1}$ exists and

$$\left\| (\lambda e - a)^{-1} \right\|_\Gamma \leq \left(\inf \left\{ |\lambda - z| : z \in W(a) \right\} \right)^{-1} \leq \frac{1}{|\lambda| - M}.$$

Then, following the proof of Theorem 2.3.7, we get that the map $\lambda \mapsto (\lambda e - a)^{-1}$ is analytic at ∞. Therefore, by Theorem 2.3.7(1), $\infty \notin \sigma_\Gamma(a)$, hence $a \in \mathcal{A}[C_\Gamma]$. $\square$

For a pseudo-complete locally convex algebra $\mathcal{A}$, the algebra $\mathcal{A}[C_\Gamma]$ is, by the definition of pseudo-completeness, a Banach algebra. Therefore, for $a \in \mathcal{A}[C_\Gamma]$ we can define $\exp(a) \in \mathcal{A}[C_\Gamma]$ by means of the usual power series, i.e., $\exp(a) = \sum_{n=0}^{\infty} \frac{a^n}{n!}$, where the series converges with respect to $\| \cdot \|_\Gamma$. The use of the exponential function will give us, at a first stage, a closer connection between $W(a)$ and $V(\mathcal{A}[C_\Gamma], a)$ (see Proposition 7.5.19) than that we have already obtained at Proposition 7.5.8. First we need the following definition and a preparatory result.

Definition 7.5.15 Let $(\mathcal{A}, \Gamma)$ be a locally convex algebra.

(i) An element $a \in \mathcal{A}$ is said to be *dissipative* if $\mathrm{Re}\lambda \leq 0$, *for every* $\lambda \in W(a)$.
(ii) We say that the resolvent of $a \in \mathcal{A}$ *has first order decay in the right half-plane*, if for $\lambda \in \mathbb{C}$ and $\mathrm{Re}\lambda > 0$, then the element $(\lambda e - a)^{-1}$ exists and $\left\| (\lambda e - a)^{-1} \right\|_\Gamma \leq \frac{1}{\mathrm{Re}\lambda}$.

Remark 7.5.16 We observe that if a is a dissipative element of a complete locally convex algebra $(\mathcal{A}, \Gamma)$, then the resolvent of a has first order decay in the right half-plane. Indeed, if $\lambda \in \mathbb{C}$ and $\mathrm{Re}\lambda > 0$, then $\lambda \notin \overline{W(a)}^*$. Hence, by Theorem 7.5.14, we have that

$$\left\| (\lambda e - a)^{-1} \right\|_\Gamma \leq \frac{1}{\inf\{|\lambda - \mu| : \mu \in W(a)\}} \leq \frac{1}{\mathrm{Re}\lambda},$$

where the second inequality follows from the fact that $\mathrm{Re}\lambda \leq \mathrm{Re}\lambda - \mathrm{Re}\mu$, for every $\mu \in W(a)$.

As we shall see in Proposition 7.5.21, *the converse of the implication of the previous paragraph holds for a pseudo-complete and hypocontinuous locally convex algebra*.

Proposition 7.5.17 *Let $(\mathcal{A}, \Gamma)$ be a pseudo-complete locally convex algebra and $a \in \mathcal{A}[C_\Gamma]$, such that its resolvent has first order decay in the right half-plane. Then, $\| \exp(a) \|_\Gamma \leq 1$.*

Proof Consider $a_n = n^2 (ne - a)^{-1} - ne$, for $n \in \mathbb{N}$. Note that $(ne - a)^{-1}$ exists due to the assumption of the resolvent of a having first order decay in the right half-plane and moreover $\|ne - a\|_\Gamma \leq \frac{1}{n}$. Then, we have that

$$
\begin{aligned}
\|a_n - a\|_\Gamma &= \|n^2(ne - a)^{-1} - ne - a\|_\Gamma \\
&= \|(ne - a)^{-1}(n^2e - (ne - a)(ne + a))\|_\Gamma \\
&= \|a^2(ne - a)^{-1}\|_\Gamma \\
&\leq \frac{1}{n}\|a^2\|_\Gamma \to 0.
\end{aligned}
$$

Hence, $a_n \to a$ with respect to $\| \cdot \|_\Gamma$. Then,

$$
\begin{aligned}
\|\exp(a_n - a) - e\|_\Gamma &= \lim_{n \to \infty} \left\| \sum_{k=1}^{n} \frac{(a_n - a)^k}{k!} \right\|_\Gamma \\
&\leq \lim_{n \to \infty} \sum_{k=1}^{n} \frac{\|a_n - a\|_\Gamma^k}{k!} \\
&= \exp(\|a_n - a\|_\Gamma) - 1 \to 0.
\end{aligned}
$$

Therefore,

$$
\begin{aligned}
\|\exp(a_n) - \exp(a)\|_\Gamma &= \|\exp(a_n - a + a) - \exp(a)\|_\Gamma \\
&= \|(\exp(a_n - a) - e)\exp(a)\|_\Gamma \\
&\leq \|\exp(a_n - a) - e\|_\Gamma \|\exp(a)\|_\Gamma \to 0,
\end{aligned}
$$

where the second equality is due to the fact that a_n and a commute. Moreover, we have that

$$
\begin{aligned}
\|\exp(a_n)\|_\Gamma &= \|\exp(n^2(ne - a)^{-1} - ne)\|_\Gamma \\
&= \|\exp(n^2(ne - a)^{-1})\exp(-ne)\|_\Gamma \\
&\leq e^{-n} \exp\|n^2(ne - a)^{-1}\|_\Gamma \\
&\leq e^{n(n\|(ne-a)^{-1}\|_\Gamma - 1)} \leq 1.
\end{aligned}
$$

Therefore, $\|\exp(a)\|_\Gamma \leq 1$. $\square$

The following Proposition 7.5.19 provides us with a clear view of the way the sets $W(a)$, $V(\mathcal{A}[C_\Gamma], a)$ fit together, where $(\mathcal{A}, \Gamma)$ is a compete locally convex algebra and a is an element of $\mathcal{A}[C_\Gamma]$. Essential to the proof of Proposition 7.5.19 is the Lumer–Phillips theorem for Banach algebras with identity, which is subsequently stated. For the proof of the latter result the reader is referred to [30, Theorem 3.6].

Theorem 7.5.18 (Lumer–Phillips) *Let* $\mathcal{A}[\|\cdot\|]$ *be a Banach algebra with identity and let* $a \in \mathcal{A}$. *Then,* $V(\mathcal{A}, a) \subset \{\lambda : Re\lambda \leq 0\}$, *if and only if,* $\|\exp(ta)\| \leq 1$, *for all* $t \geq 0$.

Proposition 7.5.19 *Let* $(\mathcal{A}, \Gamma)$ *be a complete locally convex algebra and let* $a \in \mathcal{A}[C_\Gamma]$. *Then,* $V(\mathcal{A}[C_\Gamma], a) \subset \overline{co}\, W(a)$ *(the latter indicates the closed convex hull of* $W(a)$*).*

Proof If we suppose that $W(a) \subset \{\lambda : Re\lambda \leq 0\}$, then for every $t > 0$ the element ta is dissipative. Then, by Remark 7.5.16, we deduce that ta has resolvent with first order decay in the right half-plane. Therefore, by Proposition 7.5.17, $\|\exp(ta)\|_\Gamma \leq 1$. Then, it follows by Theorem 7.5.18 that $V(\mathcal{A}[C_\Gamma], a) \subset \{\lambda : Re\lambda \leq 0\}$.

Aiming to obtain a contradiction, suppose that there exists $\lambda \in V(\mathcal{A}[C_\Gamma], a)$, such that $\lambda \notin \overline{co}\, W(a)$.

Then, $\{\lambda\}$ and $\overline{co}\, W(a)$ are disjoint closed convex subsets of $\mathbb{C}$. Thus, by the Hahn–Banach separation theorem (see, for instance, [36, Theorem 1.7, p. 7]), there is a linear functional $x^* : \mathbb{C} \to \mathbb{R}$ and $\alpha \in \mathbb{R}$, such that

$$\{\lambda\} \subset \{\mu \in \mathbb{C} : x^*(\mu) > \alpha\} \text{ and } \overline{co}\, W(a) \subset \{\mu \in \mathbb{C} : x^*(\mu) < \alpha\}. \quad (7.5.6)$$

For x^*, there exist $\alpha_1, \alpha_2 \in \mathbb{R}$, such that $x^*(\mu) = \alpha_1 Re(\mu) + \alpha_2 Im(\mu)$, for all μ in $\mathbb{C}$.

Therefore, we have that

$$W(a) \subset \overline{co}\, W(a) \subset \{\mu \in \mathbb{C} : x^*(\mu) < \alpha\} \subset \{\mu \in \mathbb{C} : \alpha_1 Re(\mu) + \alpha_2 Im(\mu) < \alpha\}.$$

Let $\alpha_1 \neq 0$. Then, after some calculations we have that

$$W(a) \subset \{\mu \in \mathbb{C} : \alpha_1 Re(\mu) - \alpha + \alpha_2 Im(\mu) < 0\}$$
$$= \left\{\mu \in \mathbb{C} : \alpha_1 Re\left(\mu - \frac{\alpha}{\alpha_1}\right) + \alpha_2 Im\left(\mu - \frac{\alpha}{\alpha_1}\right) < 0\right\} \quad (7.5.7)$$
$$= \left\{\mu \in \mathbb{C} : Re\left(\alpha_1\left(\mu - \frac{\alpha}{\alpha_1}\right)\right) - Re\left(i\alpha_2\left(\mu - \frac{\alpha}{\alpha_1}\right)\right) < 0\right\}.$$

Consider $z \in W\left(\alpha_1 a - \alpha e - i\alpha_2(a - \frac{\alpha}{\alpha_1}e)\right)$. For brevity, in what follows let us denote $\alpha_1 a - \alpha e - i\alpha_2(a - \frac{\alpha}{\alpha_1}e)$ by b. Recall that by the definition of the numerical range, $W(b) = W^1(T_b) \bigcup W^2(T_b')$, where

$$W^1(T_b) = \bigcup_{p \in \Gamma} W^1_p(T_b) \text{ and } W^2(T_b) = \bigcup_{q \in \Gamma'} W^1_q(T_b').$$

Let $z \in W_p^1(T_b)$, for some $p \in \Gamma$. Then, $z = f(T_b x) = f(bx)$, for some $(x, f) \in \mathcal{A} \times \mathcal{A}'$, such that $f(x) = 1 = p(x)$ and $|f(y)| \le p(y)$, for all $y \in \mathcal{A}$. Therefore,

$$
\begin{aligned}
z &= f\left(\left(\alpha_1 a - \alpha e - i\alpha_2\left(a - \frac{\alpha}{\alpha_1}e\right)\right)x\right) \\
&= \alpha_1 f(ax) - \alpha - i\alpha_2\left(f(ax) - \frac{\alpha}{\alpha_1}\right).
\end{aligned}
\tag{7.5.8}
$$

In the previous string of relations, we note that since $f(x) = 1 = p(x)$ and $|f(y)| \le p(y)$, for all $y \in \mathcal{A}$, then $f(ax) \in W_p^1(a)$, hence $f(ax) \in W(a)$. Now, as a consequence of (7.5.7) and (7.5.8), $\mathrm{Re}(z) < 0$.

Similarly, it can be seen that if $z \in W_q^1(T_b')$, for some $q \in \Gamma'$, then $\mathrm{Re}(z) < 0$. Moreover, if $\alpha_1 = 0$, analogous considerations as these above lead to the fact that if $z \in W(b)$, then $\mathrm{Re}(z) < 0$.

So, we conclude that $W(b) \subset \{\mu \in \mathbb{C} : \mathrm{Re}(\mu) < 0\}$. Therefore, by the first paragraph of the present proof we take that

$$
\begin{aligned}
V\left(\mathcal{A}[C_\Gamma], b\right) &\equiv V\left(\mathcal{A}[C_\Gamma], \alpha_1 a - \alpha e - i\alpha_2\left(a - \frac{\alpha}{\alpha_1}e\right)\right) \\
&\subset \{\mu \in \mathbb{C} : \mathrm{Re}(\mu) \le 0\}.
\end{aligned}
\tag{7.5.9}
$$

Recall that by our initial assumption $\lambda = f(a)$, for some $f \in \mathcal{A}'$, such that $f(e) = 1 = \|f\|$. Therefore,

$$
\begin{aligned}
\alpha_1 \lambda - \alpha - i\alpha_2\left(\lambda - \frac{\alpha}{\alpha_1}\right) &= f\left(\left(\alpha_1 a - \alpha e - i\alpha_2\left(a - \frac{\alpha}{\alpha_1}e\right)x\right)\right) \\
&\in V\left(\mathcal{A}[C_\Gamma], \alpha_1 a - \alpha e - i\alpha_2\left(a - \frac{\alpha}{\alpha_1}e\right)\right).
\end{aligned}
$$

So, by (7.5.9), we conclude that

$$
\mathrm{Re}\left(\alpha_1 \lambda - \alpha - i\alpha_2\left(\lambda - \frac{\alpha}{\alpha_1}\right)\right) \le 0
$$

$$
\Leftrightarrow \mathrm{Re}\left(\left(\alpha_1(\mathrm{Re}(\lambda) + i\mathrm{Im}(\lambda))\right)\right) - \alpha - i\alpha_2\left(\mathrm{Re}(\lambda) + i\mathrm{Im}(\lambda)\right) \le 0
$$

$$
\Leftrightarrow \alpha_1\mathrm{Re}(\lambda) + \alpha_2\mathrm{Im}(\lambda) \le \alpha,
$$

i.e., $x^*(\lambda) \le \alpha$, a contradiction to the first part of (7.5.6). Thus, the result follows. □

By Proposition 7.5.8 and Proposition 7.5.19, we have that if $\mathcal{A}$ is a complete locally convex algebra and $a \in \mathcal{A}[C_\Gamma]$, then $\overline{\mathrm{co}}\, W(a) = V(\mathcal{A}[C_\Gamma], a)$.

Corollary 7.5.20 *Let $(\mathcal{A}, \Gamma)$ be a complete locally convex algebra and let $a \in \mathcal{A}$, such that $W(a) = \{0\}$. Then, $a = 0$.*

Proof Since $W(a) = \{0\}$, $W(a)$ is certainly bounded and hence $a \in \mathcal{A}[C_\Gamma]$ (for this implication see the proof of Theorem 7.5.14). So, by Proposition 7.5.19, we have that $V(\mathcal{A}[C_\Gamma], a) = \{0\}$. It is a well known fact in the theory of normed algebras that for a normed algebra $B[\|\cdot\|]$ and for $b \in B$, we have that

$$\frac{1}{\exp(1)}\|b\| \leq \sup\{|\lambda| : \lambda \in V(B, a)\}$$

(see [30, Theorem 4.1]). From the latter result and the fact that $V(\mathcal{A}[C_\Gamma], a) = \{0\}$, we obtain $a = 0$. $\qquad\square$

Proposition 7.5.21 *Let $(\mathcal{A}, \Gamma)$ be a pseudo-complete, hypocontinuous locally convex algebra. Let $a \in \mathcal{A}$ have resolvent with first order decay in the right half-plane. Then a is dissipative.*

Proof Let $a_n = n^2(ne - a)^{-1} - ne$, $n \in \mathbb{N}$. Then, $a_n - a = a^2(ne - a)^{-1}$ (see the calculations at the beginning of the proof in Proposition 7.5.17). For an arbitrary, fixed seminorm $p \in \Gamma$, there is a seminorm $p_o \in \Gamma$ and positive constants C_{a^2}, C_{p_o}, such that

$$
\begin{aligned}
p(a_n - a) &= & p\big(a^2(ne - a)^{-1}\big) & \\
&\leq & C_{a^2}\, q\big((ne - a)^{-1}\big) & \\
&\leq & C_{a^2}C_{p_o}\, \|(ne - a)^{-1}\|_\Gamma & \qquad (7.5.10)\\
&\leq & C_{a^2}C_{p_o}\frac{1}{n} \to 0, &
\end{aligned}
$$

where the first inequality is due to the separate continuity of multiplication (with a^2 fixed), the second inequality is due to the fact that the induced topology of $\mathcal{A}$ on $\mathcal{A}[C_\Gamma]$ is weaker than $\|\cdot\|_\Gamma$ (see the comments after Definition 7.5.7), while the third inequality is due to the assumption of the resolvent of a having first order decay in the right half-plane. By the above inequalities we deduce that $a_n \to a$, with respect to Γ. Following the arguments of the last paragraph of the proof of Proposition 7.5.17, we have that $\|\exp(ta_n)\|_\Gamma \leq 1$, for every $t > 0$ and for every $n \in \mathbb{N}$. Hence, by Theorem 7.5.18, $V(\mathcal{A}[C_\Gamma], a_n) \subset \{\lambda : \operatorname{Re}\lambda \leq 0\}$, $n \in \mathbb{N}$. Therefore, since

$$\overline{\operatorname{co}}W(a_n) = V\big(\mathcal{A}[C_\Gamma], a_n\big), \quad \text{we conclude that} \quad W(a_n) \subset \{\lambda : \operatorname{Re}\lambda \leq 0\},\ n \in \mathbb{N}.$$

So, by Proposition 7.5.6, $W(a) \subset \{\lambda : \operatorname{Re}\lambda \leq 0\}$, i.e., a is dissipative. $\qquad\square$

The exponential function plays a key role in the theory of the numerical range as this is developed in the setting of a Banach algebra. The next result records the definition of an exponential function in the realm of a hypocontinuous complete locally convex algebra. For the proof of Theorem 7.5.22 the reader is referred to [156, Theorem 5.8].

Theorem 7.5.22 *Let $(\mathcal{A}, \Gamma)$ be a hypocontinuous complete locally convex algebra and let $a \in \mathcal{A}$ be dissipative. Then, for $t \geq 0$, there is an element $\exp(ta) \in \mathcal{A}$, such that $\| \exp(ta) \|_\Gamma \leq 1$. Moreover, if $t_1, t_2 \geq 0$, then*

$$\exp\big((t_1 + t_2)a\big) = \exp(t_1 a)\exp(t_2 a) \text{ and } \exp(0a) = e.$$

Even more, for $t \geq 0$ there exists $b_t \in \frac{1}{2}e C_\Gamma$ and $c_t \in C_\Gamma$, such that

$$\exp(ta) = e + ta + t^2 a^2 b_t \text{ and } \exp(ta) = e + ta + \frac{1}{2}t^2 a^2 + t^3 a^3 c_t.$$

Corollary 7.5.23 *Let $(\mathcal{A}, \Gamma)$ be a hypocontinuous complete locally convex algebra and let $h \in \mathcal{A}$ be hermitian. Then, if $t \in \mathbb{R}$, $\| \exp(ith) \|_\Gamma = 1$.*

Proof To begin with, we note that according to Theorem 7.5.22, in order for $\exp(ith)$ to be meaningful, for $t \geq 0$, ih should be dissipative. That this is the case follows immediately from the fact that h is hermitian. Likewise, $-ih$ is dissipative. Then, it follows from Theorem 7.5.22 that, for every $t \in \mathbb{R}$, the elements $\exp(ith), \exp(-ith)$ are well defined elements in $\mathcal{A}[C_\Gamma]$ and that both $\| \exp(ith) \|_\Gamma$ and $\| \exp(-ith) \|_\Gamma$ are less than or equal to 1. Moreover,

$$e = \exp(0) = \exp(ith - ith) = \exp(ith)\exp(-ith).$$

Thus, we conclude

$$1 = \|e\|_\Gamma = \| \exp(ith)\exp(-ith) \|_\Gamma \leq \| \exp(ith) \|_\Gamma \| \exp(-ith) \|_\Gamma \leq 1.$$

Therefore, $\| \exp(ith) \|_\Gamma \| \exp(-ith) \|_\Gamma = 1$. So, by $\| \exp(-ith) \|_\Gamma \leq 1$, we have that

$$\| \exp(-ith) \|_\Gamma \leq \| \exp(ith) \|_\Gamma \| \exp(-ith) \|_\Gamma \Leftrightarrow \| \exp(ith) \|_\Gamma \geq 1.$$

Hence, the result follows.   □

Theorem 7.5.24 *Let $(\mathcal{A}, \Gamma)$ be a hypocontinuous complete locally convex algebra. If $h, k \in \mathfrak{H}$, then $i(hk - kh) \in \mathfrak{H}$.*

Proof For $t \in \mathbb{R}$, using similar arguments as in the proof of Corollary 7.5.23, we can show that

$$\| \exp(itk)\exp(ith)\exp(-itk)\exp(-ith) \|_\Gamma = 1.$$

By Theorem 7.5.22, there exists $c_t \in C_\Gamma$, such that

$$\exp(ith) = e + ith + \frac{1}{2}(ith)^2 + (ith)^3 c_t,$$

while similar representations hold for the other terms of the above product. By expanding the product, we have that $\|e + t^2(hk - kh) + t^3 d_t\|_\Gamma = 1$, where d_t is a sum of products of elements in C_Γ and of powers of the elements h, k. Taking into account that the product of two bounded subsets of a hypocontinuous algebra is bounded (see the paragraph preceding Proposition 7.5.6), we can deduce that the set $\{d_t : t \in \mathbb{R}\}$ is a bounded subset of $\mathcal{A}$.

If $p \in \Gamma$ and $x \in \mathcal{A}$ with $p(x) = 1$, we have that

$$p\Big((e + t^2(hk - kh))x\Big) - t^3 p(d_t x) \leq p\Big((e + t^2(hk - kh) + t^3 d_t)x\Big)$$

$$\leq \|e + t^2(hk - kh) + t^3 d_t\|_\Gamma \, p(x) = 1,$$

which implies

$$\frac{1}{t^2}\Big(p\big((e + t^2(hk - kh))x\big) - 1\Big) \leq t\, p(d_t x), \text{ hence}$$

$$\limsup_{t \to 0+} \frac{1}{t^2}\Big(p\big((e + t^2(hk - kh))x\big) - 1\Big) \leq \limsup_{t \to 0+} t\, p(d_t x) = 0,$$

where the last equality follows from the fact that $\{d_t : t \in \mathbb{R}\}$ is bounded and due to separate continuity of multiplication as we argue in (7.5.10). Therefore,

$$\sup_{p(x)=1} \limsup_{t \to 0+} \frac{1}{t^2}\Big(p\big((e + t^2(hk - kh))x\big) - 1\Big) = 0.$$

Then, for every $(x, f) \in \Pi_p$, we have that

$$\mathrm{Re}\Big(f\big((hk - kh)x\big)\Big) = \frac{1}{t}\,\mathrm{Re}\Big(f\big((e + t(hk - kh))x\big) - 1\Big)$$

$$\leq \frac{1}{t}\Big(\big|f\big((e + t(hk - kh))x\big)\big| - 1\Big)$$

$$\leq \frac{1}{t}\Big(p\big((e + t(hk - kh))x\big) - 1\Big).$$

Hence, we conclude that

$$W^1(hk - kh) \subset i\mathbb{R}. \tag{7.5.11}$$

Let $q_B \in \Gamma'$ (for the definition of Γ' see discussion after Theorem 7.5.1) and $f \in \mathcal{A}'$, such that $q_B(f) = 1$. Then,

$$
\begin{aligned}
q_B\Big(\big(e' + t^2(hk - kh)'\big)f\Big) - t^3 q(d_t'f) &\leq \\
q_B\Big(\big(e' + t^2(hk - kh)' + t^3 d_t'\big)f\Big) &\leq \\
\big\| e' + t^2(hk - kh)' + t^3 d_t' \big\|_{\Gamma'} q_B(f) &= 1,
\end{aligned}
\tag{7.5.12}
$$

where the last equality is due to the fact that

$$
\big\| e' + t^2(hk - kh)' + t^3 d_t' \big\|_{\Gamma'} = \big\| e + t^2(hk - kh) + t^3 d_t \big\|_{\Gamma}
$$

(see the proof of Proposition 7.5.8). We recall here that, as proposed in Remark 7.5.11, e' (resp. d_t') denotes T_e' (resp. T_{d_t}').

We note that the set $\{d_t' : t \in \mathbb{R}\}$ is equicontinuous. Indeed, for an arbitrary 0-neighbourhood U in $\mathcal{A}'$ we can assume without loss of generality that $U = \{f \in \mathcal{A}' : q_B(f) < \varepsilon\}$, $\varepsilon > 0$ (see quotation at the beginning of the last paragraph). Consider the set $C = \{d_t : t \in \mathbb{R}\} \cdot B$, i.e., the product of the two sets. Since $\mathcal{A}$ is hypocontinuous, C is bounded. Hence, if $V = \{g \in \mathcal{A}' : q_C(g) < \varepsilon\}$, then for every $t \in \mathbb{R}$ and $g \in V$, we have that

$$
\begin{aligned}
q_B(d_t'g) &= \sup\{|d_t'g(x)| : x \in B\} \\
&= \sup\{|g(d_t x)| : x \in B\} \\
&\leq \sup\{|g(y)| : y \in C\} < \varepsilon.
\end{aligned}
$$

So, $d_t'V \subset U$ for every $t \in \mathbb{R}$, i.e., $\{d_t' : t \in \mathbb{R}\}$ is equicontinuous.

Therefore, from (7.5.12) we have that

$$
\frac{1}{t^2}\Big(q_B\big((e' + t^2(hk - kh)')f\big) - 1\Big) \leq t\, q_B(d_t'f),
$$

so that

$$
\limsup_{t \to 0+} \frac{1}{t^2}\Big(q_B\big((e' + t^2(hk - kh)')f\big) - 1\Big) \leq \limsup_{t \to 0+} t\, q_B(d_t'f) = 0,
$$

where the last equality follows since $\{d_t' : t \in \mathbb{R}\}$ is equicontinuous.

Then, for every $(f, x'') \in \Pi_q$, we have

$$\mathrm{Re}\Big(x''\big((hk - kh)'f\big)\Big) = \frac{1}{t}\,\mathrm{Re}\Big(x''\big((e' + t(hk - kh)')f\big) - 1\Big)$$

$$\leq \frac{1}{t}\Big(\big|x''\big((e' + t(hk - kh)')f\big)\big| - 1\Big)$$

$$\leq \frac{1}{t}\Big(q_B\big((e' + t(hk - kh)')f\big) - 1\Big).$$

So, we deduce

$$W^2(hk - kh) \subset i\mathbb{R}. \tag{7.5.13}$$

Thus, relations (7.5.11) and (7.5.13), show that $i(hk - kh) \in \mathfrak{H}$.   □

Theorem 7.5.26 provides a generalization of a result, which is known in the theory of numerical ranges of elements in normed algebras as Vidav's lemma (see [30, Theorem 5.10]). *The proof of Theorem 7.5.26 is based on the following complex analytical result*, whose the statement we include for convenience of the reader and for its proof we refer to [156, Proposition 6.2].

Proposition 7.5.25 *Let g be a complex-valued analytic function on $\{z \in \mathbb{C} : \mathrm{Re}z > 0\}$, continuous on the closure in $\mathbb{C}$ of the latter set. Suppose that for $z \notin \mathbb{R}$, $|g(z)| \leq \frac{1}{|\mathrm{Im}z|}$. Then, $|g(z)| \leq \frac{1}{|z|}$, for $\mathrm{Re}z > 0$.*

Theorem 7.5.26 *Let $(\mathcal{A}, \Gamma)$ be a hypocontinuous complete locally convex algebra and let $h \in \mathcal{A}$ be hermitian. Then, $\sup\big(\sigma_\Gamma(h) \cap \mathbb{R}\big) = \sup W(h)$.*

Proof We first show that if $\sigma_\Gamma(h) \cap \mathbb{R} \subset (-\infty, 0)$, then $W(h) \subset (-\infty, 0]$. For this suppose that $\sigma_\Gamma(h) \cap \mathbb{R} \subset (-\infty, 0)$ and let $f \in \mathcal{A}'$, such that $|f(a)| \leq 1$, for $a \in \mathcal{A}[C_\Gamma]$. For $z \in \mathbb{C}$, such that $\mathrm{Re}z \geq 0$, consider the function $g(z) = f\big((ze - h)^{-1}\big)$. It follows from Theorem 2.3.7 that the function g is analytic on $\{z \in \mathbb{C} : \mathrm{Re}z \geq 0\}$. Moreover, since h is hermitian, we have that $W(ih) \subset \{z \in \mathbb{C} : \mathrm{Re}z = 0\}$ and thus ih is dissipative. Hence, by Remark 7.5.16, ih has resolvent of first order decay in the right half-plane. Let $z \notin \mathbb{R}$, such that $\mathrm{Im}z < 0$. Then,

$$\big|g(z)\big| = \big|f(ze - h)^{-1}\big| \leq \big\|(ze - h)^{-1}\big\|_\Gamma \leq \frac{1}{\mathrm{Re}(iz)} = \frac{1}{-\mathrm{Im}z}.$$

Similarly, we deduce that $|g(z)| \leq \frac{1}{\mathrm{Im}z}$, in case $\mathrm{Im}z > 0$.

Therefore, the function g satisfies the conditions of Proposition 7.5.25, hence $|g(z)| \leq \frac{1}{|z|}$, for $z \in \mathbb{C}$, such that $\mathrm{Re}z > 0$. By the Hahn–Banach theorem, it follows that $\|(ze - h)^{-1}\|_\Gamma \leq \frac{1}{|z|}$, for $z \in \mathbb{C}$, such that $\mathrm{Re}z > 0$. That is, the resolvent of h has first order decay in the right half-plane. Then, by Proposition 7.5.21, h is dissipative and so $W(h) \subset (-\infty, 0]$.

Let $\lambda = \sup(\sigma_\Gamma(h) \cap \mathbb{R})$ and $\mu = \sup W(h)$. By Theorem 7.5.14, we have $\lambda \leq \mu$. Suppose that $\mu > \lambda$. Since $\lambda = \sup(\sigma_\Gamma(h) \cap \mathbb{R}) = \sup(\overline{co}(\sigma_\Gamma(h) \cap \mathbb{R}))$, by the Hahn–Banach separation theorem we have that there are α, $\alpha_1 \in \mathbb{R}$, in such a way that

$$\overline{co}(\sigma_\Gamma(h) \cap \mathbb{R}) \subset \{x \in \mathbb{R} : \alpha_1 x < \alpha\} \text{ and } \{\mu\} \subset \{x \in \mathbb{R} : \alpha_1 x > \alpha\}.$$

Then,

$$\sigma_\Gamma(\alpha_1 h - \alpha e) \cap \mathbb{R} = \{\alpha_1 \lambda_0 - \alpha : \lambda_0 \in \sigma_\Gamma(h)\} \cap \mathbb{R} \subset (-\infty, 0).$$

Therefore, $W(\alpha_1 h - \alpha e) \subset (-\infty, 0]$. Equivalently, $\alpha_1 W(h) \subset (-\infty, \alpha]$, a contradiction.   □

In order to build up to a Vidav–Palmer type theorem for a GB^*-algebra $\mathcal{A}[\tau]$, A. Wood proposed in [156, p. 260] a slight variant of the family of subsets of the given locally convex $*$-algebra than that of Allan's collection $\mathfrak{B}^*_\mathcal{A}$ (see Definition 3.3.1). In fact, for a locally convex algebra $\mathcal{A}[\tau]$ with identity e and with an involution $*$, A. Wood introduces the family $\widetilde{\mathfrak{B}}^*_\mathcal{A}$ of subsets of $\mathcal{A}$, such that

(1) each $B \in \widetilde{\mathfrak{B}}^*_\mathcal{A}$ is bounded and absolutely convex;
(2) $e \in B$, $B^2 \subset B$ and $B = B^*$.

Notice that *the only difference between the family $\widetilde{\mathfrak{B}}^*_\mathcal{A}$ and the family $\mathfrak{B}^*_\mathcal{A}$ of Definition 3.3.1 is that* the sets of the family $\widetilde{\mathfrak{B}}^*_\mathcal{A}$ are not assumed to be closed. The reason behind this choice, in this collection of subsets in $\mathcal{A}[\tau]$, can be spotted in the fact that, in Wood's course of arguments in [156], the continuity of the involution is not taken for granted (in this respect, see, for instance, the proof of Proposition 7.5.33, in which the assumption of closedness would impede us from having the same greatest member for the two topologies in question). Subsequently, the presence or absence of the assumption of continuity of the involution in a locally convex algebra provides us with certain variations in the respective definitions of a GB^*-algebra. In concrete terms, we have the following

Definition 7.5.27 A *semi-HB* algebra* is a locally convex algebra $\mathcal{A}[\tau]$ with identity e and with an involution (not necessarily continuous), such that

(1) $\widetilde{\mathfrak{B}}^*_\mathcal{A}$ has a greatest member B_0, which is τ-closed and $\mathcal{A}[B_0]$ is a Banach $*$-algebra with respect to $\|\cdot\|_{B_0}$.
(2) $(e + h^2)^{-1} \in \mathcal{A}_0$, for every $h \in H(\mathcal{A})$, i.e., for every $h \in \mathcal{A}$, such that $h^* = h$.

In case the involution is symmetric (see Definition 3.1.6), $\mathcal{A}[\tau]$ is said to be a *semi-GB*-algebra*. A semi-GB^*-algebra with a continuous involution is called a GB^*-algebra, while an HB^*-algebra is a semi-HB^*-algebra with a continuous involution.

The previous definition of a GB^*-algebra is equivalent to the definition of a GB^*-algebra of Dixon, as the latter is given in Definition 3.3.5. Indeed, on the one hand let $\mathcal{A}[\tau]$ be a GB^*-algebra according to Definition 3.3.5 and let B_0 be the greatest member of $\mathcal{B}^*_{\mathcal{A}}$. Consider $B \in \widetilde{\mathcal{B}}^*_{\mathcal{A}}$. Then, the τ-closure $\overline{B}$ of B belongs to $\mathcal{B}^*_{\mathcal{A}}$, hence $B \subset B_0$. Since $\mathcal{B}^*_{\mathcal{A}} \subset \widetilde{\mathcal{B}}^*_{\mathcal{A}}$, we have that B_0 is also the greatest member of $\widetilde{\mathcal{B}}^*_{\mathcal{A}}$. Therefore, it follows that $\mathcal{A}[\tau]$ is a GB^*-algebra according to Definition 7.5.27.

On the other hand, let $\mathcal{A}[\tau]$ be a GB^*-algebra according to Definition 7.5.27 and let B_0 be the greatest member of the family $\widetilde{\mathcal{B}}^*_{\mathcal{A}}$. Consider $B \in \mathcal{B}^*_{\mathcal{A}}$. Then, $B \in \widetilde{\mathcal{B}}^*_{\mathcal{A}}$, hence $B \subset B_0$. Since by Definition 7.5.27 B_0 is supposed to be τ-closed, we have that $B_0 \in \mathcal{B}^*_{\mathcal{A}}$. So, B_0 is the greatest member of $\mathcal{B}^*_{\mathcal{A}}$, from which it follows that $\mathcal{A}[\tau]$ is also a GB^*-algebra according to Definition 3.3.5.

Remark 7.5.28 We recall that, by Definition 3.1.3, a locally convex algebra $\mathcal{A}$ with involution is hermitian if $\sigma_{\mathcal{A}}(h) \subset \mathbb{R}$, for every $h \in H(\mathcal{A})$. If $\mathcal{A}$ is a pseudo-complete locally convex algebra with identity e, then from Proposition 3.1.5 we have that for $h \in H(\mathcal{A})$, $\sigma_{\mathcal{A}}(h) \subset \mathbb{R}$ is equivalent to $(e + h^2)^{-1} \in \mathcal{A}_0$. So, in case a semi-$HB^*$-algebra $\mathcal{A}[\tau]$ is pseudo-complete, property (2) of Definition 7.5.27 is equivalent to $\mathcal{A}[\tau]$ having hermitian involution. Even though, in general, a semi–HB^*-algebra need not be pseudo-complete, a similar to the aforementioned equivalence implied by Proposition 3.1.5 holds, as Lemma 7.5.30, below, records.

Definition 7.5.29 Let $\mathcal{A}[\tau]$ be a locally convex algebra, $a \in \mathcal{A}$ and let $(\mathcal{B})_{\mathcal{A}}$ be the family of subsets of $\mathcal{A}$ as defined just after Proposition 2.2.8. Then, for $B \in (\mathcal{B})_{\mathcal{A}}$, $\sigma_B(a)$ denotes the following subset of the extended complex plane $\mathbb{C}^*$:

$$\sigma_B(a) = \left\{ \lambda \in \mathbb{C} : \lambda e - a \text{ has no inverse in } \mathcal{A}[B] \right\}$$
$$\bigcup \left\{ \infty, \text{ if and only if, } a \notin \mathcal{A}[B] \right\}.$$

Given the fact that by Proposition 2.2.4, $\mathcal{A}_0 = \bigcup_{B \in (\mathcal{B})_{\mathcal{A}}} \mathcal{A}[B]$, we easily deduce that

$$\sigma_{\mathcal{A}}(a) = \bigcap_{B \in (\mathcal{B})_{\mathcal{A}}} \sigma_B(a), \ a \in \mathcal{A},$$

where $\sigma_{\mathcal{A}}(a)$ is the spectrum of a as defined in Definition 2.3.1.

Lemma 7.5.30 *Let $\mathcal{A}[\tau]$ be a locally convex $*$-algebra which satisfies property* (1) *of Definition 7.5.27. Then, for every $h \in H(\mathcal{A})$, $(e + h^2)^{-1}$ exists in $\mathcal{A}_0$, if and only if, $\sigma_{B_0}(h) \cap \mathbb{C} \subset \mathbb{R}$.*

Proof $\Rightarrow$ Let $(e + h^2)^{-1} \in \mathcal{A}_0$, for $h \in H(\mathcal{A})$. Then, there is $\lambda > 0$ such that the set $C := \{ (\lambda(e + h^2)^{-1})^n : n \in \mathbb{N} \}$ is bounded with respect to τ. Let B be the absolutely convex hull of $C \cup \{e\}$. Then, $B \in \widetilde{\mathcal{B}}^*_{\mathcal{A}}$. Therefore, $\lambda(e + h^2)^{-1} \in B_0$, hence $(e + h^2)^{-1} \in \mathcal{A}[B_0]$. Since,

$$\left(h(e + h^2)^{-1} \right)^2 = h^2 (e + h^2)^{-2} = (e + h^2)^{-1} - (e + h^2)^{-2},$$

we have that $\left(h(e+h^2)^{-1}\right)^2 \in A[B_0]$. So, the set

$$\left\{\left(\lambda^{-1}\left(h(e+h^2)^{-1}\right)^2\right)^n : n \in \mathbb{N}\right\} \text{ is } \|\cdot\|_{B_0}\text{-bounded for } \lambda > \left\|\left(h(e+h^2)^{-1}\right)^2\right\|_{B_0}.$$

Given that B_0 is τ-bounded, we have $\tau \prec \|\cdot\|_{B_0}$ on $A[B_0]$ (see comments after Definition 2.2.2). Therefore, $\left\{\left(\lambda^{-1}(h(e+h^2)^{-1})^2\right)^n : n \in \mathbb{N}\right\}$ is τ-bounded, i.e., $\left(h(e+h^2)^{-1}\right)^2 \in A_0$. Following the arguments of the proof of Lemma 3.1.4(ii), we then have that $h(e+h^2)^{-1} \in A_0$. Thus, by the argumentation of the first paragraph of the present proof we have $h(e+h^2)^{-1} \in A[B_0]$. Since $e+h^2 = -(ie+h)(ie-h)$, it follows that $(ie-h)$ is invertible and

$$\left(ie-h\right)^{-1} = -(ie+h)\left(e+h^2\right)^{-1} = -h\left(e+h^2\right)^{-1} - i\left(e+h^2\right)^{-1}.$$

Thus, $\left(ie-h\right)^{-1} \in A[B_0]$, i.e., $i \notin \sigma_{B_0}(h)$ (for the latter notation, see Definition 7.5.29).

For $\lambda + i\mu$, λ, $\mu \in \mathbb{R}$ and μ not zero, we have that $\mu^{-1}(h-\lambda e)$ is self-adjoint, thus from what is shown previously, $i \notin \sigma_{B_0}\left(\mu^{-1}(h-\lambda e)\right)$. So, $\lambda + i\mu \notin \sigma_{B_0}(h)$, from which the result follows.

$\Leftarrow$ Let $\sigma_{B_0}(h) \cap \mathbb{C} \subset \mathbb{R}$, for $h \in H(A)$. Therefore, $i, -i \notin \sigma_{B_0}(h)$, i.e., $(ie-h)^{-1}$, $(ie+h)^{-1} \in A[B_0]$. Thus, since $-(ie+h)(ie-h) = e+h^2$, we have that $(e+h^2)^{-1} \in A[B_0]$ and since $A[B_0] \subset A_0$, we conclude that $(e+h^2)^{-1} \in A_0$. $\square$

By Theorem 3.3.9(ii), every GB*-algebra contains a certain C*-algebra. The following result states that we have the analogous situation in a semi-HB*-algebra.

Proposition 7.5.31 *Let $A[\tau]$ be a semi-HB*-algebra. Then $A[B_0]$ is a C*-algebra with respect to $\|\cdot\|_{B_0}$.*

Proof Let h be a self-adjoint element of $A[B_0]$. By the hermiticity of involution, $(e+h^2)^{-1} \in A_0$. Therefore, by following analogous arguments as in the proof of Lemma 7.5.30, $(e+h^2)^{-1} \in A[B_0]$. So, $A[B_0]$ has hermitian involution (see Remark 7.5.28). Then, from the Shirali–Ford theorem (see [141] and/or [60, Theorem 22.23]), $A[B_0]$ has symmetric involution. The proof from this point onwards follows exactly the proof of Theorem 3.3.9(ii). $\square$

Note that in the previous Proposition 7.5.31, continuity of involution is not assumed, since the Shirali–Ford theorem, which is the key result in the proof, is formulated for a Banach algebra with a not necessarily continuous involution.

The following result, which is *due to Dixon* [47, Appendix A, Lemma A. 1], *shows that HB*-algebras and GB*-algebras coincide in the commutative setting.*

Proposition 7.5.32 (Dixon) *If $A[\tau]$ is a commutative HB*-algebra, then, $A[\tau]$ is a GB*-algebra.*

Proof Since A is commutative, $A[B_0] = A_0$. Let $h = h^* \in A$. By assumption of A being an HB*-algebra, we have that $\left(e+(e-h)^2\right)^{-1}$, $\left(e+(e+h)^2\right)^{-1} \in A[B_0]$.

So, $(e + (e - h)^2)^{-1} - (e + (e + h)^2)^{-1} \in A[B_0]$. Moreover,

$$(e + (e - h)^2)^{-1} - (e + (e + h)^2)^{-1}$$

$$= (e + (e - h)^2)^{-1}(e + (e + h)^2 - (e + (e - h)^2))(e + (e + h)^2)^{-1}$$

$$= (e + (e - h)^2)^{-1}(4h)(e + (e + h)^2)^{-1}$$

$$= 4h((e + (e + h)^2)(e + (e - h)^2))^{-1}$$

$$= 4h(4e + h^4)^{-1} = h\left(e + \frac{1}{4}h^4\right)^{-1}.$$

Therefore, $h(e + \frac{1}{4}h^4)^{-1} \in A[B_0]$, for every $h \in H(A)$. Hence,

$$h(e + h^4)^{-1} = \frac{1}{\sqrt{2}}(\sqrt{2}h)\left(e + \frac{1}{4}(\sqrt{2}h)^4\right)^{-1} \in A[B_0],$$

that is,

$$h(e + h^4)^{-1} \in A[B_0], \ \forall \, h \in H(A). \tag{7.5.14}$$

Let $x \in A$. Then, x can be written as $x = h + ik$, where $h = \frac{1}{2}(x + x^*)$ and $k = \frac{1}{2i}(x - x^*)$. Clearly $h, k \in H(A)$. Therefore, $(e + h^4)^{-1}, (e + k^4)^{-1} \in A[B_0]$. Hence, due to (7.5.14) and due to the assumption of commutativity of A, we have that

$$h_1 \equiv h(e + h^4)^{-1}(e + k^4)^{-1} \in A[B_0] \text{ and } k_1 \equiv k(e + h^4)^{-1}(e + k^4)^{-1} \in A[B_0],$$

where k_1, h_1 are self-adjoint elements in $A[B_0]$, hence there exists a self-adjoint element $l_1 \in A[B_0]$, such that $h_1^2 + k_1^2 = l_1^2$. Then,

$$(l_1(e + h^4)(e + k^4))^2 = l_1^2(e + h^4)^2(e + k^4)^2$$

$$= (h_1^2 + k_1^2)(e + h^4)^2(e + k^4)^2$$

$$= h^2 + k^2.$$

Therefore, $h^2 + k^2 = l^2$, where $l = l_1(e + h^4)(e + k^4)$ is a self-adjoint element of A. Hence,

$$e + x^*x = e + (h - ik)(h + ik) = e + h^2 + k^2 = e + l^2.$$

By assumption, $(e + l^2)^{-1} \in A[B_0]$. So, we conclude that $(e + x^*x)^{-1} \in A[B_0]$, from which it follows that $A[\tau]$ is a GB^*-algebra. $\qquad\qquad\square$

A useful "change of perspective" in terms of the topology considered, enables the passage from a semi-HB^*-algebra to an HB^*-algebra, as described in the following

Proposition 7.5.33 *Let $\mathcal{A}[\tau]$ be a semi-HB^*-algebra. Then, there is a topology τ' on $\mathcal{A}$, such that $\mathcal{A}[\tau']$ is an HB^*-algebra and moreover the greatest members of the respective collections $\mathfrak{B}^*_{\mathcal{A}[\tau]}$ and $\mathfrak{B}^*_{\mathcal{A}[\tau']}$ coincide; i.e., in symbols $B_0(\tau) = B_0(\tau')$* (in this regard, see also Corollary 6.3.7).

Proof Let $\mathcal{U}$ be a base of absolutely convex neighbourhoods of 0 for the topology τ. Then, by considering $N_U = U \cap U^*$ for each $U \in \mathcal{U}$, we get a base of neighbourhoods of 0 for a locally convex topology, τ' say, of $\mathcal{A}$. Then, it is easily seen that $\tau \prec \tau'$, the involution on $\mathcal{A}$ is τ'-continuous, B_0 is τ'-closed and B_0 is the greatest element of $\mathfrak{B}^*_{\mathcal{A}[\tau']}$. Therefore, since the completeness of $\mathcal{A}[B_0]$, and the fact that $(e + h^2)^{-1} \in \mathcal{A}_0$, for every $h \in H(\mathcal{A})$, remain valid under this change of topology (for the latter argument see Lemma 7.5.30), we conclude that $\mathcal{A}$ is an HB^*-algebra with respect to τ'.   □

▶ The next result *can be seen as "the half way" for a Vidav–Palmer type theorem for a semi-HB^*-algebra.*

Theorem 7.5.34 *Let $\mathcal{A}[\tau]$ be a hypocontinuous semi-HB^*-algebra. Then, there exists a defining family of seminorms $\widetilde{\Gamma}$ for τ, equivalent to the given one Γ, such that for every self-adjoint element h in $\mathcal{A}$, $W_{\widetilde{\Gamma}}(h) \subset \mathbb{R}$.*

Proof We first show that there is a defining family of seminorms $\widetilde{\Gamma}$ for τ, equivalent to the initial one Γ, such that $\|x\|_{B_0} = \|x\|_{\widetilde{\Gamma}}$, for every normal element x in $\mathcal{A}[B_0]$, where B_0 corresponds to the largest element of $\mathfrak{B}^*_{\mathcal{A}}$. From Lemma 7.5.10, there exists such a defining family of seminorms $\widetilde{\Gamma}$ for τ, such that $B_0 \subset C_{\widetilde{\Gamma}}$. Consequently, $\mathcal{A}[B_0] \subset \mathcal{A}[C_{\widetilde{\Gamma}}]$ and $\|x\|_{B_0} \geq \|x\|_{\widetilde{\Gamma}}$, for every $x \in \mathcal{A}[B_0]$. From Proposition 7.5.31, $\mathcal{A}[B_0]$ is a C^*-algebra. Hence, it follows from [127, Theorem (4.8.3)] that $r_{\widetilde{\Gamma}}(x) = r_{B_0}(x)$, for every $x \in \mathcal{A}[B_0]$, where $r_{\widetilde{\Gamma}}(x), r_{B_0}(x)$ denote the spectral radii of x in $\mathcal{A}[C_{\widetilde{\Gamma}}]$, $\mathcal{A}[B_0]$, respectively. If, in addition, x is a normal element in $\mathcal{A}[B_0]$, then $\|x\|_{B_0} = r_{B_0}(x)$, consequently $\|x\|_{B_0} = r_{\widetilde{\Gamma}}(x) \leq \|x\|_{\widetilde{\Gamma}}$. Therefore, we conclude that $\|x\|_{B_0} = \|x\|_{\widetilde{\Gamma}}$, for every normal element x in $\mathcal{A}[B_0]$.

Let us consider now a self-adjoint element h in $\mathcal{A}[B_0]$ and let $t \in \mathbb{R}$. Then, $e + ith$ is a normal element in $\mathcal{A}[B_0]$. Therefore, by the argument of the previous paragraph we have that $\|e + ith\|_{\widetilde{\Gamma}} = \|e + ith\|_{B_0}$. Since $\|\cdot\|_{B_0}$ is a C^*-norm on $\mathcal{A}[B_0]$ we have

$$\left\|e + ith\right\|_{\widetilde{\Gamma}}^2 = \left\|e + ith\right\|_{B_0}^2 = \left\|(e - ith)(e + ith)\right\|_{B_0}$$

$$= \left\|e + t^2 h^2\right\|_{B_0} = \left\|e + t^2 h^2\right\|_{\widetilde{\Gamma}}.$$

Therefore,

$$\lim_{t \to 0} \frac{1}{t} \left(\|e + ith\|_{\tilde{\Gamma}} - 1 \right) = \lim_{t \to 0} \frac{1}{t} \left(\|e + t^2 h^2\|_{\tilde{\Gamma}}^{\frac{1}{2}} - 1 \right) = 0.$$

Thus, from [30, Lemma 5.2] it follows that $V(\mathcal{A}[C_{\tilde{\Gamma}}], h) \subset \mathbb{R}$. Hence, by Proposition 7.5.8, we have that $W_{\tilde{\Gamma}}(h) \subset \mathbb{R}$.

From Proposition 7.5.33, there is a topology τ' finer than the topology τ and such that $\mathcal{A}[\tau']$ is an HB^*-algebra. Let h be a self-adjoint element in $\mathcal{A}$ and let C_h be a maximal commutative $*$-subalgebra of $\mathcal{A}[\tau']$, which contains h. Then, by Proposition 7.5.32, $C_h[\tau']$ is a GB^*-algebra. From the proof of Theorem 4.2.11, there is a sequence $\{h_n\}_n$ of self-adjoint elements in $\mathcal{A}[B_0]$, such that $h_n \underset{\tau'}{\to} h$. Hence, $h_n \underset{\tau}{\to} h$. By the argument of the previous paragraph, $W_{\tilde{\Gamma}}(h_n) \subset \mathbb{R}$. So, from Proposition 7.5.6, we conclude that $W_{\tilde{\Gamma}}(h) \subset \mathbb{R}$.   □

We shift now our attention to the inverse direction, that is, we look at algebras that are spanned as linear spaces by their hermitian elements and investigate whether this property can characterize the nature of the algebras. To begin with, we state the following helpful

Lemma 7.5.35 *Let $(\mathcal{A}, \Gamma)$ be a complete locally convex algebra with the property that every $a \in \mathcal{A}$ is written as $a = h + ik$, where $h, k \in \mathfrak{H}$. Then, the elements h, k in the decomposition of a are uniquely determined.*

Proof Let $h, k \in \mathfrak{H}$, such that $h + ik = 0$. Then, it is straightforward to check that $W(h) = -i W(k)$. Hence, $W(h), W(k) \subset \mathbb{R} \cap i\mathbb{R} = \{0\}$, therefore $W(h) = \{0\} = W(k)$. So, by Corollary 7.5.20, we have that $h = k = 0$.   □

Based on the previous lemma, *it is meaningful to write $\mathcal{A} = \mathfrak{H} + i\mathfrak{H}$, in case $(\mathcal{A}, \Gamma)$ is a complete locally convex algebra, which is spanned, as a linear space, by its hermitian elements*. In this case, we say that $(\mathcal{A}, \Gamma)$ has *a hermitian decomposition*. The next step we are going to take is to show that $\mathcal{A}$ possesses an involution.

Proposition 7.5.36 *Let $(\mathcal{A}, \Gamma)$ be a complete locally convex algebra with a hermitian decomposition. Then, there is a map $\natural : \mathcal{A} \to \mathcal{A}$, such that*

(i) $(a^\natural)^\natural = a$, $a \in \mathcal{A}$;
(ii) $(a + b)^\natural = a^\natural + b^\natural$, $a, b \in \mathcal{A}$;
(iii) $(\lambda a)^\natural = \bar{\lambda} a^\natural$, $\lambda \in \mathbb{C}$, $a \in \mathcal{A}$.

Proof Let $a \in \mathcal{A}$. Then, there are unique hermitian elements $h, k \in \mathfrak{H}$, such that $a = h + ik$. Consider the map $\natural : \mathcal{A} \to \mathcal{A} : a^\natural = h - ik$. It is straightforward to check that the map $\natural$ enjoys the desired properties.   □

While the establishment of a vector space involution on $\mathcal{A}$ is straightforward, as shown in Proposition 7.5.36, the proof that this involution is also algebraic (i.e.,

$(ab)^\natural = b^\natural a^\natural$, $a, b \in \mathcal{A}$) requires some more effort. Towards this direction, we first have the following result.

Proposition 7.5.37 *Let $(\mathcal{A}, \Gamma)$ be a complete locally convex algebra with a hermitian decomposition. Then, $\mathcal{A}[C_\Gamma]$ is a C^*-algebra with unit ball C_Γ.*

Proof Let $a \in \mathcal{A}[C_\Gamma]$. Since $\mathcal{A}$ has a hermitian decomposition, there are unique elements $h, k \in \mathfrak{H}$, such that $a = h + ik$. We show that $h, k \in \mathcal{A}[C_\Gamma]$. The latter fact will follow, once we show that $W(h)$, $W(k)$ are bounded (for this, see the proof of Theorem 7.5.14). By Proposition 7.5.8,

$$W(a) \subset V\big(\mathcal{A}[C_\Gamma], a\big); \quad \text{moreover} \;\; V\big(\mathcal{A}[C_\Gamma], a\big) \subset \big\{\lambda : |\lambda| \leq \|a\|_\Gamma\big\},$$

where the second inclusion in the previous relation is due to the following facts: by Proposition 7.5.8, $W(a) \subset V(\mathcal{A}[C_\Gamma], a)$. The element $\lambda \in V(\mathcal{A}[C_\Gamma], a)$ can be written as $\lambda = f(a)$, for some $f \in \mathcal{A}[C_\Gamma]'$ with $\|f\| = f(e) = 1$ (see the paragraph preceding Proposition 7.5.8). Hence,

$$|\lambda| = |f(a)| \leq \|f\|\|a\|_\Gamma = \|a\|_\Gamma.$$

Therefore, $W(a) \subset \{\lambda \in \mathbb{C} : |\lambda| \leq \|a\|_\Gamma\}$.

Now from the inclusion $W(a) \subset V\big(\mathcal{A}[C_\Gamma], a\big)$ as above, $W(a)$ is bounded. Since, as can be easily checked,

$$W(h) = \mathrm{Re}(W(a)) \;\; \text{and} \;\; W(k) = \mathrm{Im}(W(a)),$$

so we have that $W(h)$, $W(k)$ are bounded. Thus, $h, k \in \mathcal{A}[C_\Gamma]$. From Proposition 7.5.19,

$$V\big(\mathcal{A}[C_\Gamma], h\big) \subset \overline{\mathrm{co}}W(h) \;\; \text{and} \;\; V\big(\mathcal{A}[C_\Gamma], k\big) \subset \overline{\mathrm{co}}W(k).$$

So, $V(\mathcal{A}[C_\Gamma], h)$, $V(\mathcal{A}[C_\Gamma], k) \subset \mathbb{R}$. Therefore, the normed algebra $\mathcal{A}[C_\Gamma]$ satisfies the conditions of the Vidav–Palmer theorem (see Theorem 7.5.1), from which follows that $\mathcal{A}[C_\Gamma]$ is a C^*-algebra with unit ball C_Γ. $\qquad\square$

In the course of proving that the involution of Proposition 7.5.36 is algebraic, it would be useful to show that for certain elements of a hypocontinuous complete locally convex algebra, the closed convex hulls of the sets of the spectrum and of the numerical range coincide, a result which clearly has its independent interest (see Proposition 7.5.39). The following rather technical result is needed for the proof of Proposition 7.5.39. The proof of Proposition 7.5.38 is based on a mere recollection of results from Chaps 1 and 2. We recall that $(\mathfrak{B})_\mathcal{A}$ denotes the family of subsets of $\mathcal{A}$ as described in the paragraph following Proposition 2.2.8.

Proposition 7.5.38 *Let $\mathcal{A}[\tau]$ be a commutative locally convex algebra and $B \in (\mathfrak{B})_A$, such that $A[B]$ is a C^*-algebra. Let $a = h + ik \in \mathcal{A}$ $(h, k \in H(\mathcal{A}))$ be such that*

$$\sigma_B(a) \cap \mathbb{C} \subset \{z : \mathrm{Re}\, z \geq 0\}, \ \sigma_B(h) \cap \mathbb{C} \subset \mathbb{R} \ \text{ and } \ \sigma_B(k) \cap \mathbb{C} \subset \mathbb{R}.$$

Then, $\sigma_B(h) \cap \mathbb{C} \subset [0, \infty)$.

Proof Let $\mathfrak{M}$ be the carrier space of $A[B]$ and let $\Sigma = \{x \in \mathcal{A} : \sigma_B(x) \neq \mathbb{C}^*\}$. By using the arguments of the proof of Proposition 2.5.4 we can easily see that each $\varphi \in \mathfrak{M}$ can be extended to a $\mathbb{C}^*$-valued function φ' on Σ, such that φ' has the properties (1)–(4) of Proposition 2.5.4. Moreover, the proof of Proposition 2.5.5 carries over with exactly the same arguments, for $A[B]$ in place of $\mathcal{A}_0$. Therefore, $\sigma_B(a) = \{\varphi'(a) : \varphi \in \mathfrak{M}\}$. By assumption,

$$a = h + ik \ \text{ and } \ \sigma_B(h) \cap \mathbb{C} \subset \mathbb{R}, \ \text{ so } \ (ie - h)^{-1}, \ (ie + h)^{-1} \in A[B].$$

Thus,

$$(e + h^2)^{-1} = -(ie - h)^{-1}(ie + h)^{-1} \in A[B].$$

An analogous argument for k shows that $(e + k^2)^{-1} \in A[B]$. Now, since $(e + h^2)^{-1}, (e + k^2)^{-1} \in A[B]$, imitating the proof of Lemma 3.4.6, we conclude that the sets

$$\{\varphi \in \mathfrak{M} : \varphi(h) = \infty\}, \ \ \{\varphi \in \mathfrak{M} : \varphi(k) = \infty\}$$

are closed, nowhere dense subsets of $\mathfrak{M}$ (note that the assumption of $A[B]$ being a C^*-algebra is crucial in this point). Hence, by considering φ in the dense subset of $\mathfrak{M}$, where both $\varphi'(h), \varphi'(k)$ are finite, we have that $\varphi'(a) = \varphi'(h) + i\varphi'(k)$. So, since

$$\varphi'(h) \in \sigma_B(h) \cap \mathbb{C} \subset \mathbb{R} \ \text{ and } \ \varphi'(k) \in \sigma_B(k) \cap \mathbb{C} \subset \mathbb{R},$$

we conclude that $\varphi'(h) = \mathrm{Re}(\varphi'(a))$, which is not negative by assumption. Thus, $\sigma_B(h) \cap \mathbb{C} \subset [0, \infty)$. $\qquad \square$

Proposition 7.5.39 *Let $(\mathcal{A}, \Gamma)$ be a hypocontinuous complete locally convex algebra with a hermitian decomposition. Let $a \in \mathcal{A}$ be an element, such that $a = h + ik$, $h, k \in \mathfrak{H}$ and $hk = kh$. Then, $\overline{co}\,(\sigma_\Gamma(a) \cap \mathbb{C}) = \overline{co}\,W(a)$.*

Proof By Theorem 7.5.14, $\sigma_\Gamma(a) \cap \mathbb{C} \subset \overline{W(a)}$. The proof can be completed by an implication of the Hahn–Banach separation theorem, once it is shown that the relation $\sigma_\Gamma(a) \cap \mathbb{C} \subset \{z : \mathrm{Re}\, z \geq 0\}$ implies that $W(a) \subset \{z : \mathrm{Re}\, z \geq 0\}$. Given the fact that h, k commute, we can suppose without loss of generality that the algebra $\mathcal{A}$ is commutative. Moreover, by Proposition 7.5.37, $\mathcal{A}[C_\Gamma]$ is a C^*-algebra. We therefore note that all the conditions of Proposition 7.5.38 are satisfied, hence $\sigma_\Gamma(h) \cap \mathbb{C} \subset [0, \infty)$. By Theorem 7.5.26, we then have that $W(h) \subset [0, \infty)$. Since $k \in \mathfrak{H}$ and thus $W(ik) \subset i\mathbb{R}$, we conclude that $W(a) \subset \{z : \mathrm{Re}z \geq 0\}$. $\quad\square$

Theorem 7.5.40 *Let* $(\mathcal{A}, \Gamma)$ *be a hypocontinuous complete locally convex algebra with a hermitian decomposition. Then, the map* $\natural$ *of Proposition 7.5.36 is an algebraic involution.*

Proof We first show that for $h \in \mathfrak{H}$, $h^2 \in \mathfrak{H}$. Indeed, for h^2, there are unique $u, v \in \mathfrak{H}$, such that $h^2 = u + iv$. Then,

$$h(u + iv) = (u + iv)h, \quad \text{so that } hu - uh = i(vh - hv).$$

By Theorem 7.5.24,

$$hu - uh, \; i(hu - uh) \in \mathfrak{H}, \quad \text{so } i(vh - hv), \; vh - hv \in \mathfrak{H}.$$

Therefore, $W(i(vh - hv)) \subset \mathbb{R} \cap i\mathbb{R} = \{0\}$, which from Corollary 7.5.20 implies that $i(hv - vh) = 0$. Then, $hu = uh$ so that $h^2 u = h(uh) = (hu)h = uh^2$. Hence, $(u + iv)u = u(u + iv)$, if and only if, $uv = vu$. By Proposition 7.5.39, $\overline{\mathrm{co}}(W(h^2)) = \overline{\mathrm{co}}(\sigma_\Gamma(h^2) \cap \mathbb{C})$. Hence,

$$\overline{\mathrm{co}}(W(h^2)) = \overline{\mathrm{co}}((\sigma_\Gamma(h) \cap \mathbb{C})^2) \subset \overline{\mathrm{co}}(\overline{W(h)}^2) \subset \mathbb{R}.$$

The last but one inclusion in the aforementioned string of relations is due to Theorem 7.5.14. Therefore, we conclude that $W(h^2) \subset \mathbb{R}$; hence $h^2 \in \mathfrak{H}$.

Now let $a, b \in \mathcal{A}$. Based on the relation $ab = \frac{1}{2}(ab + b^\natural a^\natural) - i\frac{1}{2}i(ab - b^\natural a^\natural)$, we show that $ab + b^\natural a^\natural \in \mathfrak{H}$. Indeed, let $a = h + ik$, $b = u + iv$ for unique hermitian elements h, k, v, v. Then,

$$ab + b^\natural a^\natural = (hu + uh) - (kv + vk) + i(ku - uk) + i(hv - vh).$$

By Theorem 7.5.24, $i(ku - uk)$, $i(hv - vh) \in \mathfrak{H}$. Moreover, by the argument of the previous paragraph,

$$k^2, \; v^2, \; (k + v)^2, \; u^2, \; h^2, \; (h + u)^2 \in \mathfrak{H}, \quad \text{hence } kv + vk, \; hu + uh \in \mathfrak{H}.$$

Therefore, $ab + b^\natural a^\natural \in \mathfrak{H}$. Similarly, it can be shown that $i(ab - b^\natural a^\natural) \in \mathfrak{H}$. Thus, based on the definition of the map $\natural$, we deduce that

$$(ab)^\natural = \frac{1}{2}(ab + b^\natural a^\natural) + i\frac{1}{2}i(ab - b^\natural a^\natural) = b^\natural a^\natural$$

$\quad\square$

Our next objective is to show that for a hypocontinuous complete locally convex algebra with hermitian decomposition, the family $\widetilde{\mathfrak{B}}^*_{\mathcal{A}}$ attains a greatest member. The following two results are devoted to this goal.

Theorem 7.5.41 *Let* $(\mathcal{A}, \Gamma)$ *be a commutative hypocontinuous complete locally convex algebra with a hermitian decomposition and let* $B \in (\mathfrak{B})_{\mathcal{A}}$. *Then,* $B \subset C_\Gamma$.

Proof Since $\mathcal{A}$ is commutative and complete, the family $(\mathfrak{B})_{\mathcal{A}}$ is outer-directed with respect to inclusion (see Theorem 2.2.10). Hence, there is a set $F \in (\mathfrak{B})_{\mathcal{A}}$, such that $B \cup C_\Gamma \subset F$. By Lemma 7.5.10, there is a defining family of seminorms Δ for the topology τ corresponding to the initial family Γ on $\mathcal{A}$, such that $F \subset C_\Delta$. It suffices to show that $C_\Delta \subset C_\Gamma$. By $C_\Gamma \subset C_\Delta$, we have that $\mathcal{A}[C_\Gamma] \subset \mathcal{A}[C_\Delta]$. Therefore, if $a \in \mathcal{A}$ and $\lambda \notin \sigma_\Gamma(a) \cap \mathbb{C}$, then

$$(\lambda e - a)^{-1} \in \mathcal{A}[C_\Gamma] \text{ and so } (\lambda e - a)^{-1} \in \mathcal{A}[C_\Delta].$$

Thus, $\sigma_\Delta(a) \cap \mathbb{C} \subset \sigma_\Gamma(a) \cap \mathbb{C}$. Since $\mathcal{A}$ is commutative and has a hermitian decomposition, Proposition 7.5.39 yields that $\overline{\mathrm{co}}(W_\Delta(a)) \subset \overline{\mathrm{co}}(W_\Gamma(a))$. The previous inclusion gives us, in particular, that $\mathfrak{H}_\Gamma \subset \mathfrak{H}_\Delta$.

Let $a \in \mathcal{A}[C_\Delta]$. Then, $a = h + ik$, where $h, k \in \mathfrak{H}_\Gamma$ and thus a fortiori $h, k \in \mathfrak{H}_\Delta$. By the second paragraph of the proof of Proposition 7.5.19, we have that $W_\Delta(a) \subset \{\lambda : |\lambda| \leq \|a\|_\Delta\}$. So, $W_\Delta(a)$ is bounded, hence $W_\Delta(h)$, $W_\Delta(k)$ are bounded. Thus, $h, k \in \mathcal{A}[C_\Delta]$. So, $a^\natural = h - ik \in \mathcal{A}[C_\Delta]$ and thus $\mathcal{A}[C_\Delta]$ is a Banach algebra with involution $\natural$. If h is a self-adjoint element of $\mathcal{A}[C_\Delta]$, then $h \in \mathfrak{H}_\Gamma$. Indeed, if $h = u + iv$, where $u, v \in \mathfrak{H}_\Gamma$ then, by $h = h^\natural$, we have that $v = 0$, hence $h = u \in \mathfrak{H}_\Gamma$. Now, from Theorem 7.5.22 and Corollary 7.5.23, we have that $\|\exp(ih)\|_\Delta \leq \|\exp(ih)\|_\Gamma = 1$. Since

$$\|\exp(ih)\|_\Delta \geq \|\exp(ih)\|_\Delta \|\exp(-ih)\|_\Delta \geq \|\exp(ih)\exp(-ih)\|_\Delta = 1,$$

we obtain that $\|\exp(ih)\|_\Delta = 1$. Then, from [126, Theorem 10.1] follows that $\mathcal{A}[C_\Delta]$ is a C^*-algebra with unit ball C_Δ. According to [126, Theorem 9.7], C_Δ is the closed convex hull, with respect to $\|\cdot\|_\Delta$, of the set $L = \{\exp(ia) : a \in H(\mathcal{A}[C_\Delta])\}$. Clearly $\exp(ih) \in L$ and since $\|\exp(ih)\|_\Gamma = 1$,

$$\exp(ih) \in U(\mathcal{A}[C_\Gamma]) \Rightarrow \exp(ih) \in C_\Gamma \Rightarrow L \subset C_\Gamma.$$

By the way it is defined, C_Γ is closed with respect to the topology τ of $\mathcal{A}$ and $C_\Gamma \subset C_\Delta$ by our initial assumption. Since $\tau \prec \|\cdot\|_\Delta$ on $\mathcal{A}[C_\Delta]$, C_Γ is $\|\cdot\|_\Delta$-closed. Therefore, by $L \subset C_\Gamma$, it follows that $C_\Delta \subset C_\Gamma$. Therefore, $B \subset C_\Delta \subset C_\Gamma$.
□

Theorem 7.5.42 *Let* $(\mathcal{A}, \Gamma)$ *be a hypocontinuous complete locally convex algebra with a hermitian decomposition and let* $B \in \widetilde{\mathfrak{B}}^*_{\mathcal{A}}$ *be arbitrary. Then,* $B \subset C_\Gamma$; *i.e.,* C_Γ *is a greatest member in* $\widetilde{\mathfrak{B}}^*_{\mathcal{A}}$.

Proof Let $h \in \mathfrak{H} \cap B$. Then $\overline{B}$, the closure of B in $(\mathcal{A}, \Gamma)$, belongs to $(\mathfrak{B})_{\mathcal{A}}$. Since $h \in \mathfrak{H}$, h is a self-adjoint element of $\mathcal{A}$ with respect to the involution $\natural$. If we consider a maximal commutative $\natural$-subalgebra of $\mathcal{A}$ containing h, then by Theorem 7.5.41, we have that $h \in C_\Gamma$.

Now, let $x = h + ik \in B$, where $h, k \in \mathfrak{H}$. Since $x^\natural \in B^\natural = B$, $h - ik \in B$ ($B^\natural$ denotes the set of all $x^\natural$ for $x \in B$). Since B is convex, $\frac{1}{2} x^\natural + \frac{1}{2} x \in B$, from which follows that $h \in B$. Similarly, since B is absorbent, $\frac{1}{i} x$, $\frac{1}{i} x^\natural \in B$, thus $\frac{1}{2}(\frac{1}{i} x) + \frac{1}{2}(-\frac{1}{i} x^\natural) = k \in \mathfrak{H} \cap B$. So, $h, k \in \mathfrak{H} \cap B$. By the argument of the previous paragraph, we then have that $x \in 2C_\Gamma$. Therefore, B is a bounded subset of $\mathcal{A}[C_\Gamma]$ with respect to $\|\cdot\|_\Gamma$, such that $B^2 \subset B$ and $B^\natural = B$. By Proposition 7.5.37, $\mathcal{A}[C_\Gamma]$ is a C^*-algebra with unit ball C_Γ. Hence, C_Γ is the largest element, which is bounded, self-adjoint (i.e., $C_\Gamma^\natural = C_\Gamma$) and submultiplicative (i.e., $C_\Gamma^2 \subset C_\Gamma$). This can be easily seen as follows: for $x \in B$, $x^\natural x \in B^\natural B = B^2 \subset B$ and by induction $(x^\natural x)^n \in B$, $n \in \mathbb{N}$. Suppose that $x \notin C_\Gamma$. Then,

$$\|x\|_\Gamma > 1 \text{ and so } \|x^\natural x\|_\Gamma = \|x\|_\Gamma^2 > 1; \text{ therefore } \left\|(x^\natural x)^{2^n}\right\|_\Gamma = \|x^\natural x\|_\Gamma^{2^n} \underset{n \to \infty}{\to} \infty,$$

which is a contradiction given that $(x^\natural x)^{2^n} \in B$ and B is a bounded subset of $\mathcal{A}[C_\Gamma]$. $\qquad \square$

The following Theorem 7.5.43 *is a Vidav-Palmer type theorem for semi-HB*-algebras.*

Theorem 7.5.43 (Wood) *Let $\mathcal{A}[\tau]$ be a hypocontinuous complete locally convex algebra. Then, $\mathcal{A}[\tau]$ is a semi-HB*-algebra, if and only if, there is a defining family of seminorms Γ for τ, such that $(\mathcal{A}, \Gamma)$ has a hermitian decomposition.*

Proof $\Rightarrow$ By Theorem 7.5.34, there is a defining family of seminorms $\widetilde{\Gamma}$ for τ, equivalent to the initial family Γ, such that $W_{\widetilde{\Gamma}}(h) \subset \mathbb{R}$, for every $h \in H(\mathcal{A})$. It follows then that $H(\mathcal{A}) \subset \mathfrak{H}_{\widetilde{\Gamma}}$. Hence, since every $a \in \mathcal{A}$ can be expressed as a linear combination of elements in $H(\mathcal{A})$, namely $a = \frac{1}{2}(a + a^*) + i \frac{1}{2i}(a - a^*)$, we have that $\mathcal{A} = \mathfrak{H}_{\widetilde{\Gamma}} + i \mathfrak{H}_{\widetilde{\Gamma}}$. The uniqueness of the latter decomposition of $(\mathcal{A}, \widetilde{\Gamma})$ follows from Lemma 7.5.35.

$\Leftarrow$ Let Γ be a defining family of seminorms for τ, with respect to which $(\mathcal{A}, \Gamma)$ has a hermitian decomposition. Then, by Theorems 7.5.40, 7.5.42 and Proposition 7.5.37, $\mathcal{A}$ has an involution $\natural$, C_Γ is the greatest member of the family $\widetilde{\mathfrak{B}}_{\mathcal{A}}^*$, with respect to inclusion, and $\mathcal{A}[C_\Gamma]$ is a C^*-algebra. Let $h \in \mathfrak{H}_\Gamma$. Then, since h clearly satisfies the condition of Proposition 7.5.39, we have that

$$\sigma_\Gamma(h) \cap \mathbb{C} \subset \overline{\mathrm{co}}(\sigma_\Gamma(h) \cap \mathbb{C}) = \overline{\mathrm{co}}(W_\Gamma(h)) \subset \mathbb{R}.$$

Therefore, by Lemma 7.5.30, the involution $\natural$ is such that $(e + h^2)^{-1} \in \mathcal{A}_0$, for every h in $\mathcal{A}$, with $h = h^\natural$. Thus, we conclude that $\mathcal{A}[\tau]$ is a semi-HB*-algebra. $\qquad \square$

Corollary 7.5.44 *Let* $\mathcal{A}[\tau]$ *be a hypocontinuous complete semi-HB*-algebra. Then, there is a defining family of seminorms* $\widetilde{\Gamma}$ *for* τ, *equivalent to the given* Γ, *with respect to which* $\mathfrak{H}_{\widetilde{\Gamma}} = H(\mathcal{A})$.

Proof By the first part of the proof of Theorem 7.5.43 we have that there is a defining family of seminorms $\widetilde{\Gamma}$, equivalent to the given Γ, such that $H(\mathcal{A}) \subset \mathfrak{H}_{\widetilde{\Gamma}}$. On the other hand, let $h \in \mathfrak{H}_{\widetilde{\Gamma}}$. Since

$$h = \frac{1}{2}\left(h + h^*\right) + i\frac{1}{2i}\left(h - h^*\right) \text{ and } \frac{1}{2}\left(h + h^*\right), \ \frac{1}{2i}\left(h - h^*\right) \in H(\mathcal{A}) \subset \mathfrak{H}_{\widetilde{\Gamma}},$$

we have that

$$h - \frac{1}{2}\left(h + h^*\right) \in \mathfrak{H}_{\widetilde{\Gamma}} \cap i\mathfrak{H}_{\widetilde{\Gamma}}, \text{ hence } W_{\widetilde{\Gamma}}\left(h - \frac{1}{2}\left(h + h^*\right)\right) \in \mathbb{R} \cap i\mathbb{R}.$$

So, by Corollary 7.5.20, $h = \frac{1}{2}(h + h^*)$ equivalently $h = h^*$, i.e., $h \in H(\mathcal{A})$. $\qquad\square$

Theorem 7.5.43 can be further enhanced. Indeed, the next three lemmas culminate in Proposition 7.5.48, which shows that every complete hypocontinuous semi-HB^*-algebra is actually a semi-GB^*-algebra.

We note that, in accordance with the notation introduced in Chap. 3 (see comments after Definition 3.1.2), a self-adjoint element h of a semi-HB^*-algebra $\mathcal{A}$ is called *positive*, denoted by $h \geq 0$, if $\sigma_{\mathcal{A}}(h) \cap \mathbb{C} \subset [0, \infty)$.

Lemma 7.5.45 *Let* $\mathcal{A}[\tau]$ *be a hypocontinuous, complete semi-HB*-algebra and let* $h, k \in \mathcal{A}$ *with* $h, k \geq 0$. *Then,* $h + k \geq 0$.

Proof By Corollary 7.5.44, there is a defining family of seminorms $\widetilde{\Gamma}$ for τ, such that $\mathfrak{H}_{\widetilde{\Gamma}} = H(\mathcal{A})$. The assumption of positivity for h, k and Theorem 7.5.26, lead us to the fact that $W_{\widetilde{\Gamma}}(h)$ and $W_{\widetilde{\Gamma}}(k)$ are contained in $[0, \infty)$. So, $W_{\widetilde{\Gamma}}(h+k) \subset [0, \infty)$. Therefore, with the aid of Proposition 7.5.39, we have that

$$\sigma_{\mathcal{A}}(h + k) \cap \mathbb{C} = \sigma_{\widetilde{\Gamma}}(h + k) \cap \mathbb{C}$$
$$\subset \overline{\text{co}}(\sigma_{\widetilde{\Gamma}}(h + k) \cap \mathbb{C})$$
$$= \overline{\text{co}} W_{\widetilde{\Gamma}}(h + k) \subset \mathbb{R}.$$

Hence, we conclude that $h + k \geq 0$. $\qquad\square$

In the proof of the following result, as far as quasi-invertible elements are concerned, the reader is referred to the second last paragraph before Definition 2.1.1.

Lemma 7.5.46 *Let* $\mathcal{A}[\tau]$ *be a locally convex algebra with identity and let* $x, \ y \in \mathcal{A}$. *If* $\lambda \in \mathbb{C}^*(\lambda \neq 0)$. *Then* $\lambda \in \sigma_{\mathcal{A}}(xy)$, *if and only if,* $\lambda \in \sigma_{\mathcal{A}}(yx)$.

Proof We first show that $\beta(xy) = \beta(yx)$: let $\lambda > 0$, such that $(\lambda^{-1}xy)^n \to 0$, for $n \to \infty$. By induction, $(\lambda^{-1}xy)^n = \lambda^{-n}x(yx)^{n-1}y$, $n \in \mathbb{N}$. So, by continuity of multiplication we have that $\lambda^{-n}(yx)^{n+1} \to 0$, for $n \to \infty$. Thus, by Proposition 2.2.14 (3), we have that $\beta(xy) \geq \beta(yx)$. The reverse inequality is derived by symmetrical arguments.

Let $\infty \in \sigma_A(xy)$. That is, $xy \notin A_0$. By Proposition 2.2.14 (2) we equivalently have that $\beta(xy) = \infty$. By the previously established equality, $\beta(yx) = \infty$. Thus, $yx \notin A_0$, hence $\infty \in \sigma_A(yx)$.

Let $\lambda \in \mathbb{C} \setminus \{0\}$, such that $\lambda \notin \sigma_A(xy)$. Then, $(\lambda e - xy)^{-1} \in A_0$. Therefore, since $e - \lambda^{-1}xy$ is invertible, we have that $\lambda^{-1}xy$ is quasi-invertible, with quasi-inverse $(\lambda^{-1}xy)^\circ = e - (e - \lambda^{-1}xy)^{-1}$. By the following equality,

$$\lambda^{-1}yx \circ \left(\lambda^{-1}y(\lambda^{-1}xy)^\circ x - \lambda^{-1}yx\right) = \lambda^{-1}y\left(\lambda^{-1}xy \circ (\lambda^{-1}xy)^\circ\right)x,$$

we then have that $\lambda^{-1}yx$ is quasi-invertible, with quasi-inverse

$$(\lambda^{-1}yx)^\circ = \lambda^{-1}y(\lambda^{-1}xy)^\circ x - \lambda^{-1}yx.$$

Hence, $e - \lambda^{-1}yx$ is invertible with inverse

$$(e - \lambda^{-1}yx)^{-1} = e - (\lambda^{-1}yx)^\circ = e + \lambda^{-1}y(e - \lambda^{-1}xy)^{-1}x.$$

Furthermore, we have

$$\begin{aligned}
\beta\left(\lambda^{-1}y(e - \lambda^{-1}xy)^{-1}x\right) &= \beta\left(\lambda^{-1}xy(e - \lambda^{-1}xy)^{-1}\right) \\
&= \beta\left((e - \lambda^{-1}xy - e)(e - \lambda^{-1}xy)^{-1}\right) \\
&= \beta\left(e - (e - \lambda^{-1}xy)^{-1}\right) \\
&\leq \beta(e) + \beta\left((e - \lambda^{-1}xy)^{-1}\right) < \infty,
\end{aligned}$$

where the one to last inequality is due to Proposition 2.2.15. Therefore, $\lambda^{-1}y(e - \lambda^{-1}xy)^{-1}x \in A_0$. We conclude that $(e - \lambda^{-1}yx)^{-1} \in A_0$, i.e., $\lambda \notin \sigma_A(yx)$. By exchanging the roles of x and y, we have that for $\lambda \in \mathbb{C} \setminus \{0\}$, $\lambda \in \sigma_A(xy)$, if and only if, $\lambda \in \sigma_A(yx)$.   □

Lemma 7.5.47 Let $A[\tau]$ be a complete hypocontinuous semi-HB*-algebra, and $x \in A$. If $xx^* \leq 0$, then $xx^* = 0$.

Proof We consider a defining family of seminorms Γ for τ, such that $\mathfrak{H}_\Gamma = H(A)$, as is provided by Corollary 7.5.44. Let $x = h + ik$, where $h, k \in H(A)$. Then,

$$xx^* = 2(h^2 + k^2) - x^*x. \tag{7.5.15}$$

By using similar arguments to the proof of Proposition 2.5.5, we can see that $\sigma_A(h^2) \cap \mathbb{C}$ and $\sigma_A(k^2) \cap \mathbb{C}$ are contained in $[0, \infty)$. Hence, $h^2, k^2 \geq 0$. Moreover, by Lemma 7.5.46 and from the assumption '$xx^* \leq 0$', we have that $-x^*x \geq 0$. Therefore, Lemma 7.5.45 and (7.5.15) imply $xx^* \geq 0$. So, we have that $xx^* \leq 0$ and

$xx^* \geq 0$, from which it follows that $\sigma_A(xx^*) \subset \{0, \infty\}$. By considering a maximal commutative *-subalgebra of A that contains xx^*, we may suppose without loss of generality that A is commutative. By Theorem 3.4.9 and Lemma 3.4.6, the Gelfand transform, $\widehat{xx^*}$, of xx^* is a continuous $\mathbb{C}^*$-valued function on the carrier space of $A_0 = A[B_0]$, that has as possible values the points 0 and ∞; moreover, it takes the value ∞ on a nowhere dense subset. Therefore, we conclude that $\widehat{xx^*} = 0$, hence $xx^* = 0$. $\qquad\qquad\square$

Proposition 7.5.48 *Let $A[\tau]$ be a complete hypocontinuous semi-HB*-algebra. Then, $A[\tau]$ is a semi-GB*-algebra.*

Proof Let $x \in A$. The element x^*x is contained in a commutative semi-HB^*-algebra, say C. By Proposition 7.5.32, C is a semi-GB^*-algebra. Hence, from Theorem 4.1.4(ii), there exist $h, \ k \geq 0$, such that $x^*x = h - k$ and $hk = kh = 0$. Since $(xk)^*(xk) = k(x^*x)k = -k^3 \leq 0$, Lemma 7.5.47 gives that $(xk)^*(xk) = 0$. Thus, $k = 0$ and so $x^*x = h \geq 0$. Thus, A is symmetric (see the discussion before Definition 3.1.3), i.e., A is a semi-GB^*-algebra. $\qquad\qquad\square$

Based on the aforementioned result, Theorem 7.5.43 can take the following improved form. In particular, *Theorem 7.5.49 is a Vidav-Palmer type theorem for semi-GB*-algebras.*

Theorem 7.5.49 (Wood) *Let $A[\tau]$ be a complete hypocontinuous locally convex algebra. Then, $A[\tau]$ is a semi-GB*-algebra, if and only if, there is a defining family of seminorms Γ for τ, with respect to which (A, Γ) has a hermitian decomposition.*

Corollary 7.5.50 *Every hypocontinuous complete GB*-algebra has a defining family of seminorms, with respect to which the algebra has a hermitian decomposition.*

In the case where $A[\tau]$ is metrizable, we can deduce a more general "if and only if" argument than that of Theorem 7.5.49, as the next result informs us, *which namely is a Vidav–Palmer type theorem for GB*-algebras.*

Theorem 7.5.51 (Wood) *Let $A[\tau]$ be a Fréchet locally convex algebra. Then, $A[\tau]$ is a GB*-algebra, if and only if, there exists a defining family of seminorms Γ with respect to which (A, Γ) has a hermitian decomposition.*

Proof $\Rightarrow$ Since $A[\tau]$ is Fréchet, the multiplication of $A[\tau]$ is jointly continuous (see comments following Definition 3.1.1). This results in $A[\tau]$ being hypocontinuous. Indeed, let U be a 0-neighbourhood and B a bounded subset in $A[\tau]$. By joint continuity of multiplication, there is a 0-neighbourhood V, such that $V V \subset U$. Since B is bounded, there is $\varepsilon > 0$, such that $B \subset \varepsilon V$. Hence, for the 0-neighbourhood $\frac{1}{\varepsilon} V$, we have that

$$B\left(\frac{1}{\varepsilon}V\right) \subset \varepsilon V\left(\frac{1}{\varepsilon}V\right) = V V \subset U \ \text{ and similarly } \ \left(\frac{1}{\varepsilon}V\right)B \subset U,$$

i.e., $A[\tau]$ is hypocontinuous. Then, the proof of this direction follows immediately from Theorem 7.5.49.

$\Leftarrow$ suppose that there exists a defining family of seminorms Γ for τ, with respect to which $(\mathcal{A}, \Gamma)$ has a hermitian decomposition. By Theorems 7.5.43 and 7.5.49, $\mathcal{A}[\tau]$ is a semi-GB^*-algebra with respect to the involution $\natural$. Thus, what remains to be shown in order for $\mathcal{A}[\tau]$ to be a GB^*-algebra is the continuity of the involution $\natural$. We first note that by the definition of $\natural$ it is straightforward that $H(\mathcal{A}) = \mathfrak{H}_\Gamma$. Since $\mathcal{A}[\tau]$ is Fréchet and $\natural$ is conjugate linear we can apply the closed graph theorem [74, p. 301, Theorem 4]. Therefore, the continuity of the map $\natural$ will follow once it has a closed graph. So, let (x_n) be a sequence in $\mathcal{A}[\tau]$, such that $x_n \to x$ and $x_n^\natural \to y$, where $x, y \in \mathcal{A}$. Then,

$$x + y = \lim_n \left(x_n + x_n^\natural\right) \in \overline{H(\mathcal{A})} = \overline{\mathfrak{H}_\Gamma} = \mathfrak{H}_\Gamma = H(\mathcal{A}),$$

where the next to last equality follows from Proposition 7.5.6. Moreover,

$$i\left(x - y\right) = \lim_n i\left(x_n - x_n^\natural\right) \in \overline{H(\mathcal{A})} = \overline{\mathfrak{H}_\Gamma} = \mathfrak{H}_\Gamma = H(\mathcal{A}).$$

Therefore, we have that

$$(x + y)^\natural = x + y \quad \text{and} \quad i\left(x - y\right) = \left(i(x - y)\right)^\natural,$$

from which we obtain $y = x^\natural$. This shows that $\natural$ has a closed graph and so the proof is complete. $\square$

We remark that for the left-hand side direction ($\Leftarrow$) in the proof of the previous Theorem 7.5.51, the hermitian decomposition together with the metrizability of the algebra guarantee the continuity of the involution $\natural$, a property which is encoded in the "DNA" of a GB^*-algebra.

Notes The results of this section are due to A.W. Wood and can be found in [156]. Proposition 7.5.32 is due to P.G. Dixon and comes from [47, Appendix A, Lemma A.1]. We remark that, the concept of a GB^*-algebra, defined by A.W. Wood (Definition 7.5.27), for the needs of the Vidav–Palmer Theorem in the context of GB^*-algebras (see, for example, Theorem 7.5.51), is equivalent to that of a GB^*-algebra of P.G. Dixon as given by Definition 3.3.5.

Chapter 8
Applications II: Tensor Products

Tensor products of GB^*-algebras were investigated for the first time in [64]. As we have noticed, GB^*-algebras being algebras of unbounded operators are, in particular, important for mathematical physics and quantum mechanics. In this aspect, we want to emphasize that there is a physical justification for using tensor products. For example, tensor products are used to describe two quantum systems as one joint system (see [2]), while the physical significance of tensor products always depends on the applications, which may involve wave functions, spin states, oscillators etc.; see e.g., [29, 70]. Other results on topological tensor products of unbounded operator algebras can be found in [1, 65, 71].

8.1 Prerequisites: Examples

▶ *We emphasize that throughout this chapter we deal with locally convex (∗-)algebras with jointly continuous multiplication.*

Let us now recall some standard definitions concerning topological tensor products.

Definition 8.1.1 ([60, Definition 29.5]) Let $\mathcal{A}_1[\tau_1]$, $\mathcal{A}_2[\tau_2]$ be locally convex ∗-algebras. Let $\mathcal{A}_1 \otimes \mathcal{A}_2$ be their (algebraic) tensor product ∗-algebra. A topology τ on $\mathcal{A}_1 \otimes \mathcal{A}_2$ is called *∗-admissible* (with the tensor product ∗-algebra structure of $\mathcal{A}_1 \otimes \mathcal{A}_2$), if the following conditions are satisfied:

(1) $\mathcal{A}_1 \otimes \mathcal{A}_2$ endowed with τ is a locally convex ∗-algebra, with jointly continuous multiplication, denoted by $\mathcal{A}_1 \underset{\tau}{\otimes} \mathcal{A}_2$ and its completion by $\mathcal{A}_1 \widehat{\underset{\tau}{\otimes}} \mathcal{A}_2$,.

(2) The tensor map $\otimes : \mathcal{A}_1 \times \mathcal{A}_2 \to \mathcal{A}_1 \underset{\tau}{\otimes} \mathcal{A}_2 : (x, y) \mapsto x \otimes y$ is continuous.

© The Author(s), under exclusive license to Springer Nature Switzerland AG 2022
M. Fragoulopoulou et al., *Generalized B*-Algebras and Applications*, Lecture Notes
in Mathematics 2298, https://doi.org/10.1007/978-3-030-96433-7_8

(3) For any equicontinuous subsets M, N of the duals A_1^*, A_2^* of A_1, A_2 respectively, the set $M \otimes N = \{f \otimes g : f \in M, g \in N\}$ is an equicontinuous subset of the dual of $A_1 \underset{\tau}{\otimes} A_2$.

When A_1, A_2 have no involution, then τ is called a *compatible* topology on $A_1 \otimes A_2$, see [111, p. 375, Definition 3.1]. Namely, *a tensor topology τ on $A_1 \otimes A_2$ is $*$-admissible* (resp. *compatible*), if and only if,

$$\varepsilon \prec \tau \prec \pi \tag{8.1.1}$$

(see [60, p. 370, discussion before 29.7]), where ε, π stand for the *biprojective* (equivalently *injective*, cf. also [111]), respectively *projective locally convex tensor topologies*. Suppose that $A_1[\tau_1]$, $A_2[\tau_2]$ are locally convex $(*-)$algebras, with corresponding defining families of $(*-)$seminorms, given by $\Gamma_1 = \{p\}$, respectively by $\Gamma_2 = \{q\}$. Then, the defining families of $(*-)$seminorms for ε, π, denoted by $\Gamma_{\otimes_\varepsilon}$, respectively $\Gamma_{\otimes_\pi}$ are given, as

$$\varepsilon_{p,q}(z) := \sup\left\{\left|\sum_{i=1}^{n} f(x_i)g(y_i)\right| : f \in U_p^\circ,\, g \in U_q^\circ\right\},$$

$$z = \sum_{n=1}^{n} x_i \otimes y_i \in A_1 \underset{\varepsilon}{\otimes} A_2,$$

$$\pi_{p,q}(z) := \inf\left\{\sum_{n=1}^{n} p(x_i)q(y_i) : z = \sum_{n=1}^{n} x_i \otimes y_i\right\},\quad z \in A_1 \underset{\pi}{\otimes} A_2,$$

for all $(p, q) \in \Gamma_1 \times \Gamma_2$, where infimum (in the last equality) is taken over all representations $\sum_{n=1}^{n} x_i \otimes y_i$ of z in $A_1 \underset{\pi}{\otimes} A_2$. Note that U_k°, $k = p$, q, is the polar of the closed unit semiball $U_k := \{x_i \in A_i : k(x_i) \leq 1\}$, $k = p$, q in A_i, $i = 1, 2$. In particular, one has that

$$\varepsilon_{p,q}(x \otimes y) = p(x)q(y) = \pi_{p,q}(x \otimes y), \;\forall\, x \in A_1, y \in A_2.$$

Clearly, ε, π are Hausdorff, compatible (resp. $*$-admissible) tensor topologies on $A_1 \otimes A_2$.

In the case when $A_i[\tau_i]$, $i = 1, 2$ are normed $(*-)$algebras, the tensor topologies ε, π are respectively denoted by λ and γ and the corresponding $(*-)$norms by $\|\cdot\|_\lambda$ and $\|\cdot\|_\gamma$.

Coming back to $A_i[\tau_i]$, $i = 1, 2$, as C^*-convex algebras, observe that the *injective*, respectively *projective C^*-convex tensor topologies* α and ω on $A_1 \otimes A_2$ [60, Section 31] are analogous to the minimal (resp. maximal) C^*-crossnorms max and min on the tensor product of two C^*-algebras [144, p. 203]. Both are Hausdorff $*$-admissible tensor topologies. If $A_1[\tau_1]$, $A_2[\tau_2]$ are pro-C^*-algebras, let $\Gamma_1 = \{p\}$ and $\Gamma_2 = \{q\}$ be defining families of C^*-seminorms for τ_1, τ_2 respectively. Denote by $\Gamma_{\otimes_\alpha} = \{\alpha_{p,q}\}$, $\Gamma_{\otimes_\omega} = \{\omega_{p,q}\}$ the families of C^*-seminorms defining the C^*-convex tensor topologies α and ω, respectively. Particularly, let $R_k(\mathcal{A}_i)$ and $R_{p,q}(\mathcal{A}_1 \widehat{\otimes}_\pi \mathcal{A}_2)$, $k = p, q$, $i = 1, 2$, stand for the sets of continuous Hilbert space representations μ_i on $\mathcal{A}_i[\tau_i]$, $i = 1, 2$ and $\mu_1 \otimes \mu_2$ on $\mathcal{A}_1 \widehat{\otimes}_\pi \mathcal{A}_2$, whose continuity is respectively given by the inequalities

$$\|\mu_1(x)\| \le p(x), \ \|\mu_2(y)\| \le q(y) \text{ and } \|(\mu_1 \otimes \mu_2)(z)\| \le \pi_{p,q}(z),$$

for all $x \in \mathcal{A}_1$, $y \in \mathcal{A}_2$ and $z \in \mathcal{A}_1 \otimes_\pi \mathcal{A}_2$. Then,

$$\alpha_{p,q}(z) := \sup \left\{ \|(\mu_1 \otimes \mu_2)(z)\| : (\mu_1, \mu_2) \in R_p(\mathcal{A}_1) \times R_q(\mathcal{A}_2) \right\}, \ \forall \, z \in \mathcal{A}_1 \widehat{\otimes}_\pi \mathcal{A}_2$$

$$\text{and } \omega_{p,q}(z) := \sup \left\{ \|\mu(z)\| : \mu \in R_{p,q}(\mathcal{A}_1 \widehat{\otimes}_\pi \mathcal{A}_2) \right\}, \ \forall \, z \in \mathcal{A}_1 \widehat{\otimes}_\pi \mathcal{A}_2.$$

In this respect, we have that (see [60, Proposition 31.3])

$$\varepsilon \prec \alpha \prec \omega \prec \pi \text{ and} \tag{8.1.2}$$

$$\alpha_{p,q}(x \otimes y) = p(x)q(y) = \omega_{p,q}(x \otimes y),$$

for every elementary tensor $x \otimes y$ in $\mathcal{A}_1 \otimes \mathcal{A}_2$ and any p, q.

We proceed now with some examples of topological tensor product $*$-algebras that are, in particular, GB^*-algebras.

Example 8.1.2 Let $\mathcal{A}_1[\tau_1]$, $\mathcal{A}_2[\tau_2]$ be two pro-C^*-algebras (hence GB^*-algebras). Consider the C^*-convex algebras $\mathcal{A}_1 \underset{\alpha}{\otimes} \mathcal{A}_2$, $\mathcal{A}_1 \underset{\omega}{\otimes} \mathcal{A}_2$. Then, their completions, denoted respectively by $\mathcal{A}_1 \widehat{\otimes}_\alpha \mathcal{A}_2$, $\mathcal{A}_1 \widehat{\otimes}_\omega \mathcal{A}_2$ are pro-C^*-algebras, therefore GB^*-algebras too [60, p. 386].

Example 3.3.16(6) also serves as a topological tensor product that, in particular, is a GB^* algebra.

In order to proceed to our third example we need some prerequisites, which we first exhibit. Let X be a locally compact space and $\mathcal{A}[\tau]$ a complete GB^*-algebra with a natural family $\{p_\nu\}_{\nu \in \Lambda}$ of $*$-seminorms (see Remark 3.5.7(4) and take, for instance, $\mathcal{A}[\tau]$ to be a C^*-like locally convex $*$-algebra; in this respect, cf. also Corollary 3.5.4 and Proposition 3.5.8). Consider the locally convex $*$-algebra $\mathcal{C}(X, \mathcal{A})$ of all $\mathcal{A}$-valued continuous functions on X, with the compact-open topology. According to our conventions, $\mathcal{A}[\tau]$ as a locally convex $*$-algebra has a jointly continuous multiplication, therefore so does $\mathcal{C}(X, \mathcal{A})$ too. It is known that

$$\mathcal{C}(X, \mathcal{A}) = \mathcal{C}_c(X) \widehat{\otimes}_\varepsilon \mathcal{A},$$

up to an isomorphism of locally convex $*$-algebras [111, p. 391, Theorem 1.1], where $\mathcal{C}_c(X)$ is the pro-C^*-algebra of all complex-valued continuous functions on X, under the topology "c" of uniform convergence on the compacta of X. As we have already seen, $\mathcal{C}_c(X)$ as a pro-C^*-algebra is a GB^*-algebra. If $\mathcal{K}$ is the family of all compact subsets of X, the topology "c" is determined by the C^*-seminorms

$$q_K(f) = \sup\{|f(x)| : x \in K\}, \ f \in \mathcal{C}_c(X), \ \forall \ K \in \mathcal{K}.$$

Suppose now that τ is defined by a directed family, $(p_\nu)_{\nu \in \Lambda}$, of $*$-seminorms; then the functions

$$q_{K,\nu}(f) = \sup\{p_\nu(f(x)) : x \in K\}, \ f \in \mathcal{C}(X, \mathcal{A}), \ K \in \mathcal{K}, \ \nu \in \Lambda,$$

are $*$-seminorms determining the compact-open topology on $\mathcal{C}(X, \mathcal{A})$.

With X as before, let $\mathcal{C}_b(X)$ denote the C^*-algebra of all bounded continuous complex-valued functions on X.

Lemma 8.1.3 *Let X be a locally compact space and $\mathcal{A}[\tau]$ a GB^*-algebra with a natural family $\{p_\nu\}_{\nu \in \Lambda}$ of $*$-seminorms. Let $\mathcal{C}(X, \mathcal{A})$, $\mathcal{C}_b(X)$ be as above, and $A_0 = \mathcal{C}_b(X)\widehat{\otimes}_\lambda \mathcal{A}[B_0]$, where λ is the normed injective tensor topology on $\mathcal{C}_b(X) \otimes \mathcal{A}[B_0]$. Then,*

(1) $\varepsilon\restriction_{A_0} \prec \lambda$.
(2) *The C^*-algebra $\mathcal{C}_b(X) \widehat{\otimes}_{\max} \mathcal{A}[B_0]$ is ε-dense in $\mathcal{C}_c(X)\widehat{\otimes}_\varepsilon \mathcal{A}$.*

Proof

(1) A defining family of seminorms for ε is denoted by $(\varepsilon_{K,\nu})_{(K,\nu)\in\mathcal{K}\times\Lambda}$, where for every element $z = \sum_{i=1}^n f_i \otimes x_i \in \mathcal{C}_c(X) \otimes \mathcal{A}$, we have

$$\varepsilon_{K,\nu}(z) := \sup\left\{\left|\sum_{i=1}^n \phi(f_i)y'(x_i)\right| : \phi \in U_K^\circ, \ y' \in U_\nu^\circ\right\}. \tag{8.1.3}$$

Note that U_K° is the polar of the closed unit semiball $U_K := \{f \in \mathcal{C}_c(X) : q_K(f) \leq 1\}$ in $\mathcal{C}_c(X)$ and U_ν° the polar of the closed unit semiball $U_\nu := \{x \in \mathcal{A} : p_\nu(x) \leq 1\}$ in $\mathcal{A}[\tau]$. The normed injective tensor topology λ on $\mathcal{C}_b(X) \otimes \mathcal{A}[B_0]$ is determined by the vector space norm $\|\cdot\|_\lambda$ (see also long discussion after (8.1.1)), where for every $z = \sum_{i=1}^n f_i \otimes x_i \in \mathcal{C}_b(X) \otimes \mathcal{A}[B_0]$,

$$\|z\|_\lambda := \sup\left\{\left|\sum_{i=1}^n \phi(f_i)y'(x_i)\right| : \phi \in U_{\mathcal{C}_b(X)}^\circ, \ y' \in U_{\mathcal{A}[B_0]}^\circ\right\},$$

with $U_{\mathcal{C}_b(X)}^\circ$ and $U_{\mathcal{A}_b}^\circ$ standing for the polars of the closed unit balls $U(\mathcal{C}_b(X))$ and $U(\mathcal{A}[B_0])$ of the C^*-algebras $\mathcal{C}_b(X)$ and $\mathcal{A}[B_0]$, respectively.

We shall show that for each $(K, \nu) \in \mathcal{K} \times \Lambda$,

$$\varepsilon_{K,\nu}(z) \leq \|z\|_\lambda, \ \forall z \in A_0,$$

so the claim (1) will have been proved. According to (8.1.3) it suffices to show that

$$U(\mathcal{C}_b(X)) \subset U_K \text{ and } U(\mathcal{A}[B_0]) \subset U_\nu, \ \forall (K, \nu) \in \mathcal{K} \times \Lambda.$$

Since $\|f\|_\infty = \sup\{q_K(f) : K \in \mathcal{K}\}$, the first inclusion is obvious. Let now $y \in U(\mathcal{A}[B_0])$. Then, $\sup\{p_\nu(y) : \nu \in \Lambda\} \leq 1$, which implies that $y \in U_\nu$, for all $\nu \in \Lambda$.

(2) Since the C^*-algebra $\mathcal{C}_b(X)$ is commutative, the normed injective tensor topology λ coincides on $\mathcal{C}_b(X) \otimes \mathcal{A}[B_0]$ with the C^*-tensor topology max [144, p. 215, Lemma 4.18]. Denote by $\|\cdot\|_{\max}$ the C^*-crossnorm corresponding to max. From (1), $\varepsilon\restriction_{A_0} \prec \max$. Hence, $A_0 = \mathcal{C}_b(X) \widehat{\underset{\max}{\otimes}} \mathcal{A}[B_0]$ can be continuously embedded into $\mathcal{C}(X, \mathcal{A}) = \mathcal{C}_c(X)\widehat{\underset{\varepsilon}{\otimes}}\mathcal{A}$. Moreover, by the continuity of the tensor map $\otimes$, one gets that

$$\mathcal{C}_c(X) \underset{\varepsilon}{\otimes} \mathcal{A} \subset \overline{\mathcal{C}_b(X) \underset{\max}{\otimes} \mathcal{A}[B_0]}^\varepsilon \subset \overline{\mathcal{C}_b(X) \widehat{\underset{\max}{\otimes}} \mathcal{A}[B_0]}^\varepsilon \subset \mathcal{C}_c(X)\widehat{\underset{\varepsilon}{\otimes}}\mathcal{A},$$

where the first inclusion follows from the fact that $\mathcal{C}_b(X)$, $\mathcal{A}[B_0]$ are sequentially dense in $\mathcal{C}_c(X)$, $\mathcal{A}[\tau]$ respectively, according to Theorem 4.2.11.

□

Example 8.1.4 Let $\mathcal{A}[\tau]$ be a complete GB*-algebra. We shall show that $\mathcal{C}(X, \mathcal{A}) = \mathcal{C}_c(X)\widehat{\underset{\varepsilon}{\otimes}}\mathcal{A}$ is a *GB*-algebra. We consider the C^*-algebra $A_0 = \mathcal{C}_b(X) \widehat{\underset{\lambda=\max}{\otimes}} \mathcal{A}[B_0]$. Then, from Lemma 8.1.3, $\varepsilon\restriction_{A_0} \prec \lambda = \max$ and $\mathcal{C}_c(X)\widehat{\underset{\varepsilon}{\otimes}}\mathcal{A} = \widetilde{A_0}[\varepsilon]$, where $A_0[\varepsilon]$ has a jointly continuous multiplication (see [111, p. 391, Theorem 1.1], as well the note ▶ at the beginning of Sect. 8.1). So from Corollary 7.1.3, it follows that $\mathcal{C}_c(X)\widehat{\underset{\varepsilon}{\otimes}}\mathcal{A}$ is a *GB*-algebra over the ε-closed unit ball $U(A_0)$ of $A_0[\|\cdot\|_{\max}]$.

8.2 𝔅*-Collections in Tensor Product Locally Convex *-Algebras

Let $\mathcal{A}[\tau]$ be a locally convex *-algebra with identity e. We denote by $\widetilde{\mathcal{A}}[\tau]$ the completion of $\mathcal{A}[\tau]$. Since, the multiplication in $\mathcal{A}[\tau]$ is jointly continuous (as we have emphasized at the beginning of Sect. 8.1), $\widetilde{\mathcal{A}}[\tau]$ is a complete locally convex *-algebra. In what follows, we shall use, for distinction, the symbol $\widetilde{\tau}$ for the topology of the complete locally convex *-algebra $\widetilde{\mathcal{A}}[\tau]$. We denote by $\mathfrak{B}^*_{\mathcal{A}}$ and $\mathfrak{B}^*_{\widetilde{\mathcal{A}}}$ the respective $\mathfrak{B}^*$-collections of Definition 3.3.1, in $\mathcal{A}[\tau]$, respectively $\widetilde{\mathcal{A}}[\tau]$.

Proposition 8.2.1 *Let* $\mathcal{A}[\tau]$ *be a locally convex* $*$*-algebra. Then, the following hold:*

(1) *For every* $B \in \mathfrak{B}^*_{\widetilde{\mathcal{A}}}$, *the set* $B \cap \mathcal{A} \in \mathfrak{B}^*_{\mathcal{A}}$.

(2) *If* $B \in \mathfrak{B}^*_{\mathcal{A}}$ *and* $\overline{B}^{\widetilde{\tau}}$ *is the closure of* B *in* $\widetilde{\mathcal{A}}[\tau]$*, then* $\overline{B}^{\widetilde{\tau}} \in \mathfrak{B}^*_{\widetilde{\mathcal{A}}}$ *and* $B = \overline{B}^{\widetilde{\tau}} \cap \mathcal{A}$.

Proof It follows easily by the very definitions.   □

Let now $\mathcal{A}_1[\tau_1]$ and $\mathcal{A}_2[\tau_2]$ be locally convex $*$-algebras and let τ be a $*$-admissible topology on $\mathcal{A}_1 \otimes \mathcal{A}_2$ (see Definition 8.1.1). Let $B_1 \in \mathfrak{B}^*_{\mathcal{A}_1}$, $B_2 \in \mathfrak{B}^*_{\mathcal{A}_2}$ and $\Gamma(B_1 \otimes B_2)$ the absolutely convex hull of $B_1 \otimes B_2$ in $\mathcal{A}_1 \underset{\tau}{\otimes} \mathcal{A}_2$. Denote by $\overline{\Gamma(B_1 \otimes B_2)}^{\tau}$, respectively $\overline{\Gamma(B_1 \otimes B_2)}^{\widetilde{\tau}}$, the closure of $\Gamma(B_1 \otimes B_2)$ in $\mathcal{A}_1 \underset{\tau}{\otimes} \mathcal{A}_2$, respectively $\mathcal{A}_1 \widehat{\underset{\tau}{\otimes}} \mathcal{A}_2$. Then, we have the following

Theorem 8.2.2 *Let* $\mathcal{A}_1[\tau_1]$, $\mathcal{A}_2[\tau_2]$, B_1 *and* B_2 *be as in the previous paragraph. Then, the collections*

$$\left\{ \overline{\Gamma(B_1 \otimes B_2)}^{\tau} : B_i \in \mathfrak{B}^*_{\mathcal{A}_i}, i = 1, 2 \right\} \text{ and } \left\{ \overline{\Gamma(B_1 \otimes B_2)}^{\widetilde{\tau}} : B_i \in \mathfrak{B}^*_{\mathcal{A}_i}, i = 1, 2 \right\},$$

are contained in $\mathfrak{B}^*_{\mathcal{A}_1 \underset{\tau}{\otimes} \mathcal{A}_2}$ *and* $\mathfrak{B}^*_{\mathcal{A}_1 \widehat{\underset{\tau}{\otimes}} \mathcal{A}_2}$ *respectively.*

Proof The absolutely convex hull of the set $B_1 \otimes B_2$ is given by

$$\Gamma(B_1 \otimes B_2) := \left\{ \sum_{\text{finite}} \lambda_i (x_i \otimes y_i) : \lambda_i \in \mathbb{C}, \sum_{\text{finite}} |\lambda_i| \leq 1 \text{ and } x_i \otimes y_i \in B_1 \otimes B_2 \right\}.$$

Using the very definitions, the continuity of the tensor map $\otimes$ and the fact that the $*$-admissible topology τ fulfills the inequality $\varepsilon \prec \tau \prec \pi$, we conclude easily that the collection $\{\Gamma(B_1 \otimes B_2) : B_i \in \mathfrak{B}^*_{\mathcal{A}_i}, i = 1, 2\}$ consists of absolutely convex, bounded subsets of $\mathcal{A}_1 \underset{\tau}{\otimes} \mathcal{A}_2$, which contain the identity and are symmetric and idempotent. Due to the continuity of the algebraic operations of the locally convex $*$-algebras under consideration, all of the preceding properties pass to the τ-(resp. $\widetilde{\tau}$-) closure of the sets $\Gamma(B_1 \otimes B_2)$, as before. With regard to this, recall also that the closure of a bounded subset of a topological vector space is bounded. Hence,

$$\left\{ \overline{\Gamma(B_1 \otimes B_2)}^{\tau} : B_i \in \mathfrak{B}^*_{\mathcal{A}_i}, i = 1, 2 \right\} \subset \mathfrak{B}^*_{\mathcal{A}_1 \underset{\tau}{\otimes} \mathcal{A}_2}, \text{ resp.}$$

$$\left\{ \overline{\Gamma(B_1 \otimes B_2)}^{\widetilde{\tau}} : B_i \in \mathfrak{B}^*_{\mathcal{A}_i}, i = 1, 2 \right\} \subset \mathfrak{B}^*_{\mathcal{A}_1 \widehat{\underset{\tau}{\otimes}} \mathcal{A}_2}.$$

□

Now from the investigation done in Example 8.1.4, we are led to the formulation of the following

Theorem 8.2.3 *Let $\mathcal{A}_1[\tau_1]$, $\mathcal{A}_2[\tau_2]$ be GB*-algebras and B_1^0, B_2^0 maximal members in $\mathfrak{B}^*_{\mathcal{A}_1}$, $\mathfrak{B}^*_{\mathcal{A}_2}$, respectively. Let τ be a $*$-admissible topology on $\mathcal{A}_1 \otimes \mathcal{A}_2$. The following statements are equivalent:*

(i) *$\mathcal{A}_1 \widehat{\otimes}_\tau \mathcal{A}_2$ is a GB*-algebra.*

(ii) *There is a C*-crossnorm $\| \cdot \|$ on $\mathcal{A}_1[B_1^0] \otimes \mathcal{A}_2[B_2^0]$, such that $A_0 := \mathcal{A}_1[B_1^0] \widehat{\otimes}_{\|\cdot\|} \mathcal{A}_2[B_2^0]$ is a C*-algebra contained in $\mathcal{A}_1 \widehat{\otimes}_\tau \mathcal{A}_2$ and $\tau \prec \| \cdot \|$ on A_0.*

Proof (ii) $\Rightarrow$ (i) From our hypotheses the following facts are true: (1) $\mathcal{A}_i[B_i^0]$, $i = 1, 2$ are dense in $\mathcal{A}_i[\tau_i]$, $i = 1, 2$ (see Theorem 4.2.11), (2) the tensor map $\otimes : \mathcal{A}_1[\tau_1] \times \mathcal{A}_2[\tau_2] \to \mathcal{A}_1 \otimes_\tau \mathcal{A}_2$ is continuous and (3) $\tau \prec \| \cdot \|$ on A_0. Thus, first

$$\mathcal{A}_1 \otimes \mathcal{A}_2 \subset \overline{\mathcal{A}_1[B_1^0] \otimes \mathcal{A}_2[B_2^0]}^\tau \subset \overline{\mathcal{A}_1[B_1^0] \widehat{\otimes}_{\|\cdot\|} \mathcal{A}_2[B_2^0]}^\tau \subset \mathcal{A}_1 \widehat{\otimes}_\tau \mathcal{A}_2,$$

therefore A_0 is τ-dense in $\mathcal{A}_1 \widehat{\otimes}_\tau \mathcal{A}_2$. Secondly (i) follows by applying Theorem 7.1.2.

(i) $\Rightarrow$ (ii) Suppose that $\mathcal{A}_1 \widehat{\otimes}_\tau \mathcal{A}_2$ is a GB*-algebra. Then, $(\mathcal{A}_1 \widehat{\otimes}_\tau \mathcal{A}_2)[B_0]$ is a C*-algebra with respect to the gauge function $\| \cdot \|_{B_0}$, where B_0 is the maximal element in $\mathfrak{B}^*_{\mathcal{A}_1 \widehat{\otimes}_\tau \mathcal{A}_2}$ (see Theorem 3.3.9(2)). If B_1^0, B_2^0 are maximal elements in $\mathfrak{B}^*_{\mathcal{A}_1}$, $\mathfrak{B}^*_{\mathcal{A}_2}$ respectively, then it follows from Theorem 8.2.2 that

$$\mathcal{A}_1[B_1^0] \otimes \mathcal{A}_2[B_2^0] \subset (\mathcal{A}_1 \widehat{\otimes}_\tau \mathcal{A}_2)[\overline{\Gamma(B_1^0 \otimes B_2^0)}^{\widetilde{\tau}}] \subset (\mathcal{A}_1 \widehat{\otimes}_\tau \mathcal{A}_2)[B_0].$$

So we may restrict $\| \cdot \|_{B_0}$ to $\mathcal{A}_1[B_1^0] \otimes \mathcal{A}_2[B_2^0]$ and obtain

$$A_0 := \mathcal{A}_1[B_1^0] \widehat{\otimes}_{\|\cdot\|_{B_0}} \mathcal{A}_2[B_2^0] \subset (\mathcal{A}_1 \widehat{\otimes}_\tau \mathcal{A}_2)[B_0]. \tag{8.2.4}$$

Thus, A_0 is a C*-algebra and since $\mathcal{A}_1 \widehat{\otimes}_\tau \mathcal{A}_2$ is a GB*-algebra over B_0, we have $\tau \prec \| \cdot \|_{B_0}$ on $(\mathcal{A}_1 \widehat{\otimes}_\tau \mathcal{A}_2)[B_0]$, hence on A_0 as well. Consequently, the proof is completed by considering $\| \cdot \| = \| \cdot \|_{B_0}$. $\qquad\square$

The following example demonstrates that not every complete tensor product of GB*-algebras under some $*$-admissible topology is necessarily a GB*-algebra.

Example 8.2.4 Let $\mathcal{A}_1$ and $\mathcal{A}_2$ be noncommutative C*-algebras. Consider the injective tensor norm $\| \cdot \|_\lambda$ on $\mathcal{A}_1 \otimes \mathcal{A}_2$. Then, $\| \cdot \|_\lambda$ is not a C*-norm on $\mathcal{A}_1 \otimes \mathcal{A}_2$ (see

[144, Theorems IV.4.14 and IV.4.19]). Therefore, $\mathcal{A}_1 \widehat{\otimes}_\lambda \mathcal{A}_2$ is not a C^*-algebra, but a Banach $*$-algebra. Hence, by Corollary 3.3.11(i), $\mathcal{A}_1 \widehat{\otimes}_\lambda \mathcal{A}_2$ is not a GB^*-algebra.

Thus, we are led to the following

Definition 8.2.5 *When the equivalent conditions of Theorem 8.2.3 hold, the locally convex $*$-algebra $\mathcal{A}_1 \widehat{\otimes}_\tau \mathcal{A}_2$ will be called* a tensor product GB^*-algebra of the GB^*-algebras $\mathcal{A}_1[\tau_1]$ and $\mathcal{A}_2[\tau_2]$.

The results of Sect. 8.2 lead to the following

Question It would be interesting and very helpful to know whether there are conditions under which the structure of a tensor product GB^*-algebra $\mathcal{A}_1 \widehat{\otimes}_\tau \mathcal{A}_2$ could be described by the collection

$$\left\{ \overline{\Gamma(B_1 \otimes B_2)}^{\widetilde{\tau}} : B_i \in \mathfrak{B}^*_{\mathcal{A}_i}, i = 1, 2 \right\}.$$

Maybe this would contribute to a relationship, like the one in Example 8.4.2(3), between $M_n(\mathcal{A})_b$ and $M_n(\mathcal{C}) \underset{max}{\widehat{\otimes}} \mathcal{A}_b$, which is nicer than the one we have up to now for the C^*-algebras

$$\mathcal{A}_1[B_1^0] \underset{\|\cdot\|}{\widehat{\otimes}} \mathcal{A}_2[B_2^0] \quad \text{and} \quad \left(\mathcal{A}_1 \widehat{\otimes}_\tau \mathcal{A}_2 \right)[B_0],$$

whose τ-closures coincide in $\mathcal{A}_1 \widehat{\otimes}_\tau \mathcal{A}_2$, as proof of Theorem 8.2.3 shows (see also (8.3.5)).

8.3   Tensor Product GB*-Algebras

In this section we verify Definition 8.2.5 in some cases. The reader should not expect an analogy of GB^*-tensor products with the C^*-tensor products. For instance, a GB^*-algebra attains unbounded $*$-representations (see Theorem 6.3.5), so to define the C^*-crossnorms max and min on the algebraic tensor product of two such algebras, seems not attainable. On the other hand, the biprojective tensor locally convex topology ε plays, in a way, the role of max (and min) in the GB^*-case, as Theorem 8.3.2 and Corollary 8.3.6, of this section, show. For instance, suppose that $\mathcal{A}_1[\tau_1]$, $\mathcal{A}_2[\tau_2]$ in Theorem 8.3.2 are pro-C^*-algebras (hence GB^*-algebras), with one of them being commutative. Then, the assumptions of Theorem 8.3.2 are satisfied and $\mathcal{A}_1 \widehat{\otimes}_\varepsilon \mathcal{A}_2$ is a pro-C^*-algebra (hence a GB^*-algebra), with $\varepsilon = \omega = \alpha$ (see (8.1.2) and [60, Corollary 31.16]). A crucial property for the last equalities is the commutativity of one of the pro-C^*-algebras $\mathcal{A}_1[\tau_1]$, $\mathcal{A}_2[\tau_2]$; the respective classical C^*-algebra result, in this case, is Lemma 4.18, p. 215, in [144].

Furthermore, the fact that commutativity appears for one of the factors of a GB^*-tensor product in the results of this section, is not surprising, since the definition of a GB^*-algebra requires symmetry (see Definition 3.3.2). More precisely, when a GB^*-algebra $\mathcal{A}[\tau]$ is also m-convex, then symmetry means that every element of the form $e + x^*x$, $x \in \mathcal{A}$, is invertible in $\mathcal{A}$ and to pass symmetry to the topological tensor product of two complete symmetric m^*-convex algebras, you need one of them to be commutative [60, Corollary 34.15] (the corresponding Banach algebra result is due to K.B. Laursen [109, Theorem III.3, p. 65]).

Yet, as we shall see, the C^*-crossnorms max, min are essentially used in the tensor product of the C^*-algebras generated by the maximal elements of the $\mathfrak{B}^*$-collections for the GB^*-algebras participating in a GB^*-tensor product algebra (concerning these C^*-algebras, see also Remark 3.5.7 and the discussion before it, as well as Remark 3.5.5(4)). Recall that given a GB^*-algebra $\mathcal{A}[\tau]$, the C^*-algebra $\mathcal{A}[B_0]$ generated by the maximal element B_0 of the collection $\mathfrak{B}^*_{\mathcal{A}}$ is the pillar on which the study of the structure of a GB^*-algebra is based.

Lemma 8.3.1 that follows is required in the proof of Theorem 8.3.2. Before we proceed to its statement, we set some notation that we shall need for it. Let $\mathcal{A}_1[\tau_1]$, $\mathcal{A}_2[\tau_2]$ be two locally convex $*$-algebras and τ a $*$-admissible topology on $\mathcal{A}_1 \otimes \mathcal{A}_2$ (see Definition 8.1.1). Suppose that $\Gamma_1 = \{p\}$, $\Gamma_2 = \{q\}$ and $\Gamma_{\otimes_\tau} = \{\tau_{p,q}\}$ are the defining families of $*$-seminorms for the topologies τ_1, τ_2, τ respectively.

To avoid indices in the preceding families Γ_1, Γ_2 and $\Gamma_{\otimes_\tau}$, we abuse the symbol $\mathcal{D}(p_\Lambda)$ as in (3.5.24) and put

$$\mathcal{D}(\Gamma_1) := \left\{ x \in \mathcal{A}_1 : \sup_{p \in \Gamma_1} p(x) < \infty \right\},$$

$$\mathcal{D}(\Gamma_2) := \left\{ y \in \mathcal{A}_2 : \sup_{q \in \Gamma_2} q(y) < \infty \right\},$$

$$\mathcal{D}(\Gamma_{\otimes_\tau}) := \left\{ z \in \mathcal{A}_1 \widehat{\otimes}_\tau \mathcal{A}_2 : \sup_{(p,q) \in \Gamma_1 \times \Gamma_2} \tau_{p,q}(z) < \infty \right\}.$$

In this regard, we have

Lemma 8.3.1 *Let $\mathcal{A}_1[\tau_1]$ and $\mathcal{A}_2[\tau_2]$ be locally convex $*$-algebras and τ a $*$-admissible topology on $\mathcal{A}_1 \otimes \mathcal{A}_2$. Let Γ_1, Γ_2, $\Gamma_{\otimes_\tau}$ be as before. Suppose that for each p, q, the $*$-seminorm $\tau_{p,q}$ is a cross-seminorm, i.e., $\tau_{p,q}(x \otimes y) = p(x)q(y)$, for all $x \in \mathcal{A}_1$ and $y \in \mathcal{A}_2$. Then,*

$$\mathcal{D}(\Gamma_1) \underset{\gamma}{\otimes} \mathcal{D}(\Gamma_2) \subseteq \mathcal{D}(\Gamma_{\otimes_\tau}),$$

where γ is the normed projective tensor topology. .

Proof According to our modified notation just before Lemma 8.3.1, we use in the place of p_Λ as in (3.5.24), the symbols p_{Γ_i}, $i = 1, 2$ and $p_{\Gamma_{\otimes_\tau}}$. Namely, $p_{\Gamma_1}(x) :=$

$\sup_{p \in \Gamma_1} p(x)$, $x \in \mathcal{D}(\Gamma_1)$, $pr_{\Gamma_2}(y) := \sup_{q \in \Gamma_2} q(y)$, $y \in \mathcal{D}(\Gamma_2)$ and $pr_{\Gamma_{\otimes_\tau}}(z) :=$
$\sup_{(p,q) \in \Gamma_1 \times \Gamma_2} \tau_{p,q}(z)$, $z \in \mathcal{D}(\Gamma_{\otimes_\tau})$, respectively.

Let now $z = \sum_{i=1}^{n} x_i \otimes y_i \in \mathcal{D}(\Gamma_1) \otimes \mathcal{D}(\Gamma_2)$. Then, we have

$$pr_{\Gamma_{\otimes_\tau}}(z) := \sup_{(p,q) \in \Gamma_1 \times \Gamma_2} \tau_{p,q}(z) = \sup_{(p,q) \in \Gamma_1 \times \Gamma_2} \tau_{p,q} \left(\sum_{i=1}^{n} x_i \otimes y_i \right)$$

$$\leq \sup_{(p,q) \in \Gamma_1 \times \Gamma_2} \sum_{i=1}^{n} \tau_{p,q}(x_i \otimes y_i)$$

$$\leq \sum_{i=1}^{n} \sup_{p \in \Gamma_1} p(x_i) \sup_{q \in \Gamma_2} q(y_i) = \sum_{i=1}^{n} pr_{\Gamma_1}(x_i) pr_{\Gamma_2}(y_i)| < \infty.$$

Taking now infimum with respect to all representations of z in $\mathcal{D}(\Gamma_1) \otimes \mathcal{D}(\Gamma_2)$, we obtain

$$pr_{\Gamma_{\otimes_\tau}}(z) := \sup_{(p,q) \in \Gamma_1 \times \Gamma_2} \tau_{p,q}(z) \leq \|z\|_\gamma, \ \forall \, z \in \mathcal{D}(\Gamma_1) \otimes \mathcal{D}(\Gamma_2),$$

where the mentioned infimum is denoted by $\| \cdot \|_\gamma$ (see also long discussion after (8.1.1)) and reads as follows

$$\|z\|_\gamma := \inf \left\{ \sum_{i=1}^{n} pr_{\Gamma_1}(x_i) pr_{\Gamma_2}(y_i), \ z \in \mathcal{D}(\Gamma_1) \otimes \mathcal{D}(\Gamma_2) \right\}.$$

The normed projective tensor topology γ on $\mathcal{D}(\Gamma_1) \otimes \mathcal{D}(\Gamma_2)$ is generated by $\| \cdot \|_\gamma$. It follows that $\mathcal{D}(\Gamma_1) \underset{\gamma}{\otimes} \mathcal{D}(\Gamma_2)$ is embedded continuously in $\mathcal{D}(\Gamma_{\otimes_\tau})$ □

Theorem 8.3.2 *Let $\mathcal{A}_1[\tau_1]$ and $\mathcal{A}_2[\tau_2]$ be GB^*-algebras, with a natural family of $*$-seminorms (take, for instance, $\mathcal{A}_1[\tau_1]$ and $\mathcal{A}_2[\tau_2]$ to be C^*- like locally convex $*$- algebras; see Corollary 3.5.4 and Remark 3.5.7(4)). If either $\mathcal{A}_1$ or $\mathcal{A}_2$ is commutative, then $\mathcal{A}_1 \widehat{\underset{\varepsilon}{\otimes}} \mathcal{A}_2$ is a tensor product GB^*-algebra.*

Proof Since one of the $\mathcal{A}_1$, $\mathcal{A}_2$ is commutative, $\mathcal{A}_1[B_1^0] \widehat{\underset{\lambda}{\otimes}} \mathcal{A}_2[B_2^0]$ is a C^*-algebra [144, Theorem IV.4.14]. Suppose that $\mathcal{A}_1[B_1^0] \widehat{\underset{\lambda}{\otimes}} \mathcal{A}_2[B_2^0]$ can be algebraically embedded into $\mathcal{A}_1 \widehat{\underset{\varepsilon}{\otimes}} \mathcal{A}_2$. By the same proof as that of Lemma 8.1.3(1), $\varepsilon \restriction_{\mathcal{A}_1[B_1^0] \widehat{\underset{\lambda}{\otimes}} \mathcal{A}_2[B_2^0]} \prec \lambda$, therefore by Theorem 8.2.3, $\mathcal{A}_1 \widehat{\underset{\varepsilon}{\otimes}} \mathcal{A}_2$ is a tensor product GB^*-algebra.

So all that remains is to prove that $\mathcal{A}_1[B_1^0]\widehat{\otimes}_\lambda\mathcal{A}_2[B_2^0]$ can be algebraically embedded into $\mathcal{A}_1\widehat{\otimes}_\varepsilon\mathcal{A}_2$. By Lemma 8.3.1, $\mathcal{A}_1[B_1^0]\otimes\mathcal{A}_2[B_2^0]\subset\mathcal{D}(\Gamma_{\otimes_\varepsilon})$, where

$$\mathcal{D}(\Gamma_{\otimes_\varepsilon})=\left\{z\in\mathcal{A}_1\widehat{\otimes}_\varepsilon\mathcal{A}_2:\sup_{(p,q)\in\Gamma_1\times\Gamma_2}\varepsilon_{p,q}(z)<\infty\right\}\text{ and}$$

$$p_{\Gamma_{\otimes_\varepsilon}}(z):=\sup_{(p,q)\in\Gamma_1\times\Gamma_2}\varepsilon_{p,q}(z),\ z\in\mathcal{D}(\Gamma_{\otimes_\varepsilon}).$$

We may restrict $p_{\Gamma_{\otimes_\varepsilon}}$ to $\mathcal{A}_1[B_1^0]\otimes\mathcal{A}_2[B_2^0]$. Let $z_0=\sum_{i=1}^n x_i\otimes y_i\in\mathcal{A}_1[B_1^0]\otimes\mathcal{A}_2[B_2^0]$. Then,

$$
\begin{aligned}
p_{\Gamma_{\otimes_\varepsilon}}(z_0)&=\sup_{(p,q)\in\Gamma_1\times\Gamma_2}\varepsilon_{p,q}(z_0)\\
&=\sup_{(p,q)\in\Gamma_1\times\Gamma_2}\sup\left\{\left|\sum_{i=1}^n x'(x_i)y'(y_i)\right|:x'\in U_p^\circ,\ y'\in U_q^\circ\right\}\\
&=\sup\left\{\left|\sum_{i=1}^n x'(x_i)y'(y_i)\right|:x'\in U_{\mathcal{A}_1[B_1^0]}^\circ,\ y'\in U_{\mathcal{A}_2[B_2^0]}^\circ\right\}\\
&=\|z_0\|_\lambda,
\end{aligned}
$$

where $U_{\mathcal{A}_1[B_1^0]}^\circ$, $U_{\mathcal{A}_2[B_2^0]}^\circ$ are the polars of the closed unit balls $U(\mathcal{A}_1[B_1^0])$, $U(\mathcal{A}_2[B_2^0])$ of the C^*-algebras $\mathcal{A}_1[B_1^0]$, $\mathcal{A}_2[B_2^0]$, respectively.

Since $\mathcal{A}_1\widehat{\otimes}_\varepsilon\mathcal{A}_2$ is complete, it follows from the same proof as in [60, Theorem 10.23] that $\mathcal{D}(\Gamma_{\otimes_\varepsilon})$ is complete with respect to $p_{\Gamma_{\otimes_\varepsilon}}$. Hence,

$$\mathcal{A}_1[B_1^0]\widehat{\otimes}_\lambda\mathcal{A}_2[B_2^0]=\mathcal{A}_1[B_1^0]\ \widehat{\otimes}_{p_{\Gamma_{\otimes_\varepsilon}}}\ \mathcal{A}_2[B_2^0]\subset\mathcal{A}_1\widehat{\otimes}_\varepsilon\mathcal{A}_2.$$

$\square$

As in the case of Theorem 8.2.3, for the GB^*-algebras $\mathcal{A}_1[\tau_1]$ and $\mathcal{A}_2[\tau_2]$ of Theorem 8.3.2, one concludes that

$$\overline{\mathcal{A}_1[B_1^0]\ \widehat{\underset{\lambda=\max}{\otimes}}\ \mathcal{A}_2[B_2^0]}^{\ \varepsilon}=\overline{(\mathcal{A}_1\widehat{\otimes}_\varepsilon\mathcal{A}_2)[B_0]}^{\ \varepsilon}=\mathcal{A}_1\widehat{\otimes}_\varepsilon\mathcal{A}_2.\qquad(8.3.5)$$

Corollary 8.3.3 *Let $\mathcal{A}_1[\tau_1]$ and $\mathcal{A}_2[\tau_2]$ be C^*-like locally convex $*$-algebras, with either $\mathcal{A}_1$, or $\mathcal{A}_2$ being commutative. Then, the following statements are equivalent:*

(i) *$\mathcal{A}_1\widehat{\otimes}_\varepsilon\mathcal{A}_2$ admits a C^*-like family of seminorms defining the topology ε.*

(ii) *$\mathcal{A}_1\widehat{\otimes}_\varepsilon\mathcal{A}_2$ is a C^*-like locally convex $*$-algebra.*

Proof It is obvious that (ii) $\Rightarrow$ (i). So we prove that (i) $\Rightarrow$ (ii). Since $\mathcal{A}_1[\tau_1]$ and $\mathcal{A}_2[\tau_2]$ are C^*-like locally convex $*$-algebras, they are also GB^*-algebras and $\mathcal{A}_1[B_1^0] = (\mathcal{A}_1)_b$, $\mathcal{A}_2[B_2^0] = (\mathcal{A}_2)_b$, according to Theorem 3.5.3 and Remark 3.5.7(3). Thus, by Theorem 8.3.2, $\mathcal{A}_1 \widehat{\otimes}_\varepsilon \mathcal{A}_2$ is a tensor product GB^*-algebra. Hence, from the proof (i) $\Rightarrow$ (ii) of Theorem 8.2.3, the relation (8.2.4) holds, with B_0 the maximal element in $\mathfrak{B}^*_{\mathcal{A}_1 \widehat{\otimes}_\varepsilon \mathcal{A}_2}$. Since either $\mathcal{A}_1[B_1^0]$ or $\mathcal{A}_2[B_2^0]$ is commutative,

we conclude that

$$\| \cdot \|_{B_0} = \| \cdot \|_\lambda = p_{\Gamma_{\otimes\varepsilon}},$$

on $A_0 := \mathcal{A}_1[B_1^0] \otimes \mathcal{A}_2[B_2^0]$, (see [144, p. 215, Lemma 4.18] and the proof of Theorem 8.3.2). It follows that B_0 coincides with the unit ball $U(p_{\Gamma_{\otimes\varepsilon}})$ of the C^*-algebra $\mathcal{D}(\Gamma_{\otimes\varepsilon})$ (for the last two symbols, see discussion after (3.5.24), before Lemma 3.5.2 and the beginning of the proof of Lemma 8.3.1). So from (i) and Theorem 3.5.3, we obtain (ii). $\square$

Example 8.3.4 Consider the locally convex $*$-algebra $C(X, \mathcal{A}) \cong C_c(X) \widehat{\otimes}_\varepsilon \mathcal{A}$ of Example 8.1.4, where X is a locally compact space and $\mathcal{A}$ a C^*-like locally convex $*$-algebra. Recall that if $\{p_\nu\}_{\nu \in \Lambda}$ is a C^*-like family of seminorms for $\mathcal{A}$, a defining family of $*$-seminorms for $C(X, \mathcal{A})$ is given by the functions

$$q_{K,\nu}(f) := \sup\{p_\nu(f(x)), x \in K\}, \ \forall \ f \in C(X, \mathcal{A}),$$

for all $\nu \in \Lambda$ and $K \subset X$, compact. It is easily verified that, for every compact subset K of X and every $\nu \in \Lambda$, there exists $\nu' \in \Lambda$ such that

$$q_{K,\nu}(fg) \leq q_{K,\nu'}(f)q_{K,\nu'}(g), \ q_{K,\nu}(f)^2 \leq q_{K,\nu'}(f^*f) \ \text{and} \ q_{K,\nu}(f^*) \leq q_{K,\nu'}(f),$$

for all $f, g \in C(X, \mathcal{A})$, i.e., the family of seminorms $(q_{K,\nu})_{\nu \in \Lambda}$ is a C^*-like family of seminorms defining the topology of $C(X, \mathcal{A})$. Hence, by Corollary 8.3.3, $C(X, \mathcal{A})$ is a tensor product C^*-like locally convex $*$-algebra.

In [59, p. 241, Corollary] it is proved that if $\mathcal{A}[\tau]$ is a locally convex algebra with jointly continuous multiplication and an identity element e, then there is a family of seminorms, say $\{q\}$, defining the topology τ of $\mathcal{A}$, such that $q(e) = 1$, for every q.

Suppose now that $\mathcal{A}[\tau]$ has a continuous involution $*$. Then, a family of $*$-seminorms $\{q'\}$ is defined on $\mathcal{A}$ that preserves involution, identity and defines the topology τ. Indeed, using the preceding family of seminorms $\{q\}$, we put $q'(a) := \max\{q(a), q(a^*)\}$, $a \in \mathcal{A}$. It follows that for every q', $q'(a^*) = q'(a)$, for all $a \in \mathcal{A}$ and $q'(e) = 1$. So the aforementioned result in [59], for the $*$-case, reads as follows:

Lemma 8.3.5 *Let* $\mathcal{A}[\tau]$ *be a locally convex* $*$-*algebra with identity* e *and jointly continuous multiplication. Then,* τ *is defined by a family* $\Gamma = \{q'\}$ *of* $*$-*seminorms, such that* $q'(e) = 1$, *for every* $q' \in \Gamma$.

Corollary 8.3.6 *Let $A_1[\tau_1]$ and $A_2[\tau_2]$ be complete locally convex $*$-algebras with identities e_1, e_2, respectively. Consider the following statements:*

(i) *$A_1 \widehat{\otimes}_\varepsilon A_2$ is a tensor product GB*-algebra.*

(ii) *$A_1[\tau_1]$ and $A_2[\tau_2]$ are GB*-algebras.*

Then (i) *implies* (ii). *If $A_1[\tau_1]$ and $A_2[\tau_2]$ are GB*-algebras with a natural family of $*$-seminorms and either A_1 or A_2 is commutative, then* (ii) *implies* (i) *too.*

Proof (i) $\Rightarrow$ (ii) Let $\Gamma_1 = \{p\}$ (resp. $\Gamma_2 = \{q\}$) be a family of $*$-seminorms defining the topology τ_1 on A_1 (resp. τ_2 on A_2). Denote by $\Gamma_{p,q} = \{\varepsilon_{p,q}\}$ the family of $*$-seminorms defining the topology ε on $A_1 \otimes A_2$. Then, $\varepsilon_{p,q}(x \otimes y) = p(x)q(y)$ for all $x \in A_1$ and $y \in A_2$. Applying Lemma 8.3.5, we may suppose that the families of $*$-seminorms Γ_1, respectively Γ_2, preserve the corresponding identities. Thus, one has

$$\varepsilon_{p,q}(x \otimes e_2) = p(x)q(e_2) = p(x), \ \forall \, x \in A_1.$$

Consider now the embedding

$$A_1 \hookrightarrow A_1 \widehat{\otimes}_\varepsilon A_2 : x \mapsto x \otimes e_2.$$

Then $A_1[\tau_1]$ becomes a closed $*$-subalgebra of $A_1 \widehat{\otimes}_\varepsilon A_2$ with $e_1 \otimes e_2 \in A_1$. Hence, by Proposition 3.3.19, $A_1[\tau_1]$ is a GB^*-algebra. Similarly, $A_2[\tau_2]$ is a GB^*-algebra.
  (ii) $\Rightarrow$ (i) It follows from Theorem 8.3.2. □

Remark 8.3.7 Note that for $A_i[\tau_i]$, $i = 1, 2$, as before and τ a $*$-admissible topology on $A_1 \otimes A_2$, such that $A_1 \widehat{\otimes}_\tau A_2$ is a tensor product GB^*-algebra, one concludes from the proof of Corollary 8.3.6, that $A_1[\tau_1]$ and $A_2[\tau_2]$ are GB^*-algebras too.

8.4 GB*-Nuclearity

The concept of a nuclear topological vector space was first given by A. Grothendieck and much of the theory of these types of spaces was developed by him and published in [69]. The nuclearity of a C^*-algebra was first discussed under the term "Property T" by M. Takesaki, in 1964 (see, for instance, [144, p. 204, Vol. III]). The present term is due to E. Effros and C. Lance [55]. Nuclear C^*-algebras are developed in various directions and are very important in the C^*-algebra theory; see also [39, 41, 101, 102] and the literature therein. For the non-normed case, see [26, 124].

Throughout this section, given two GB^*-algebras $\mathcal{A}[\tau_i]$, $i = 1, 2$ and their tensor product GB^*-algebra $\mathcal{A}_1 \widehat{\underset{\tau}{\otimes}} \mathcal{A}_2$, we shall denote by B_0^i and B_0 the maximal elements in $\mathfrak{B}^*_{\mathcal{A}_i}$, $i = 1, 2$ and $\mathfrak{B}^*_{\mathcal{A}_1 \widehat{\underset{\tau}{\otimes}} \mathcal{A}_2}$ respectively.

Our definition of GB^*-nuclearity that follows is based on the definition of nuclearity given by S.J. Bhatt and D.J. Karia for pro-C^*-algebras in [26, Theorem 4.5].

Definition 8.4.1 Let $\mathcal{A}[\tau]$ be a GB^*-algebra with B_0 the greatest member in $\mathfrak{B}^*_{\mathcal{A}}$. We say that $\mathcal{A}[\tau]$ is *nuclear* if the C^*-subalgebra $\mathcal{A}[B_0]$ of $\mathcal{A}[\tau]$ is nuclear.

Examples 8.4.2

(1) Every nuclear pro-C^*-algebra $\mathcal{A}[\tau]$ is a nuclear GB^*- algebra. Indeed, by Bhatt and Karia [26, Theorem 4.5], a pro-C^*-algebra $\mathcal{A}[\tau]$ is nuclear, if and only if, its bounded part $\mathcal{A}_b$ (which is a C^*-algebra, see Remark 3.5.7(3)) is nuclear. Moreover, $\mathcal{A}[\tau]$ is a GB^*-algebra with B_0 the closed unit ball of $\mathcal{A}_b$, therefore $\mathcal{A}_b = \mathcal{A}[B_0]$ (see Example 3.3.16(1)), whence the conclusion.

(2) Every commutative GB^*-algebra is nuclear. In particular, the Arens algebra $L^\omega[0, 1]$ (see Example 3.3.16(5)) is a nuclear GB^*-algebra, with $(L^\omega[0, 1])[B_0] = L^\infty[0, 1]$.

(3) Let $\mathcal{A}[\tau]$ be a commutative C^*-like locally convex $*$-algebra and $M_n(\mathcal{A})$ all $n \times n$ matrices with entries from $\mathcal{A}$. The assumption for $\mathcal{A}[\tau]$ implies that this is a GB^*-algebra with $\mathcal{A}[B_0] = \mathcal{A}_b$, B_0 being the maximal element in $\mathfrak{B}^*_{\mathcal{A}}$. Then, since $M_n(\mathbb{C})$ is nuclear, all C^*-crossnorms coincide on $M_n(\mathbb{C}) \otimes \mathcal{A}_b$ and are also equal to the injective norm λ since $\mathcal{A}_b$ is commutative (see [144, Theorem IV.4.14]). Thus, Theorem 8.3.2 implies that $M_n(\mathcal{A}) \cong M_n(\mathbb{C}) \widehat{\underset{\varepsilon}{\otimes}} \mathcal{A}$ is a GB^*-algebra. Let us now consider the structure of $M_n(\mathcal{A})$ as a locally convex $*$-algebra through that of $\mathcal{A}[\tau]$. If τ is defined by the family $\Gamma = \{p\}$ of seminorms, then a defining family of seminorms, say $\{r_{\tilde{p},\tilde{q}}\}$, $p, q \in \Gamma$, for $M_n(\mathcal{A})$, is given as follows: Let $M := \mathcal{A}^n$ be the finitely generated free (left) $\mathcal{A}$-module, corresponding to $\mathcal{A}$. Then, each $x \in M_n(\mathcal{A})$ defines a continuous $\mathcal{A}$-linear operator T_x on M given by

$$T_x(a) := \left(\sum_{i=1}^{n} x_{1i} x_i, \ldots, \sum_{i=1}^{n} x_{ni} x_i \right),$$

where x_{ij} are the entries of the matrix x and $a = \sum_{i=1}^{n} x_i e_i$ is an element of M, with $x_i \in \mathcal{A}$ and $e_i := (\delta_{ij})_{1 \leq j \leq n} \in \mathcal{A}^n$, such that $\delta_{ii} = e$ (the identity of $\mathcal{A}$) and $\delta_{ij} = 0$, for $i \neq j$. Furthermore, since Γ is a C^*-like family, for every $p \in \Gamma$, there is $q \in \Gamma$, such that $p(xy) \leq q(x)q(y)$, for any $x, y \in \mathcal{A}$. For such

p, q as before, we put

$$r_{\tilde{p},\tilde{q}}(x) := \sup\left\{\tilde{p}(T_x(a)) : \tilde{q}(a) \leq 1\right\}, \quad x \in M_n(\mathcal{A}), \quad \text{where}$$

$$\tilde{p}(T_x(a)) := \sum_{j=1}^{n} p\left(\sum_{i=1}^{n} x_{ji} x_i\right) \text{ and } \tilde{q}(a) := \sum_{i=1}^{n} q(x_i), \quad \forall\, a \in M$$

(see also [60, pp. 109–110, (6)]). An easy calculation shows that $r_{\tilde{p},\tilde{q}}(x) < \infty$, for all $x \in M_n(\mathcal{A})$. Also, we can easily verify that $M_n(\mathcal{A})_b = M_n(\mathcal{A}_b) = M_n(\mathbb{C}) \underset{\max}{\widehat{\otimes}} \mathcal{A}_b$, where the last equality results from the nuclearity of $M_n(\mathbb{C})$. At the same time, we conclude that $M_n(\mathcal{A})_b$ is nuclear, therefore what remains to be proved is that $M_n(\mathcal{A})[B_0] = M_n(\mathcal{A})_b$. This will follow by showing that $M_n(\mathcal{A})$ accepts a C^*-like family of seminorms defining its topology, so that by Corollary 8.3.3, $M_n(\mathcal{A})$ will be a C^*-like locally convex $*$-algebra. This demands a series of technical long computations, which are based on the same method, with obvious modifications, as the one applied in [110, Section 2, pp. 463 - 467] for showing that: if $\mathcal{B}[\tau']$ is a pro-C^*-algebra and $M_n(\mathcal{B})$ the algebra of all $n \times n$ matrices with entries in $\mathcal{B}$, then $M_n(\mathcal{B})$ also becomes a pro-C^*-algebra (ibid., p. 466, Corollary 2.1). For this, we use the form of the seminorms $\{r_{\tilde{p},\tilde{q}}\}$ defined above and the fact that $\Gamma = \{p\}$ is a C^*-like family of seminorms defining the topology τ of our commutative C^*-like locally convex $*$-algebra $\mathcal{A}$.

Note that we can also define a GB^*-algebra to be nuclear by assuming no identity; this is, for instance, required for the two-sided ideal I in the following

Theorem 8.4.3 *Let $\mathcal{A}[\tau]$ be a GB^*-algebra. Let I be a closed two-sided ideal in $\mathcal{A}[\tau]$. Then, $\mathcal{A}[\tau]$ is nuclear, if and only if, I and $\mathcal{A}/I$ are nuclear.*

Proof By Theorem 7.4.8 we have that I is a $*$-ideal, therefore a GB^*-subalgebra of $\mathcal{A}[\tau]$ by Proposition 3.3.19. Applying again Theorem 7.4.8, we conclude that $\mathcal{A}/I$ is a GB^*-algebra with the quotient topology and $(\mathcal{A}/I)[B_I^0] = \mathcal{A}[B_0]/(I \cap \mathcal{A}[B_0])$, where B_I^0, B_0 are the maximal elements in $\mathfrak{B}_{\mathcal{A}/I}^*$, $\mathfrak{B}_{\mathcal{A}}^*$, respectively.

Now, $\mathcal{A}[\tau]$ is nuclear, if and only if, $\mathcal{A}[B_0]$ is nuclear, if and only if, (see [39, Corollary 4]) $I \cap \mathcal{A}[B_0]$ and $\mathcal{A}[B_0]/(I \cap \mathcal{A}[B_0])$ are nuclear, if and only if, (from the previous paragraph), $I \cap \mathcal{A}[B_0]$ and $(\mathcal{A}/I)[B_I^0]$ are nuclear. But, the GB^*-structure of I is determined by the collection $\mathfrak{B}_I^* = I \cap \mathfrak{B}_{\mathcal{A}}^*$, so that $I[I \cap B_0] = I \cap \mathcal{A}[B_0]$. Thus, we have proved that the GB^*-algebra $\mathcal{A}[\tau]$ is nuclear, if and only if, I and $\mathcal{A}/I$ are nuclear GB^*-algebras. □

Applying Theorem 8.4.3 and the open mapping theorem for Fréchet spaces, we obtain the following

Corollary 8.4.4 *Let $\mathcal{A}_1[\tau_1]$, $\mathcal{A}_2[\tau_2]$ be Fréchet GB^*-algebras. If $\phi : \mathcal{A}_1[\tau_1] \to \mathcal{A}_2[\tau_2]$ is a continuous surjective $*$-homomorphism, then $\mathcal{A}_1[\tau_1]$ is nuclear, if and only if, $\ker(\phi)$ and $\mathcal{A}_2[\tau_2]$ are nuclear.*

Theorem 8.4.5 *Let $\mathcal{A}[\tau]$ be a GB*-algebra. Then, the following are equivalent:*

(1) *$\mathcal{A}[\tau]$ is nuclear.*
(2) *For every C*-algebra $\mathcal{B}$, $\mathcal{A}[B_0] \widehat{\underset{\max}{\otimes}} \mathcal{B} = \mathcal{A}[B_0] \widehat{\underset{\min}{\otimes}} \mathcal{B}$, with max = min.*
(3) *For every pro-C*-algebra $\mathcal{B}[\tau_\mathcal{B}]$, $\mathcal{A}[B_0]\widehat{\underset{\alpha}{\otimes}}\mathcal{B} = \mathcal{A}[B_0]\widehat{\underset{\omega}{\otimes}}\mathcal{B}$, with $\alpha = \omega$, where*
   α, ω are, the corresponding to the classical C*-cross-norms $\| \cdot \|_{\min}$, $\| \cdot \|_{\max}$, in the context of pro-C*-algebras (see (8.1.2)).

Proof The equivalence of (1), (2) follows easily from the very definitions.
    (2) $\Rightarrow$ (3) The topology $\tau_\mathcal{B}$ is determined by an upwards directed family of C^*−seminorms, say, $\{q\}$. Then, $\mathcal{B}[\tau_\mathcal{B}] = \varprojlim \mathcal{B}_q$, where each $\mathcal{B}_q = \mathcal{B}/\ker(q)$, is a C^*−algebra [60, pp. 15–16]. Thus, from (2), we have that $\mathcal{A}[B_0] \widehat{\underset{\max}{\otimes}} \mathcal{B}_q = \mathcal{A}[B_0] \widehat{\underset{\min}{\otimes}} \mathcal{B}_q$, with max = min, for each C^*−seminorm q. The assertion now follows from [60, Corollary 31.11].
    (3) $\Rightarrow$ (2) It is obvious.   □

Proposition 8.4.6 *Let $\mathcal{A}_i[\tau_i]$, $i = 1, 2$, be nuclear GB*-algebras. Suppose that $\mathcal{A}_1\widehat{\underset{\tau}{\otimes}}\mathcal{A}_2$ is the tensor product GB*-algebra of $\mathcal{A}_i[\tau_i]$, $i = 1, 2$, such that $(\mathcal{A}_1\widehat{\underset{\tau}{\otimes}}\mathcal{A}_2)[B_0] = \mathcal{A}_1[B_0^1] \widehat{\underset{\|\cdot\|_{B_0}}{\otimes}} \mathcal{A}_2[B_0^2]$. Then, $\mathcal{A}_1\widehat{\underset{\tau}{\otimes}}\mathcal{A}_2$ is nuclear too.*

Proof It is straightforward from the nuclearity of the C^*−algebras $\mathcal{A}_i[B_0^i]$, $i = 1, 2$ and the fact that max = $\| \cdot \|_{B_0}$ = min (see [102, p. 389]).   □

    Under the basis put in this section, it would be interesting to see how far GB*-nuclearity goes. A **natural question** is, for instance, under what conditions a subalgebra of a nuclear GB*-algebra is also nuclear.
    A further investigation of GB*-nuclearity was started by the third author in [152], where a more detailed study of completely positive maps of GB*-algebras and their relationship to GB*-nuclearity is carried out.

8.5   Some Applications on Tensor Product GB*-Algebras

For the terminology used in what follows, cf. the discussion in Chap. 5, before Definition 5.1.1 and after Proposition 5.1.2.
    Let $\mathcal{A}$ be a $*$-algebra and p an unbounded C^*-(semi)norm of $\mathcal{A}$ with domain $\mathcal{D}(p)$ (see discussion before Definition 3.5.1). Denote by $\mathcal{A}_p$ the C^*-algebra completion of the pre-C^*-algebra $(\mathcal{D}(p)/N_p)[\| \cdot \|_p]$ under the C^*-norm $\|x + N_p\|_p := p(x)$, $x \in \mathcal{D}(p)$, induced by p, where $N_p = \{x \in \mathcal{D}(p) : p(x) = 0\}$. For the concepts of a *w-semifinite* unbounded C^*-seminorm and of a *well-behaved* $*$-representation of a $*$-algebra, that we use right after, we refer the reader to [24].
    Let now $\mathcal{A}[\tau]$ be a GB*-algebra, π a $*$-representation of $\mathcal{A}$ in a Hilbert space $\mathcal{H}$, with domain $\mathcal{D}(\pi)$, and $\mathcal{I}_b^\mathcal{A} = \{x \in B(\mathcal{A}) : ax \in B(\mathcal{A}), \forall a \in \mathcal{A}\}$. For the

*-subalgebra $B(\mathcal{A})$, see Remark 3.5.5(4). For every $x \in \mathcal{I}_b^{\mathcal{A}}$, define a C^*-seminorm on $\pi(\mathcal{A})$ as follows

$$p_x(\pi(a)) = \|\overline{\pi(x^*ax)}\|, \ \forall \, a \in \mathcal{A}.$$

The locally convex topology induced by the family of C^*-seminorms $\{p_x \ : \ x \in \mathcal{I}_b^{\mathcal{A}}\}$, we denote with τ_{lu} [22, p. 96].

A *-representation π of a *-algebra $\mathcal{A}$ is said to be *uniformly nondegenerate* if $[\pi(\mathcal{I}_b^{\mathcal{A}})\mathcal{D}(\pi)] = \mathcal{H}_\pi$, where $[K]$ means the closed linear span of a subset K of a Hilbert space $\mathcal{H}$.

In this regard, we now have the following

Proposition 8.5.1 *Let $\mathcal{A}_1[\tau_1]$ and $\mathcal{A}_2[\tau_2]$ be metrizable C^*-like locally convex *-algebras with an identity element. Suppose that either $\mathcal{A}_1$ or $\mathcal{A}_2$ is commutative. Then, every uniformly nondegenerate *-representation π of the tensor product GB^*-algebra $\mathcal{A}_1\widehat{\otimes}_\varepsilon\mathcal{A}_2$ into the image $Im(\pi)[\tau_{lu}]$ of π, under the topology τ_{lu}, is continuous. The restrictions π_1, π_2 of π to $\mathcal{A}_1[\tau_1]$, $\mathcal{A}_2[\tau_2]$, respectively, are also continuous.*

Proof From Theorem 3.5.3, every C^*-like locally convex *-algebra is a GB^*-algebra, therefore Theorem 8.3.2 yields that $\mathcal{A}_1\widehat{\otimes}_\varepsilon\mathcal{A}_2$ is a GB^*-algebra. The latter being Fréchet is quasi-complete and bornological [74, p. 222, Proposition 3] (for the preceding terminology see also Definition 4.3.14(2) and the discussion after it). So the result follows from [22, Proposition 4.8].   □

Notice that if π is an unbounded *-representation of a Fréchet *-algebra $\mathcal{A}[\tau]$, then there exist two topologies on $Im(\pi)$ making π continuous [136, Theorem 3.6.8]. These topologies $\tau_i, i = 1, 2$, do not necessarily have the property that $Im(\pi)[\tau_i]$ is complete [136, Proposition 3.3.19(i)].

Proposition 8.5.2 *Let $\mathcal{A}_1[\tau_1]$ and $\mathcal{A}_2[\tau_2]$ be metrizable C^*-like locally convex *-algebras with an identity element. Suppose that either $\mathcal{A}_1$ or $\mathcal{A}_2$ is commutative. Then, every positive linear functional on $\mathcal{A}_1\widehat{\otimes}_\varepsilon\mathcal{A}_2$ is continuous.*

Proof As in the proof of Proposition 8.5.1, $\mathcal{A}_1\widehat{\otimes}_\varepsilon\mathcal{A}_2$ is a quasi-complete and bornological GB^*-algebra, so the result follows from Corollary 7.2.2.   □

A consequence of Theorem 8.3.2 and of the proofs of Proposition 8.5.1 and Proposition 8.5.2 is the following

Corollary 8.5.3 *Let $\mathcal{A}$ be a Fréchet GB^*-algebra with $\mathcal{A}[B_0] = \mathcal{D}(p_\Lambda)$ (see (3) and (4) in Remark 3.5.7). Then, we have that*

(1) *every uniformly nondegenerate *-representation*

$$\pi : \mathcal{C}(\mathbb{R}, \mathcal{A}) \cong C_c(\mathbb{R})\widehat{\otimes}_\varepsilon\mathcal{A} \to Im(\pi)[\tau_{lu}]$$

is continuous;

(2) *every positive linear functional on* $\mathcal{C}(\mathbb{R}, \mathcal{A}) \cong \mathcal{C}_c(\mathbb{R}) \widehat{\otimes}_\varepsilon \mathcal{A}$ *is continuous.*

Proposition 8.5.4 *Let* $\mathcal{A}[\tau]$ *be a metrizable* C^*-*like locally convex* $*$-*algebra with* $B_0 = U(\mathcal{A}_b)$ *(the closed unit ball of* $\mathcal{A}_b$) *and* $\mathcal{I}_b^{\mathcal{A}} \neq \{0\}$. *Consider the Fréchet* GB^*-*algebra* $\mathcal{C}(\mathbb{R}, \mathcal{A})$ *of all* $\mathcal{A}$-*valued continuous functions on the real line* $\mathbb{R}$. *Then,* $\mathcal{C}(\mathbb{R}, \mathcal{A}) \cong \mathcal{C}_c(\mathbb{R}) \widehat{\otimes}_\varepsilon \mathcal{A}$ *attains a well-behaved* $*$-*representation.*

Proof By Theorem 3.5.3 and Proposition 3.5.8, we have

$$\mathcal{A}_b[\| \cdot \|_b] = \mathcal{A}[B_0][\| \cdot \|_{B_0}] = B(\mathcal{A})[\| \cdot \|_{B_0}]. \qquad (8.5.6)$$

Hence, $\mathcal{A}_b[\| \cdot \|_b]$ is a C^*-algebra, therefore hermitian (see discussion after Definition 3.1.2). In the case of the Fréchet pro-C^*-algebra (hence Fréchet GB^*-algebra) $\mathcal{C}_c(\mathbb{R})$ we have that $\mathcal{C}_c(\mathbb{R})[\| \cdot \|_b] = \mathcal{C}_b(\mathbb{R})$, the C^*-algebra of all bounded continuous functions on $\mathbb{R}$, and $\mathcal{I}_b^{\mathcal{C}_c(\mathbb{R})} = \{f \in \mathcal{C}(\mathbb{R}) : supp(f) \text{ is compact}\}$. Consequently, $\mathcal{I}_b^{\mathcal{C}_c(\mathbb{R})} \neq \{0\} = \ker(\| \cdot \|_\infty)$ (see also [22, Example 3.9(3)]). We can now apply [61, Theorem 5.2(2) and Application (1) in p. 269] to obtain a well-behaved $*$-representation of the tensor product GB^*-algebra $\mathcal{C}(\mathbb{R}, \mathcal{A})$ (Example 8.1.4). □

An example of a GB^*-algebra $\mathcal{A}[\tau]$ with the property $\mathcal{I}_b^{\mathcal{A}} \neq \{0\}$ can be found in [25, Example 6.4]. In general, for $i = 1, 2$ and any metrizable C^*-like locally convex $*$-algebras $\mathcal{A}_i[\tau_i]$, such that at least one of them is commutative and $\mathcal{I}_b^{\mathcal{A}_i} \neq \{0\}$, we obtain that the tensor product GB^*-algebra $\mathcal{A}_1 \widehat{\otimes}_\varepsilon \mathcal{A}_2$ (cf. Theorem 8.3.2) accepts a well-behaved $*$-representation.

Notes Lemma 8.3.5 is a $*$-version of a result of A. Fernández and V. Müller, which can be found in [59, p. 241, Corollary]. Definition 8.4.1 is an analogue of a definition of nuclearity for pro-C^*-algebras given by S.J. Bhatt and D.J. Karia in [26, Theorem 4.5]. All other results are due to the first three authors of this book and are contained in [64].

Bibliography

1. M.-S. Adamo, M. Fragoulopoulou, Tensor products of normed and Banach quasi *-algebras. J. Math. Anal. Appl. **490**, 1–27 (2020)
2. D. Aerts, I. Daubechies, Physical justification for using the tensor product to describe two quantum systems as one joint system. Helvetica Phys. Acta **51**, 661–675 (1978)
3. G.R. Allan, A note on B^*-algebras. Proc. Camb. Philos. Soc. **61**, 29–32 (1965)
4. G.R. Allan, A spectral theory for locally convex algebras. Proc. Lond. Math. Soc. (3) **15**, 399–421 (1965)
5. G.R. Allan, On a class of locally convex algebras. Proc. Lond. Math. Soc. (3) **17**, 91–114 (1967)
6. J. Alcantara, J. Yngvason, Algebraic quantum field theory and non commutative moment problems. I, Ann. Inst. Henri Poincaré **48**, 147–159 (1988)
7. J.-P. Antoine, A. Inoue, C. Trapani, *Partial *-Algebras and their Operator Realizations* (Kluwer Academic Publ., Dordrecht, 2002)
8. C. Apostol, b^*-algebras and their representation. J. Lond. Math. Soc. **3**, 30–38 (1971)
9. R. Arens, The space L^ω and convex topological rings. Bull. Am. Math. Soc. **52**, 931–935 (1946)
10. R. Arens, Linear topological division algebras. Bull. Am. Math. Soc. **53**, 623–630 (1947)
11. R. Arens, A generalization of normed rings. Pac. J. Math. **2**, 455–471 (1952)
12. F. Bagarello, M. Fragoulopoulou, A. Inoue, C. Trapani, The completion of a C^*-algebra with a locally convex topology. J. Oper. Theory **56**, 357–376 (2006)
13. F. Bagarello, M. Fragoulopoulou, A. Inoue, C. Trapani, Structure of locally convex quasi C^*-algebras. J. Math. Soc. Jpn. **60**, 511–549 (2008)
14. F. Bagarello, M. Fragoulopoulou, A. Inoue, C. Trapani, Locally convex quasi C^*-normed algebras. J. Math. Anal. Appl. **366**, 593–606 (2010)
15. E. Beckenstein, L. Narici and C. Suffel, *Topological Algebras* (North-Holland, Amsterdam, 1977)
16. S.K. Berberian, The maximal ring of quotients of a finite von Neumann algebra. Rocky Mountain J. Math. **12**, 149–164 (1982)
17. S.J. Bhatt, A note on generalized B^*-algebras. I, J. Indian Math. Soc. **43**, 253–257 (1979)
18. S.J. Bhatt, A note on generalized B^*-algebras. II, J. Indian Math. Soc. **44**(1980), 285–290
19. S.J. Bhatt, On the dual of a generalized B^*-algebra. Math. Rep. Acad. Sci. Can. **4**, 3–9 (1982)
20. S.J. Bhatt, Generalized B^*-algebras, in *Topics in Banach Algebras and Topological Algebras* (Sardar Patel University Press, 1984), pp. 39–84
21. S.J. Bhatt, Irreducible representations of a class of unbounded operator algebras. Jpn. J. Math. No. 1 **11**, 103–108 (1985)

© The Author(s), under exclusive license to Springer Nature Switzerland AG 2022
M. Fragoulopoulou et al., *Generalized B*-Algebras and Applications*, Lecture Notes in Mathematics 2298, https://doi.org/10.1007/978-3-030-96433-7

22. S.J. Bhatt, M. Fragoulopoulou and A. Inoue, Existence of well-behaved ∗-representations of locally convex ∗-algebras. Math. Nachr. **279**, 86–100 (2006)
23. S.J. Bhatt, M. Fragoulopoulou, A. Inoue, Existence of spectral well-behaved ∗-representations. J. Math. Anal. Appl. **317**, 475–495 (2006)
24. S.J. Bhatt, A. Inoue, H. Ogi, Unbounded C^*-seminorms and unbounded C^*-spectral algebras. J. Oper. Theory **45**, 53–80 (2001)
25. S.J. Bhatt, A. Inoue, K-D. Kürsten, Well-behaved unbounded operator representations and unbounded C^*-seminorms. J. Math. Soc. Jpn. **56**, 417–445 (2004)
26. S.J. Bhatt, D.J. Karia, Complete positivity, tensor products and C^*-nuclearity for inverse limits of C^*-algebras. Proc. Indian Acad. Sci. Math. Sci. **101**, 149–167 (1991)
27. S.J. Bhatt, D.J. Karia, Uniqueness of the uniform norm with an application to topological algebras. Proc. Am. Math. Soc. **116**, 499–503 (1993)
28. G. Birkhoff, *Lattice Theory*, 2nd edn, vol. 25 (Amer. Math. Soc., Colloquium Publications, New York, 1948)
29. A. Böhm, *Quantum Mechanics* (Springer, New York, 1979)
30. F.F. Bonsall, J. Duncan, *Numerical Ranges of Operators on Normed Spaces and Elements of Normed Algebras*. London Math. Soc. Lectures Notes Series 2, (Cambridge University Press, Cambridge, 1971)
31. F.F. Bonsall, J. Duncan, *Numerical Ranges II*. London Math. Soc. Lectures Notes Series 10 (Cambridge University Press, Cambridge, 1973)
32. F.F. Bonsall, J. Duncan, *Complete Normed Algebras* (Springer, Berlin, 1973)
33. H.J. Borchers, On structure of the algebra of field operators. Nuovo Cimento (10) **24**, 214–236 (1962)
34. H.J. Borchers, J. Yngvason, On the algebra of field operators. The weak commutant and integral decompositions of states. Commun. Math. Phys. **42**, 231–252.
35. N. Bourbaki, *Espaces Vectoriels Topologiques*, Chap. I et II (Hermann, Paris, 1966)
36. H. Brezis, *Functional Analysis, Sobolev Spaces and Partial Differential Equations* (Springer, New York, 2010)
37. R.M. Brooks, On Locally m-convex ∗-algebras. Pac. J. Math. **25**, 5–23 (1967)
38. V.I. Chilin, B.S. Zakirov, Abstract characterization of EW^*-algebras. Funct. Anal. Appl. **25**, 76–78 (1991) (Russian). English translation: **25**, 63–64 (1991)
39. M.-D. Choi, E.G. Effros, Nuclear C^*-algebras and approximation property. Am. J. Math. **100**, 61–79 (1978)
40. C.L. Crowther, *Aspects of Duality Theory for Spaces of Measurable Operators*, PhD Thesis, University of Cape Town, South Africa, 1997
41. M. Dadarlat, S. Eilers, On the classification of nuclear C^*-algebras. Proc. Lond. Math. Soc. **85**, 168–210 (2002)
42. H.G. Dales, Automatic continuity: a Survey. Bull. Lond. Math. Soc. **10**, 129–183 (1978)
43. H.G. Dales, *Banach Algebras and Automatic Continuity* (Clarendon Press, Oxford, 2000)
44. H. Danrun, Z. Dianzhou, Joint spectrum and unbounded operator algebras. Acta Math. Sin. **3**, 260–269 (1986)
45. J. Dixmier, C^*-*Algebras* (North-Holland, Amsterdam, 1977)
46. J. Dixmier, *Von Neumann Algebras* (North-Holland, Amsterdam, 1981)
47. P.G. Dixon, *Generalized B^*-algebras*, PhD Thesis, University of Cambridge, 1970
48. P.G. Dixon, Generalized B^*-algebras. Proc. Lond. Math. Soc. **21**, 693–715 (1970)
49. P.G. Dixon, Unbounded operator algebras. Proc. Lond. Math. Soc. **23**, 53–69 (1971)
50. P.G. Dixon, Generalized B^*-algebras II. J. Lond. Math. Soc. **5**, 159–165 (1972)
51. P.G. Dixon, Automatic continuity of positive functionals on topological involution algebras. Bull. Austr. Math. Soc. **23**, 265–281 (1981)
52. R.S. Doran, V.A. Belfi, *Characterizations of C^*- Algebras* (Marcel Dekker, New York, 1986)
53. D.A. Dubin, M.A. Hennings, T.B. Smith, *Mathematical Aspects of Weyl Quantization and Phase* (World Scientific, Singapore, 2000)

54. M. Dubois-Viollette, A generalization of the classical moment problem on ∗-algebras with applications to relativistic quantum filed theory I, II. Commun. Math. Phys. **43**, 225–254 (1975), resp. Commun. Math. Phys. **54**, 151–172 (1977)

55. E. Effros and C. Lance, Tensor products of operator algebras. Adv. Math. **25**, 1–34 (1977)

56. S.J.L. van Eijndhoven, P. Kruszyński, GB^*-algebras associated with inductive limits of Hilbert spaces. Stud. Math. **85**, 107–123 (1987)

57. G.G. Emch, *Mathematical and Conceptual Foundations of 20th-Century Physics* (North-Holland, Amsterdam, 1984)

58. R.C. Fabec, *Fundamentals of Infinite Dimentional Representation Theory*. Monographs and Surveys in Pure and Applied Mathematics, vol. 114 (Chapman and Hall/CRC, 2000)

59. A. Fernández, V. Müller, Renormalizations of Banach and locally convex algebras. Stud. Math. **96**, 237–242 (1990)

60. M. Fragoulopoulou, *Topological Algebras with Involution* (North-Holland, Amsterdam, 2005)

61. M. Fragoulopoulou, A. Inoue, Unbounded ∗-representations of tensor product locally convex ∗-algebras induced by unbounded C^*-seminorms, Stud. Math. **183**, 259–271 (2007)

62. M. Fragoulopoulou, A. Inoue, K.-D. Kürsten, On the completion of a C^*-normed algebra under a locally convex algebra topology. Contemp. Math. **427**, 155–166 (2007)

63. M. Fragoulopoulou, A. Inoue, K.-D. Kürsten, Old and new results on GB^*-algebras. Banach Center Publ. **91**, 169–178 (2010)

64. M. Fragoulopoulou, A. Inoue, M. Weigt, Tensor products of generalized B^*-algebras. J. Math. Anal. Appl. **420**, 1787–1802 (2014)

65. M. Fragoulopoulou, A. Inoue, M. Weigt, Tensor products of unbounded operator algebras. Rocky Mountain J. Math. **44**, 895–912 (2014)

66. M. Fragoulopoulou, C. Trapani, *Locally Convex Quasi ∗-Algebras and their Representations*. Lecture Notes in Mathematics, vol. 2257 (Springer Cham, 2020)

67. L. Gillman, M. Jerison, *Rings of Continuous Functions* (Van Nostrand, 1960)

68. L. Gillman, M. Henriksen and M. Jerison, On a theorem of Gelfand and Kolmogoroff concerning maximal ideals in rings of continuous functions. Proc. Amer. Math. Soc. **5**, 447–455 (1954)

69. A. Grothendieck, Produits tensoriels topologiques et espaces nucléaires. Mem. Am. Math. Soc. No 16 (1955)

70. S.J. Gustafson, I.M. Sigal, *Mathematical Concepts of Quantum Mechanics*, 2nd edn. (Springer, Berlin, 2006)

71. W.-D. Heinrichs, Topological tensor products of unbounded operator algebras on Fréchet domains. Kyoto Univ. **33**, 241–255 (1997)

72. A.Ya. Helemskii, *Banach and Locally Convex Algebras* (Oxford Science Publ., Oxford, 1993)

73. A.Ya. Helemskii, *Lectures and Exercises on Functional Analysis*. Translations of Math. Monographs, vol. 233 (Amer. Math. Soc., Providence, 2006)

74. J. Horváth, *Topological Vector Spaces and Distributions* (Addison-Wesley, Reading, 1966)

75. A. Inoue, Locally C^*-algebra. Mem. Faculty Sci. Kyushu Univ. (Ser. A) **25**, 197–235 (1971)

76. A. Inoue, Representations of GB^*-algebras I. Fukuoka Univ. Sci. Rep. **5**, 63–78 (1975)

77. A. Inoue, Representations of GB^*-algebras II. Fukuoka Univ. Sci. Rep. **5**, 21–41 (1975)

78. A. Inoue, On a class of unbounded operator algebras. Pac. J. Math. **65**, 77–95 (1976)

79. A. Inoue, On a class of unbounded operator algebras II. Pac. J. Math. **66**, 411–431 (1976)

80. A. Inoue, Unbounded Hilbert algebras as locally convex ∗-algebras. Math. Rep. Col. Gen. Educ. Kyushu Univ. **10**, 113–129 (1976)

81. A. Inoue, On a class of unbounded operator algebras III. Pac. J. Math. **69**, 105–115 (1977)

82. A. Inoue, On a class of unbounded operator algebras IV. J. Math. Anal. Appl. **64**, 334–347 (1978)

83. A. Inoue, Unbounded generalizations of standard von Neumann algebras. Rep. Math. Phys **13**, 25–35 (1978)

84. A. Inoue, Unbounded generalizations of left Hilbert algebras. I, II, J. Funct. Anal. **34**, 339–362 (1979), ibid., **35**, 230–250 (1980)

85. A. Inoue, *Tomita–Takesaki Theory in Algebras of Unbounded Operators*. Lecture Notes in Mathematics, vol. 1699 (Springer, Berlin, 1998)
86. A. Inoue, *Tomita's Lectures on Observable Algebras in Hilbert Space*. Lecture Notes in Mathematics, vol. 2285 (Springer Nature, 2021)
87. A. Inoue, K. Kuriyama, S. Ôta, Topologies on unbounded operator algebras. Mem. Faculty Sci. Kyushu Univ. Ser. A **33**, 335–375 (1974)
88. A. Inoue, K.-D. Kürsten, On C^*-like locally convex $*$-algebras. Math. Nachr. **235**, 51–58 (2002)
89. A. Inoue, K. Takesue, Self-adjoint representations of polynomial algebras. Trans. Am. Math. Soc. **280**, 393–400 (1983)
90. G. Jameson, *Ordered Linear Spaces*. Lecture Notes in Math., vol. 141 (Springer, Berlin, 1970)
91. H. Jarchow, *Locally Convex Spaces* (B.G. Teubner, Stuttgart, 1981)
92. J.P. Jurzuk, Simple facts about algebras of unbounded operators. J. Funct. Anal. **21**, 469–482 (1976)
93. J.P. Jurzuk, Topological aspects of algebras of unbounded operators. J. Funct. Anal. **24**, 397–425 (1977)
94. J.P. Jurzuk, Unbounded operator algebras and DF-spaces. Publ. RIMS Kyoto Univ. **17**, 755–776 (1981)
95. R.V. Kadison, J.R. Ringrose, *Fundamentals of the Theory of Operator Algebras*. Volume I, Elementary Theory, Volume II, Advanced Theory (Academic Press, London, 1983 and 1986)
96. I. Kaplansky, Topological rings. Am. J. Math. **59**, 153–183 (1947)
97. J.L. Kelley, *General Topology* (Springer, New York, 1955)
98. W. Kunze, Zur algebraischen Struktur der GC^*-Algebren. Math. Nachr. **88**, 7–11 (1979)
99. W. Kunze, Halbordnung und topologie in GC^*-Algebren. Wiss. Z. KMU Leipzig, Math. Naturwiss. R. **31**, 55–62 (1982)
100. K. Kuratowski, *Topology*, Vol. I (Academic Press, London, 1966)
101. C. Lance, On nuclear C^*-algebras. J. Func. Anal. **12**, 157–176 (1973)
102. C. Lance, Tensor products and nuclear C^*-algebras. Proc. Symp. Pure Math. **38**, 379–399 (1982)
103. G. Lassner, Topological algebras of operators. Rep. Math. Phys. **3**, 279–293 (1972)
104. G. Lassner, Uber Realisierungen gewisser $*$-Algebren. Math. Nachr. **52**, 161–166 (1972)
105. G. Lassner, Mathematische beschreibung von observalben-zustandssystemen. Wiss. Z. KMU Leipzig, Math. Naturwiss. R. **22**, 103–138 (1973)
106. G. Lassner, Topological algebras and their applications in quantum statistics. Wiss. Z. KMU Leipzig, Math. Naturwiss. R. **30**, 572–595 (1981)
107. G. Lassner, Algebras of unbounded operators and quantum dynamics. Phys. A **124**, 471–480 (1984)
108. G. Lassner, Topological algebras and quantum spaces, in *Proc. of a Fest-Colloquium in Honour of Prof. A. Mallios, Sept. 1999*, ed. by P. Strantzalos, M. Fragoulopoulou (Athens Univ. Publications, Athens, 2002), pp. 39–53
109. K.B. Laursen, *Tensor products of Banach $*$-algebras with involution*, Ph.D. Thesis, Univ. of Minnesota, 1967
110. A. Mallios, Hermitian K-theory over topological $*$-algebras. J. Math. Anal. Appl. **106**, 454–539 (1985)
111. A. Mallios, *Topological Algebras. Selected Topics* (North-Holland, Amsterdam, 1986)
112. E.A. Michael, Locally multiplicatively-convex topological algebras. Mem. Am. Math. Soc. No 11 (1952)
113. R. Meise, D. Vogt, *Introduction to Functional Analysis* (Clarendon Press, Oxford, 1997)
114. G.J. Murphy, *C^*-Algebras and Operator Theory* (Academic Press, London, 1990)
115. M.A. Naimark, *Normed Rings* (Noordhoff, Groningen, 1960)
116. T. Ogasawara, A theorem on operator algebras. J. Sci. Hiroshima Univ. Ser. A **18**, 307–309 (1955)
117. T. Ogasawara, K. Yoshinaga, A noncommutative theory of integration for operators. J. Sci. Hiroshima Univ. **18**, 311–347 (1955)

118. M. Oudadess, Ogasawara's commutativity condition in GB^*-algebras. Preprint
119. R. Pallu de la Barrière, Algèbres unitaires et espaces d' Ambrose. Ann. Éc. Norm. Sup. **79**, 381–401 (1953)
120. T.W. Palmer, *Banach Algebras and the General Theory of $*$-Algebras*, vol. 1 and 2. Encyclopedia Math. Appl. 49 and 79 (Cambridge University Press, Cambridge, 1994 and 2001)
121. A.B. Patel, A joint spectral theorem for unbounded normal operators. J. Aust. Math. Soc. **34**, 203–213 (1983)
122. G.K. Pedersen, C^*-*Algebras and Their Automorphism Groups* (Academic Press, London, 1979)
123. P. Pérez Carreras, J. Bonet, *Barrelled Locally Convex Spaces* (North-Holland, Amsterdam, 1987)
124. N.C. Phillips, Inverse limits of C^*-algebras. J. Oper. Theory **19**, 150–195 (1988)
125. R.T. Powers, Self-adjoint algebras of unbounded operators. Commun. Math. Phys. **21**, 85–124 (1971)
126. V. Pták, Banach algebras with involution. Manuscripta Math. **6**, 245–290 (1972)
127. C.E Rickart, *General Theory of Banach Algebras* (R.E. Krieger Rubl. Co., Huntington, New York, 1974)
128. A.P. Robertson, W. Robertson, *Topological Vector Spaces* (Cambridge University Press, Cambridge, 1973)
129. W. Rudin, *Functional Analysis* (McCraw-Hill, New York, 1973)
130. S. Sakai, C^*-*Algebras and W^*-Algebras* (Springer, Berlin, 1971)
131. H.H. Schaefer, *Topological Vector Spaces* (Springer, Berlin, 1970)
132. K. Schmüdgen, Über LMC^*-Algebren. Math. Nachr. **68**, 167–182 (1975)
133. K. Schmüdgen, The order structure of topological $*$-algebras of unbounded operators I. Rep. Math. Phys. **7**, 215–227 (1975)
134. K. Schmüdgen, Uniform topologies and strong operator topologies on polynomial algebras and on the algebra of C.C.R.. Rep. Math. Phys. **10**, 369–384 (1976)
135. K. Schmüdgen, Lokal multiplikativ konvexe O_p^*-algebren. Math. Nach. **85**, 161–170 (1978)
136. K. Schmüdgen, *Unbounded Operator Algebras and Representation Theory* (Birkhäuser, Basel, 1990)
137. K. Schmüdgen, On well-behaved unbounded representations of $*$-algebras. J. Oper. Theory **48**, 487–502 (2002)
138. K. Schmüdgen, *An Invitation to Unbounded Representations of $*$-Algebras on Hilbert Space.* Graduate Texts in Mathematics, vol. 285 (Springer, Cham, 2020)
139. K. Schmüdgen, $*$-bimodules. Proc. Amer. Math. Soc. **149**, 3923–3938 (2021)
140. I.E. Segal, A noncommutative extension of abstract integration. Ann. Math. **57**, 401–457 (1953)
141. S. Shirali, J.W.M. Ford, Symmetry in complex involutory Banach algebras. Duke Math. J. **37**, 275–280 (1970)
142. Eu. Stefanovich, *Elementary Particle Theory*, vol. 1: Quantum Mechanics, Studies in Mathematical Physics 45 (De Gruyter, Berlin, 2019)
143. D.-S. Sya, On semi-normed rings with involution. Dokl. Akad. Nauk SSSR. 1223–1225 (1959)
144. M. Takesaki, *Theory of Operator Algebras I, II, III* (Springer-Verlag, Berlin, 1979, 2003)
145. A.E. Taylor, Spectral theory of closed distributive operators. Acta Math. **84**, 189–224 (1950)
146. S. Triolo, WQ^*-algebras of measurable operators. Indian J. Pure Appl. Math. **43**, 601–617 (2012)
147. A. Uhlmann, Über die Definition der Quantenfelder nach Wightman und Haag. Wiss. Z. KMU Leipzig Math.–Naturwiss. R. **11**, 213–217 (1962)
148. L. Waelbroeck, Algèbres commutatives: Éléments réguliers. Bull. Soc. Math. Belg. **9**, 42–49 (1957)
149. S. Warner, Inductive limits of normed algebras. Trans. Am. Math. Soc. **82**, 190–216 (1956)

150. M. Weigt, *Jordan Homomorphisms and Derivations of Algebras of Measurable Operators*, PhD Thesis, University of Cape Town, South Africa, 2008
151. M. Weigt, Derivations of τ-measurable operators. Oper. Theory Adv. Appl. **195**, 273–286 (2009)
152. M. Weigt, On nuclear generalized B^*-algebras, in *Proceedings of the conference ICTAA2018, held at the Univ. of Tallinn, Estonia, January 25–28, 2018* (Est. Math. Soc. Tartu, 2018)
153. M. Weigt, I. Zarakas, Derivations of Fréchet nuclear GB^*-algebras. Bull. Aust. Math. Soc. **92**, 290–301 (2015)
154. M. Weigt, I. Zarakas, Unbounded derivations of GB^*-algebras. Oper. Theory Adv. Appl. **247**, 69–82 (2015)
155. M. Weigt, I. Zarakas, Spaciality of derivations of Fréchet GB^*-algebras. Stud. Math. **231**, 219–239 (2015)
156. A.W. Wood, Numerical range and generalized B^*-algebras. Proc. Lond. Math. Soc. **34**, 245–268 (1977)
157. B. Yood, Hilbert algebras as topological algebras. Ark. Mat. **12**, 131–151 (1974)
158. W. Żelazko, *Metric Generalizations of Banach Algebras* (Instytut Matematyczny Polskiej Akademi Nauk (Warszawa), 1965)

Index

© The Author(s), under exclusive license to Springer Nature Switzerland AG 2022
M. Fragoulopoulou et al., *Generalized B*-Algebras and Applications*, Lecture Notes in Mathematics 2298, https://doi.org/10.1007/978-3-030-96433-7

LECTURE NOTES IN MATHEMATICS 🐎 Springer

Editors in Chief: J.-M. Morel, B. Teissier;

Editorial Policy

1. Lecture Notes aim to report new developments in all areas of mathematics and their applications – quickly, informally and at a high level. Mathematical texts analysing new developments in modelling and numerical simulation are welcome.

   Manuscripts should be reasonably self-contained and rounded off. Thus they may, and often will, present not only results of the author but also related work by other people. They may be based on specialised lecture courses. Furthermore, the manuscripts should provide sufficient motivation, examples and applications. This clearly distinguishes Lecture Notes from journal articles or technical reports which normally are very concise. Articles intended for a journal but too long to be accepted by most journals, usually do not have this "lecture notes" character. For similar reasons it is unusual for doctoral theses to be accepted for the Lecture Notes series, though habilitation theses may be appropriate.

2. Besides monographs, multi-author manuscripts resulting from SUMMER SCHOOLS or similar INTENSIVE COURSES are welcome, provided their objective was held to present an active mathematical topic to an audience at the beginning or intermediate graduate level (a list of participants should be provided).

   The resulting manuscript should not be just a collection of course notes, but should require advance planning and coordination among the main lecturers. The subject matter should dictate the structure of the book. This structure should be motivated and explained in a scientific introduction, and the notation, references, index and formulation of results should be, if possible, unified by the editors. Each contribution should have an abstract and an introduction referring to the other contributions. In other words, more preparatory work must go into a multi-authored volume than simply assembling a disparate collection of papers, communicated at the event.

3. Manuscripts should be submitted either online at www.editorialmanager.com/lnm to Springer's mathematics editorial in Heidelberg, or electronically to one of the series editors. Authors should be aware that incomplete or insufficiently close-to-final manuscripts almost always result in longer refereeing times and nevertheless unclear referees' recommendations, making further refereeing of a final draft necessary. The strict minimum amount of material that will be considered should include a detailed outline describing the planned contents of each chapter, a bibliography and several sample chapters. Parallel submission of a manuscript to another publisher while under consideration for LNM is not acceptable and can lead to rejection.

4. In general, **monographs** will be sent out to at least 2 external referees for evaluation.

   A final decision to publish can be made only on the basis of the complete manuscript, however a refereeing process leading to a preliminary decision can be based on a pre-final or incomplete manuscript.

   Volume Editors of **multi-author works** are expected to arrange for the refereeing, to the usual scientific standards, of the individual contributions. If the resulting reports can be forwarded to the LNM Editorial Board, this is very helpful. If no reports are forwarded or if other questions remain unclear in respect of homogeneity etc, the series editors may wish to consult external referees for an overall evaluation of the volume.

5. Manuscripts should in general be submitted in English. Final manuscripts should contain at least 100 pages of mathematical text and should always include

   - a table of contents;
   - an informative introduction, with adequate motivation and perhaps some historical remarks: it should be accessible to a reader not intimately familiar with the topic treated;
   - a subject index: as a rule this is genuinely helpful for the reader.
   - For evaluation purposes, manuscripts should be submitted as pdf files.

6. Careful preparation of the manuscripts will help keep production time short besides ensuring satisfactory appearance of the finished book in print and online. After acceptance of the manuscript authors will be asked to prepare the final LaTeX source files (see LaTeX templates online: https://www.springer.com/gb/authors-editors/book-authors-editors/manuscriptpreparation/5636) plus the corresponding pdf- or zipped ps-file. The LaTeX source files are essential for producing the full-text online version of the book, see http://link.springer.com/bookseries/304 for the existing online volumes of LNM). The technical production of a Lecture Notes volume takes approximately 12 weeks. Additional instructions, if necessary, are available on request from lnm@springer.com.

7. Authors receive a total of 30 free copies of their volume and free access to their book on SpringerLink, but no royalties. They are entitled to a discount of 33.3 % on the price of Springer books purchased for their personal use, if ordering directly from Springer.

8. Commitment to publish is made by a *Publishing Agreement*; contributing authors of multiauthor books are requested to sign a *Consent to Publish form*. Springer-Verlag registers the copyright for each volume. Authors are free to reuse material contained in their LNM volumes in later publications: a brief written (or e-mail) request for formal permission is sufficient.

Addresses:
Professor Jean-Michel Morel, CMLA, École Normale Supérieure de Cachan, France
E-mail: moreljeanmichel@gmail.com

Professor Bernard Teissier, Equipe Géométrie et Dynamique,
Institut de Mathématiques de Jussieu – Paris Rive Gauche, Paris, France
E-mail: bernard.teissier@imj-prg.fr

Springer: Ute McCrory, Mathematics, Heidelberg, Germany,
E-mail: lnm@springer.com

Printed in the United States
by Baker & Taylor Publisher Services

Printed in the United States
by Baker & Taylor Publisher Services